Introduction to Electric Energy Devices

F. Robert Bergseth

Professor Emeritus of Electrical Engineering
University of Washington

Subrahmanyam S. Venkata

Professor of Electrical Engineering
University of Washington

PRENTICE-HALL, INC., Englewood Cliffs, New Jersey 07632

Library of Congress Cataloging-in-Publication Data

BERGSETH, F. ROBERT.
   Introduction to electric energy devices.

   Includes index.
   1. Electric engineering. I. Venkata,
Subrahmanyam S. II. Title.
TK146.B44 1987   621.3   86-22605
ISBN 0-13-481383-9

Editorial/production supervision and
   interior design: *Mary Jo Stanley*
Cover design: *20/20 Services Inc.*
Manufacturing buyer: *Rhett Conklin*

Printed in the United States of America

10 9 8 7 6 5 4 3 2

ISBN   0-13-481383-9   025

PRENTICE-HALL INTERNATIONAL (UK) LIMITED, *London*
PRENTICE-HALL OF AUSTRALIA PTY. LIMITED, *Sydney*
PRENTICE-HALL CANADA INC., *Toronto*
PRENTICE-HALL HISPANOAMERICANA, S.A., *Mexico*
PRENTICE-HALL OF INDIA PRIVATE LIMITED, *New Delhi*
PRENTICE-HALL OF JAPAN, INC., *Tokyo*
PRENTICE-HALL OF SOUTHEAST ASIA PTE. LTD., *Singapore*
EDITORA PRENTICE-HALL DO BRASIL, LTDA., *Rio de Janeiro*

To
Gertrude
and
Padma

Contents

2 THREE-PHASE CIRCUITS

3 THE TRANSFORMER

6 THE INDUCTION MOTOR   211

7 THE COMMUTATOR MACHINE   245

8 ENERGY TRANSMISSION   275

Preface

The use of electricity for transmission and ultimate consumption of energy has long been one of the main concerns of electrical engineers. Although study of electrical energy supply and utilization may be seen as a specialty item within electrical engineering, most educators seem to feel that some exposure to electric power subjects is necessary for any individual who would assume professional status as an electrical engineer. The approach to this objective at this university has been to give a basic course in electric energy to all electrical engineering students, and then to follow with elective courses for those who would specialize in the area. This book was written with the needs of such a basic course in mind. The material is intended for a one-quarter or one-semester course. With four lecture hours per week in the ten-week quarter at this university, most faculty have found it necessary to treat some of the topics lightly in order to avoid being entirely superficial in the whole picture.

The students coming to study this course would normally be in their junior year, with a background of courses in electric circuits and electromagnetic fields. In spite of the background in circuits, however, it is often found that the students are weak in the "bread and butter" topics of sinusoidal steady-state circuit solutions, including polyphase systems. A review of these matters is given in the first two chapters, and this material might be given intensive coverage or reserved for student self-study, depending upon the needs of the class. Also the nonlinearity and complex magnetic structures of power apparatus make desirable the treatment of approximate methods of solving the magnetic field problems of power apparatus — the subject of *magnetic circuits*.

It is difficult to select specific topics to be included in a work of this nature, but some material should be included on the topics of transformers, synchronous machines, induction machines, commutator machines, transmission and distribution lines and power electronics. In addition to the specifics of these devices, an attempt has been made to give some notion of how these devices fit into the overall picture of electric power systems.

The treatment in most chapters has been along rather traditional lines with a physical description of the devices, the mathematical modeling and analysis, and illustrative examples. Each chapter is followed by problem sets. The problem sets are divided into two categories: Study Exercises and Homework Problems.

Complete solutions are given to the study exercises. It is hoped that the serious student will work on these problems *before* looking at the solutions. Whatever approach is used by the student, however, these problems will serve as additional illustrative examples.

Homework problems are of two types. Some problems involve routine "plugging in the numbers," whereas others may extend the methods and theory of the chapter to cases where the student will feel challenged to use ingenuity and thereby gain a broader understanding and viewpoint.

The notes on which this book is based have undergone a long period of trial and development at the University of Washington. The contributions of numerous students and faculty colleagues who have used the notes are too many to mention specifically. Our thanks nevertheless go out to all those who have helped so greatly in the preparation of this text.

Chapter I

Some Preliminary

Ideas and Review

- A few words about the nature of electric energy supply systems.
- A review of methods of solution of electric power circuits, phasors, notational methods, and phasor diagrams.
- A review of the concept of complex power; use of complex power in problem solution.
- Measurements in ac circuits.

1.1 ELECTRIC ENERGY SUPPLY AND CONSUMPTION

Engineers are involved with the use of electricity for two principal reasons: the processing and transmission of information, as with communications links and computers; and the transmission and utilization of energy in electrical form. This latter activity is the concern of this book.

We may be concerned with use of energy in electrical form because of the ease of transmission over great distances. By generating electric energy in very large power plants we achieve an economy of scale — the unit cost of electric energy goes down with increasing plant size. On the other hand, even if distances are not significant, we may need energy in electrical form because of its intrinsic properties; for example, the electrolytic processing industry, or even the convenience of energy storage in an automobile battery. The use of energy in electrical form involves various conversion devices as well as transmission and control means. The engineer needs to design and analyze the performance of the individual devices as well as analyze their interaction in a *system.*

A typical electric energy supply system involves three elements: generation equipment to convert energy from some other form, transmission and control equipment to bring the energy to the place of utilization, and conversion equipment to change the energy to the form of ultimate utilization. These interrelations are shown in symbolic form in Figure 1-1, where the various functions are grouped in blocks with samples of the types of equipment shown in each block — by no means to be taken as an exhaustive list or a hard-and-fast division of functions!

The scale of magnitude of different systems varies tremendously from hundreds of gigawatts of generator capacity, as with a large electric power utility, to a few watts or even milliwatts of the photovoltaic power supply to a remote radio transponder. The distances also vary greatly from hundreds of miles of transmission line for the large electric utility to a few hundred feet in a shipboard or aircraft system, to just a few feet from a solar converter to an electronic apparatus. The scope of the problems, and their nature, varies greatly with the type of energy supply system. In the pages that follow, only selected representative topics can be covered in a reasonable period of study.

The focus of an engineer's involvement with power apparatus and systems varies greatly also. Some engineers are concerned primarily with the internal physics and design of apparatus; others are more concerned with the utilization of the apparatus to perform a certain task. These latter engineers tend to concentrate on

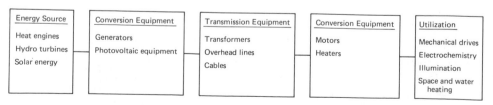

Figure 1-1 The elements of an electric energy supply and delivery system.

forming a *model* or equivalent circuit of the devices, which can be used to predict performance, either of the device singly or in combination with others to form a system. The approach of this book is to present a little of the internal physics and then concentrate on those topics necessary to form a model that will serve as a guide in the *use* of the apparatus.

The biggest portion of large energy supply systems is made up of alternating current (ac) apparatus, with a relatively small fraction of the power being converted to direct current (dc) for transmission and utilization.* We begin, then, with a review of ac circuit theory before studying devices and their interconnections.

1.2 LINEAR CIRCUIT SOLUTIONS

In circuit solution it is common to model a physical system by means of an interconnection of linear circuit elements (constant inductance, capacitance, resistance, etc.). By use of Kirchhoff's laws, the resulting equilibrium equations are found to be ordinary linear integro-differential equations with constant coefficients. Various methods of solution are straightforward and well known, as contrasted with systems described by nonlinear differential equations, where generally an explicit solution is not available.† For the linear case, the order of the system differential equations may be very high for a complex circuit and the solution to a problem may be very tedious even though the method is well known. Computer assistance in large-scale solutions is a practical necessity.

Separate components can be identified in a general linear circuit solution. There are components associated with the natural or force-free behavior of the circuits. In mathematical terms, these are called the *complementary* functions. Also there are terms that match the applied forcing functions (voltage sources in many power problems). These are known as *particular integrals*. The particular integrals have the same *form* as the driving functions (and their derivatives). Driving functions of particular interest in electric power studies are the sinusoidal (ac) driving function or the constant (dc) driving function, with which the student is familiar from previous courses. In some cases, the constant (dc) driving function is conveniently thought of as a special zero-frequency aspect of the ac sinusoidal case. The forced component (particular integral) of response to a sinusoidal ac driving function is itself a sinusoid of the same frequency, which (in theory) persists forever. This component is therefore called the *steady-state*. The natural component (complementary function) of response for practical circuits with dissipation (resistive elements) takes the form of decaying exponentials with typical time constants of less than a second. These components are therefore often called the *transient* components.

---

*As the capabilities of power electronics advance, we see an increasing number of applications where conversion from ac to dc is practiced.

†Even if the system equations are linear, but the coefficients are not constant, as in the case of rotating machines, solution may be difficult. Consideration of such cases is deferred to studies of electrical machinery.

In large-scale circuit problems it is rarely feasible even with large computers to make a completely general circuit solution involving both the natural response (complementary function) and forced response (particular integral). A modern power system involves thousands of circuit elements, and a complete classical solution would involve extraction of the roots of polynomials of degrees in the thousands! Fortunately, a complete solution is usually not necessary or desired.

The primary focus of our present work is the handling of energy in electrical form. Energy transfer is the time integral of power flow between the transfer points, and the time periods of interest are usually great with respect to the short time constants of the natural components of response. As a consequence, energy transfer or power flow is almost always concerned *only with the steady-state solution* of the circuits. For those less common cases, where the transient components are of interest, it is usually possible to simplify the system model to represent only a restricted portion of the circuit.

Although restriction of interest to the steady-state greatly simplifies problems in electric power, it must not be inferred that the problems become trivial. By means of complex number methods, the simultaneous *differential* equations become simultaneous *algebraic* equations for the steady-state solution. In a large system problem, however, there may be thousands of equations in thousands of unknowns! The immensity of these problems requires (1) computer assistance to handle the drudgery and (2) careful attention to systematic notation and methods in order to avoid ending in hopeless confusion. It is the latter point that leads us to review steady-state ac circuit solution, and to discuss some of the conventions of notation and viewpoints used in power problems, as in the following sections of this chapter.

1.3 DESCRIPTION OF A SINUSOID

At this stage, the student is familiar with the shape of a sinusoid as for example viewed on a cathode ray oscilloscope and shown in Figure 1-2. The figure is labeled in terms of a voltage as an example. The wave is characterized by a maximum value E^m and a period T. The period is the length of time taken by the quantity to pass

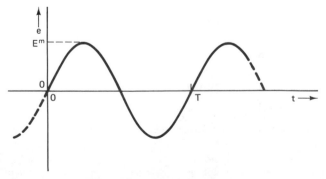

Figure 1-2 A sinusoidal voltage.

through a complete sequence of values or *cycle*. The mathematical description of the function is

$$e(t) = E^m \sin \omega t \qquad (1\text{-}1)$$

Since the argument of a sine function must be an angle, the quantity ωt is dimensionally an angle in radians and ω is an angular velocity in radians per second. In particular, a sine function passes through a complete cycle in 2π radians, hence $\omega T = 2\pi$ or $\omega = 2\pi/T$. More commonly, we observe that if T is the time per cycle, then we make $1/T$ cycles per second which is called frequency, f. A shorter term for cycles per second is *Hertz*. We then have the basic relation:

$$\omega = 2\pi f \qquad \text{radians per second}$$

It is equally common to show the sine function plotted against angle ωt, rather than time as in Figure 1-3, where the period is 2π radians.

Since problems involving the steady-state infer voltages and/or currents that have existed for some time, the origin $t = 0$ is rather arbitrary* and a sinusoidal quantity might appear as in Figure 1-4, where the sine function has been shifted by

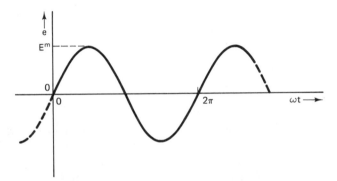

Figure 1-3   Using angle ωt as the abscissa for a sine plot.

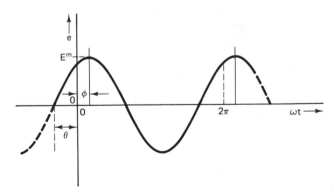

Figure 1-4   Displaced sine or cosine wave.

*Where the complete response (transient *and* steady-state) is involved, we have little choice in a practical sense in naming the origin $t = 0$; it is urged upon us by external events such as switching.

an angle θ and the equation for the wave would be

$$e(t) = E^m \sin(\omega t + \theta) \tag{1-2}$$

It should also be noted that the wave of Figure 1-4 could equally well be given in terms of a cosine function as

$$e(t) = E^m \cos(\omega t - \phi) \tag{1-3}$$

The form we use is a matter of convenience.

It might be noted that θ and ϕ in the sample are considered in radians, but calculators and students often think in terms of degrees, so we sometimes see such hybrid expressions as $E^m \sin(2\pi f t + 45°)$, which is dimensionally incorrect. No great harm is done, however, if we are conscious of what we are doing!

1.4 THE USE OF PHASORS

Let us review again the solution of a familiar circuit with the dual purpose of refreshing the memory and establishing nomenclature and viewpoints desired in this work. Consider the circuit of Figure 1-5.

A source $e(t) = E^m \sin(\omega t + \alpha)$ is impressed on a series combination with resistor, R, between a and b, and an inductor, L, between b and c. As a result, a current, i, flows. We write the Kirchhoff's voltage law equation expressing the equilibrium

$$E^m \sin(\omega t + \alpha) = Ri + L\frac{di}{dt} \tag{1-4}$$

Again, the solution consists of two parts, the natural (complementary function) component, which is transient in nature, and the forced (particular integral) component of response, which is steady-state in nature with a sinusoidal driving function. The latter component is our present subject.

One method of finding the steady-state component is the method of undetermined coefficients. We assume a solution of the form

$$i(t) = I^m \sin(\omega t + \beta)$$

where I^m and β are unknown constants and ω is the same as in Eq. (1-4). We then insert this "guess" into Eq. (1-4).

$$E^m \sin(\omega t + \alpha) = RI^m \sin(\omega t + \beta) + \omega LI^m \cos(\omega t + \beta) \tag{1-5}$$

We now solve for I^m and β by expanding by trig identities.*

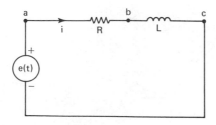

Figure 1-5 Example for review of circuit solution.

---

*It might at first seem that we have an impossible situation—one equation and two unknowns. Actually Eq. (1-5) is stronger than that, however; the two sides must be identically equal for all values of t.

We will not pursue this method further. The purpose of presenting it at all is to convince the student of the utility of the method to follow (and the student *will* surely be convinced by finishing the above illustration in trig form!).

The *phasor* method of solution avoids the cumbersome use of trig identities at the small cost of introducing complex number arithmetic into the solution. The phasor method has its roots in Euler's identity which gives

$$\varepsilon^{j\theta} = \cos\theta + j\sin\theta \qquad (1\text{-}6)$$

By aid of Eq. (1-6) we may write

$$e(t) = \operatorname{Im} E^m \varepsilon^{j(\omega t + \alpha)} \qquad (1\text{-}7)$$

where the symbol "Im" means "take the imaginary part of what follows." We then use our assumed response as

$$i(t) = \operatorname{Im} I^m \varepsilon^{j(\omega t + \beta)} \qquad (1\text{-}8)$$

From the properties of exponents

$$E^m \varepsilon^{j(\omega t + \alpha)} = \left[E^m \varepsilon^{j\alpha} \right] \varepsilon^{j\omega t}$$
$$I^m \varepsilon^{j(\omega t + \beta)} = \left\lfloor I^m \varepsilon^{j\beta} \right\rfloor \varepsilon^{j\omega t}$$

The quantities in brackets are recognized as complex numbers, which may be symbolized by a single letter as follows

$$E^m \varepsilon^{j\alpha} = \boldsymbol{E}^m \quad \text{and}$$
$$I^m \varepsilon^{j\beta} = \boldsymbol{I}^m$$

The boldface type signifies that the quantity is a complex number and, therefore, requires *two* numbers for complete specification. With the aid of the shorthand notation shown, Eqs. (1-7) and (1-8) may be written

$$e(t) = \operatorname{Im} \boldsymbol{E}^m \varepsilon^{j\omega t}$$
$$i(t) = \operatorname{Im} \boldsymbol{I}^m \varepsilon^{j\omega t}$$

When we see a quantity like $\boldsymbol{E}$ as input to a system, we must be prepared to supply the two parts of the complex number. When we see a quantity like $\boldsymbol{I}$ as the output or response to a system, we may expect a complex number in a solution. Alternate forms are the exponential (as given first), the polar, the rectangular, and the trigonometric or circular as for example:

$$\boldsymbol{E}^m = E^m \varepsilon^{j\alpha} = E^m \underline{/\alpha} = E_1^m + jE_2^m = E^m(\cos\alpha + j\sin\alpha)$$

The last form is recognized as merely using Euler's identity to replace $\varepsilon^{j\alpha}$. The components of the rectangular form are found from

$$E_1^m = E^m \cos\alpha$$
$$E_2^m = E^m \sin\alpha$$

The operation of converting from one form to another is easily done on a modern calculator and the student is well advised to practice the conversion on the particular calculator available.

Let us return to Eq. (1-4) for the illustrative *R-L* circuit. We now consider a voltage source $\boldsymbol{E}^m \varepsilon^{j\omega t}$ and assume a current $\boldsymbol{I}^m \varepsilon^{j\omega t}$, then insert into the equation and solve for $\boldsymbol{I}^m$

$$E^m \varepsilon^{j\omega t} = RI^m \varepsilon^{j\omega t} + j\omega L I^m \varepsilon^{j\omega t}$$

$$E^m = RI^m + j\omega L I^m = (R + j\omega L)I^m$$

$$I^m = \frac{E^m}{R + j\omega L} \quad \text{or} \quad = \frac{E^m \underline{/\alpha}}{R + j\omega L} \qquad (1\text{-}9)$$

Several things should be said about the process of arriving at Eq. (1-9):

a. The actual voltage applied is a real number function of time, $E^m \sin(\omega t + \alpha)$, and so it is the imaginary part of $E^m \varepsilon^{j\omega t}$ that is of interest. As a consequence, it is the imaginary part of $I^m \varepsilon^{j\omega t}$ that gives our result. If we wished to use a cosine description for the voltage and current, we would be concerned with the real part of $E^m \varepsilon^{j\omega t}$ and $I^m \varepsilon^{j\omega t}$. In practice, for power computations, we seldom wish to have a time-function expression or plot and so we need not ask whether we were thinking of the real or imaginary part of our voltage and current.

b. We have stated that the imaginary part of the response corresponds to the imaginary part of the driving voltage and similarly with respect to the real parts. The student is shown in introductory courses how this is so; it is as though the two parts lived separately in the circuit. We will not pause to prove this, but it should be pointed out that this simple result is a consequence of the *linear* equations with which we deal.

c. The complex number quantities $E^m = E^m \underline{/\alpha}$ and $I^m = I^m \underline{/\beta}$ are called *phasors* and represent the actual sinusoid in the circuit. If we are given a phasor voltage or current, we can always reconstruct the *time domain* description of the quantity. For example, if $I^m = 10\underline{/30}$, then $i(t) = \text{Im } I^m \varepsilon^{j\omega t} = 10 \sin(\omega t + 30)$. If the person using the quantities bases the use on the real part, then $i(t) = \text{Re } I^m \varepsilon^{j\omega t} = 10 \cos(\omega t + 30)$ where the symbol Re means "take the real part of what follows." For most purposes, such as finding the magnitude of a voltage or a current, the relative angular difference (known as the *phase* angle) between two quantities, or the power flow, we do not need to reconstitute the time domain functions and the question of whether the author of a work had real or imaginary part in mind is immaterial. Most people form preferences for one form or the other, however, and the *real* part designation (cosine form) is perhaps most common in texts and articles.

d. Most voltmeters and ammeters read the root-mean-square (*rms* or *effective*) value of voltage and current. Also rms values are used in power computations. As a result, rms quantities are most often used in circuit computations. One is usually justified in assuming if nothing else is said that an ac voltage or current is given as rms. For example, if we are told that the voltage of the outlet in a room is 120 volts, this is the *rms* value. For a sinusoid the maximum value is $\sqrt{2}$ times the rms or $\sqrt{2} \times 120 = 170$ volts in this example.

In connection with phasor representation, $\sqrt{2}$ is merely a scale factor multiplying the magnitude of the phasor by $\sqrt{2}$. In this case $E^m = E^m \underline{/\alpha} = \sqrt{2}E$. The quantity $E \underline{/\alpha} = E$ is an rms phasor. Because the rms values are the most common, we use the cumbersome superscript on E^m and use the plain E for the rms phasor. Note that the angles of E^m and E are the same.

If the circuit example above is treated by rms phasors, we would have as a result $\sqrt{2}I = \sqrt{2}E/(R + j\omega L)$ in Eq. (1-9) or, cancelling the common factor, $I = E/(R + j\omega L)$. This is the usual way of handling voltage and current phasors. Note that the $\sqrt{2}$ factor does not affect the denominator, *impedance,* which is discussed in Section 1.7.

1.5 REFERENCE DIRECTION AND POLARITY

The student may have noted the arrowhead and the plus $(+)$ and minus $(-)$ signs in Figure 1-5. These are references for mathematical description of voltage and current. When we relate the physical quantities of the real world to the mathematical world of numbers, we must usually set up scales of magnitude and directions — coordinate systems. For example, in describing the position of a point on a plane we often erect a set of Cartesian coordinates and *in terms of these coordinates* describe the position as, say $(3, -2)$. In terms of another coordinate system, the numbers would be different, perhaps $(-10, 13)$.

In dealing with electrical quantities, we likewise set up scales of units for magnitude and reference directions or polarity.

Electric current is defined basically as the rate of charge transfer through a surface such as the cross-sectional area of a conductor. The unit of electric charge is the *coulomb,* and we specify the positive direction of transfer by means of a reference arrow, as in Figure 1-6.

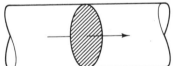

Figure 1-6   Illustration of current reference direction.

If an element of positive charge passes through the surface in the arrow direction, we record the passage as positive, whereas the opposite effect is recorded as numerically negative.* If we find that an increment of charge, Δq, passes through in a positive time increment Δt, then we define the average *current* over this time interval as $I_{ave} = \Delta q/\Delta t$ amperes. In terms of the calculus, the current right *at* a certain instant (t) is given by

$$i(t) = \lim_{\Delta t \to 0} \frac{\Delta q}{\Delta t} = \frac{dq}{dt} \text{ amperes}$$

Note that the numerical sign of Δq and therefore the current, is dependent upon the direction of the reference arrow. Reversal of the arrow direction would simply reverse the numerical sign of the current description at any particular instant of time.

In the idealized relations of circuit theory, we are concerned with conductors connecting two elements of the circuit. In such a case, we *might* consider more than one cross-sectional area of the conductor as in Figure 1-7. Here we could con-

*We could, of course, enter into a discussion of transfer by both positive and negative charge carriers but this would merely add unnecessary confusion. In circuit solution, we use positive charge as our bookkeeping quantity.

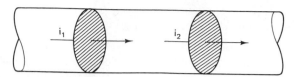

Figure 1-7   Illustrates the continuity of current in a conductor.

ceivably have two different currents, i_1 and i_2, at the two positions, *but* if such were the case, charge would accumulate (or be depleted) in the volume between the surfaces. In circuit theory, we postulate that this is *not* the case and hence $i_1 = i_2$. If we must acknowledge a charge accumulation, then the circuit theory approach is to show a capacitance for this section.

In dealing with circuit problems, we do not ordinarily concern ourselves with the cross-sectional area itself, but merely draw an arrow alongside the conductor symbol or an arrowhead on the conductor as in the example of Figure 1-5.

For a sinusoid, the sense of the time plot is reversed, depending on the reference direction, as is illustrated in Figure 1-8. Suppose the current i between two portions of a circuit is plotted as shown at the right in the time domain and, for example, described as $i(t) = 10 \sin(\omega t + 30°)$. The opposite reference choice, i', simply gives the negative of the current i, that is, $i'(t) = -10 \sin(\omega t + 30)$ or $i'(t) = 10 \sin(\omega t + 30° - 180°) = 10 \sin(\omega t - 150°)$. The phasor description corresponding would be

$$I = 7.07\underline{/30°}$$
$$I' = -7.07\underline{/30°} = 7.07\underline{/-150°}$$

Note the use of rms magnitudes in the phasor description; note also that the phasor *angle* is dependent on the reference direction chosen.

Voltage, or more precisely the difference in potential between two points, is also described in terms of references. In the case of a constant potential source (approximated by a battery as an example), we mark one terminal with a plus ($+$) sign and one with a negative ($-$) sign to indicate the terminals toward which electric charge is urged internally by the electrochemistry. In the case of time varying voltages, we also use the plus and minus signs, but the tendency of the device to

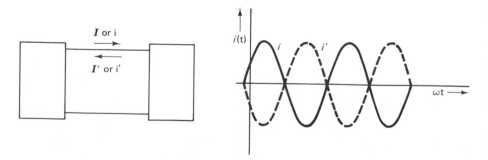

Figure 1-8   Illustration of current reference.

move charge may be first in one direction and then in the other, as in the case of an alternating voltage. The + and − signs are then references, and the voltage is described as numerically positive at an instant when the motion of positive charge from minus to plus *would* result in an increase in potential energy of the charge.*

For the ac case, the voltage description may be reversed in sign depending on the reference chosen, just as in the case of the current description. This is illustrated in Figure 1-9.

If the voltage e is given as $141.4 \sin(\omega t + 60°)$, then the same voltage in terms of the opposite polarity reference would be $e' = -141.4 \sin(\omega t + 60°)$ or $e' = 141.4 \sin(\omega t + 60° - 180°) = 141.4 \sin(\omega t - 120°)$. The phasor descriptions corresponding are $E = 100\underline{/60°}$ and $E' = -100\underline{/60°} = 100\underline{/-120°}$.

The choices of current reference and/or voltage polarity reference is of course arbitrary and usually made to suit the convenience of the problem at hand. Although the choice is arbitrary, it is essential that all mathematical work be done *consistent with the choice* or else incorrect conclusions will be drawn.

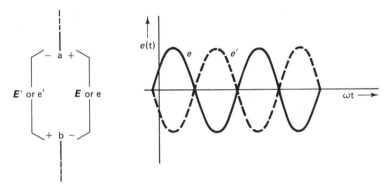

Figure 1-9   Illustration of voltage polarity reference.

1.6 SUBSCRIPT NOTATION

The use of plus and minus signs to describe voltage reference polarities is rather basic in terms of understanding, but such use tends to clutter the circuit diagram for large-scale problems. In many cases we append a subscript to the voltage symbol, which not only tells *what* voltage or potential difference is involved but also gives a reference polarity indication. Such subscripts are associated with particular points or *nodes* in the circuit.

A single subscript, such as E_a, infers that we are describing the potential of point or node a with respect to some reference node; in many cases in power apparatus studies the reference is the earth. In terms of plus and minus signs, such a notational convention is equivalent to putting a plus sign on the node and a minus sign on the reference.

*It is important to note the words like *would* or *tendency* in mentioning charge movement. There may not be any charge movement (zero current) or movement may be opposite to the tendency.

We also often find it convenient to use *double subscripts* in describing voltages in a circuit. The double subscripts tell us two things: (1) We identify the points or nodes *between* which we are describing the voltage, and (2) we give the polarity reference by agreeing that the first subscript bears a plus sign of reference and the second a minus sign.* As an example, the voltage E of Figure 1-9 might be described as E_{ab} and the voltage E' as E_{ba}. Note that $E_{ab} = -E_{ba}$.

It is well to note that the single- and double-subscript conventions may be related by an equation such as

$$E_{ab} = E_a - E_b$$

The use of double-subscript voltage notation is very convenient where voltages are to be added. Figure 1-10 shows an almost trivial example, where we may write the summation equation for overall voltage by merely tracing the sequence of the subscripts:

$$e_{ac} = e_{ab} + e_{bc} \tag{1-9}$$

Double subscripts are also sometimes used for currents to give reference directions such as i_{ab}. There is sometimes a possible ambiguity as to which direction is followed around a loop between a and b, but this is not often troublesome if other points are identified around the loop.

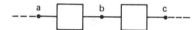

Figure 1-10 The application of double-subscript notation for voltages.

1.7 IMPEDANCE AND ADMITTANCE

The *R-L* circuit example of Figure 1-5 should have served its purpose of reviewing the relation between the circuit differential equation and the steady-state sinusoidal (or ac) solution in the form of phasor representation. In actual practice, we seldom write this differential equation if only the steady-state ac solution is of interest. Instead we find the relation between phasors E and I (voltage *across* and current *through*) the common circuit elements and then derive the overall circuit relations by using Kirchhoff's laws in combination with the building block relations. In each case we find that the voltage and current are related by a *complex number constant of proportionality* when the voltage and current are described in phasor form. The relation may be given either in the form $E = ZI$ or $I = YE$, where Z is called *impedance* and has the units of ohms and Y is called *admittance* and has the units of siemens. Table 1-1 summarizes the voltage-current relations for the common circuit elements, R, L, and C, in the form of time-domain relations, and phasor relations in both the impedance and admittance forms.

---

*This choice of meaning for the subscripts is sometimes referred to as a voltage *drop* notation — a positive test charge would lose potential energy in traveling from node a to node b in the example for a positive numerical value of the voltage. Other conventions are sometimes encountered.

TABLE 1-1   Steady-State E-I Relations of the Elements

R = 1/G	L	C
$e = Ri$	$e = L\dfrac{di}{dt}$	$e = \dfrac{1}{C}\displaystyle\int idt$
or	or	or
$E = RI$	$E = j\omega LI$	$E = \dfrac{1}{j\omega C}I$
		$= j\left(-\dfrac{1}{\omega C}\right)I$
$i = Ge$	$i = \dfrac{1}{L}\displaystyle\int edt$	$i = C\dfrac{de}{dt}$
or	or	or
$I = GE$	$I = \dfrac{1}{j\omega L}E$	$I = j\omega CE$
	$= j\left(-\dfrac{1}{\omega L}\right)E$	

In Table 1-1 relations are given in both the time domain and in the complex number or phasor domain. For example, if we have $e = L\,di/dt$ and we express $i(t)$ as

$$i = \operatorname{Im} I^m e^{j(\omega t + \theta)} = \operatorname{Im}[I^m e^{j\theta}]e^{j\omega t} = \operatorname{Im} \mathbf{I}^m e^{j\omega t}$$

when we take the derivative we have $e = \operatorname{Im} j\omega L I^m e^{j\omega t}$ from which we may write

$$E^m = j\omega L I^m \quad \text{or} \quad E^m/\sqrt{2} = j\omega L I^m/\sqrt{2} \quad \text{or} \quad \mathbf{E} = j\omega L \mathbf{I}$$

Rather than go through the lengthy sequence above in dealing with circuit problems we simply recognize, as in the table, that the operation $d(\ )/dt$ becomes multiplication by $j\omega$ in the phasor domain. In a similar manner, the operation of $\int (\ )/dt$ becomes division by $j\omega$. The operations will remind most students of the analogous relations in the *complex frequency*, $s = \sigma + j\omega$, of Laplace transform methods where $d(\ )/dt$ becomes multiplication by s and integration becomes division by s.

In Table 1-1, the complex number constants of proportionality (impedance or admittance) are seen to fall into the special case of *all real*, as with R or *all imaginary*, as with L and C. In the latter cases, the constants are given special names of *reactance*, X, and *susceptance*, B, for the imaginary parts of impedance and admittance respectively. The impedance of an inductor is then described by $Z_L = 0 + jX_L$, where $X_L = \omega L$ and the impedance of a capacitor is given by $Z_C = 0 + jX_C$, where $X_C = -1/\omega C$. In a similar manner, the admittance of an inductor is given by $Y_L = 0 + jB_L$, where $B_L = -1/\omega L$ and the admittance of a capacitor is given by $Y_C = 0 + jB_C$ where $B_C = \omega C$.

In the case of the resistive element, the impedance is all real and given by $Z_R = R + j0$ and the admittance by $Y_R = G + j0$, where R and G are named simply *resistance* and *conductance*.

The student is probably prepared to accept the statement that Kirchhoff's voltage law applies in phasor form the same as in terms of actual time functions.* Thus, for the R-L circuit of Figure 1-5, we have

$$e(t) = e_{ac} = e_{ab} + e_{bc}$$
$$E_{ac} = E_{ab} + E_{bc}$$
$$= RI + j\omega LI$$
$$= (R + j\omega L)I$$

and we further note that the total voltage across the series combination of R-L is related to the current by the complex number constant $Z = R + j\omega L$, as was named earlier. We also see an example of the combination of impedances in series. Formally, we state that impedances in series can be combined by addition to form a single complex number impedance relating the overall voltage to the through current.

The handling of the series R-L circuit of Figure 1-5 was shown in terms of impedance. The analogous use of admittance may be shown in terms of the circuit of Figure 1-11 (which is the *dual* of that of Figure 1-5).

Again, it must be stated that Kirchhoff's current law applies to the phasor representation of sinusoidal currents, just as it does to the time functions. For the case at hand, note that double-subscript notation for currents would not be as convenient as that for voltages in the preceding example.

$$i(t) = i_R + i_C$$
$$I = I_R + I_C$$
$$= GE + j\omega CE$$
$$= (G + j\omega C)E$$

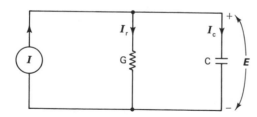

Figure 1-11   Example of use of admittance parameter.

*Consider the simple example of Figure 1-10. If the voltages are all sinusoids of the same frequency, the phasor representations would be

$$e_{ac} = \text{Im } E_{ac}\, \varepsilon^{j\omega t}$$
$$e_{ab} = \text{Im } E_{ab}\, \varepsilon^{j\omega t}$$
$$e_{bc} = \text{Im } E_{bc}\, \varepsilon^{j\omega t}$$

and Eq. (1-9) becomes

$$\text{Im } E_{ac}\, \varepsilon^{j\omega t} = \text{Im } E_{ab}\, \varepsilon^{j\omega t} + \text{Im } E_{bc}\, \varepsilon^{j\omega t}$$

and

$$E_{ac} = E_{ab} + E_{bc}$$

Thus, we might as well write a voltage relation in terms of the phasor form instead of the instantaneous form like Eq. (1-9)!

From consideration of this last equation, we see that the admittance of a parallel combination of elements is the complex number sum of the parts.

If we know the impedance or admittance relations between phasor voltage and phasor current for the common circuit elements, and if we accept the notion that Kirchhoff's laws apply to phasor voltage and current, then we may solve involved networks by writing the network equations directly in terms of the phasor representations of the voltages and currents, as was done in the preceding examples. In the case of simple two-terminal arrays of passive elements, the net relation between the voltage and the current at the terminals is expressed by a single complex number constant, the impedance or the admittance of the circuit as a whole. Examples were given in connection with Figure 1-5, where the impedance was found as $Z = R + j\omega L$ and for Figure 1-11, where the admittance was found as $Y = G + j\omega C$.

1.8 IMPEDANCE OR ADMITTANCE OF A MORE COMPLEX CIRCUIT

In the preceding section the equations $E = ZI$ and $I = YE$ were presented in terms of very simple circuits. In the case of a more complex combination of (linear) circuit elements, the ratio E/I or I/E will also be also be a complex number in the phasor approach to steady-state sinusoidal analysis, and we may speak of the impedance or admittance looking into two terminals of such a circuit. Let us use the subscript zero (0) to designate the overall voltage and current in such a situation; that is, let $E_0/I_0 = Z_0$ or $I_0/E_0 = Y_0$.

Since *impedance* is a complex number, it has a real and an imaginary component. It is usual to name the real part *resistance,* R_0, and the imaginary part *reactance,* X_0; that is, $R_0 = \mathrm{Re}\ Z_0$, $X_0 = \mathrm{Im}\ Z_0$. (The subscript 0 is appended to be able to distinguish the overall circuit resistance and reactance from that of any individual element.)

The determination of R_0 and X_0 for the circuit of Figure 1-5 follows from $Z = R + j\omega L$ to give $R_0 = \mathrm{Re}\ Z = R$ and $X_0 = \mathrm{Im}\ Z = \omega L$, a result consistent with the statement that the reactance of an inductor is simply ωL.

The apparently trivial result of $R_0 = R$ in the above is deceptive when applied to the simple series circuit. Significance of the definition is better appreciated in the case of a more complex circuit such as that of Figure 1-12, for which values of the elements are given beside the diagram. With the frequency, f, given as 60 Hz,

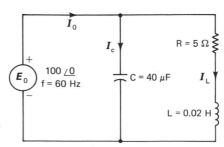

Figure 1-12   A slightly more complex circuit case.

$\omega = 2\pi f = 377$ radians per second. The capacitive reactance is given by $X_C = -10^6/\omega C = -66.3$ ohms. The current I_C is then found from $I_C = E_0/(0 + jX_C) = 100\underline{/0°}/(0 - j66.3) = 1.51\underline{/90°}$. The inductive reactance X_L is given by $X_L = \omega L = 7.54$ ohms and the current

$$I_L = E_0/(R + jX_L)$$
$$= 100\underline{/0°}/(5 + j7.54)$$
$$= 11.05\underline{/-56.5°}$$

The total current then is given by

$$I_0 = I_C + I_L$$
$$= 9.83\underline{/-51.59°}$$

The impedance looking into the circuit from the voltage source is given by

$$Z_0 = \frac{E_0}{I_0} = \frac{100\underline{/0}}{9.83\underline{/-51.59}} = 10.17\underline{/51.59} = 6.32 + j7.97$$

From this result, we have $R_0 = \text{Re } Z = 6.32$ ohms and $X_0 = \text{Im } Z = 7.97$ ohms. Note that these values cannot be associated with any particular element in the circuit.

It should also be noted that if R and L of the simple series circuit of Figure 1-5 are chosen as $R = 6.32$ and $\omega L = 7.97$, the impedance presented to the source will be the same for simple circuit which is said to be equivalent to that of Figure 1-12.*

The point of the above is that only for a series circuit (or a single element of a circuit) can the circuit's R_0 and X_0 be identified with particular elements. For some involved circuits, the net R_0 and X_0 are mathematical properties of the circuit as a whole.

As with impedance, the parameter *admittance* is often divided into real and imaginary components. The real component is named *conductance* (G_0) and the imaginary component is named *susceptance* (B_0). Thus $G_0 = \text{Re } Y$ and $B_0 = \text{Im } Y$.

For the simple parallel circuit of Figure 1-11 we have $Y = G + j\omega C$ so $G_0 = G$, and $B_0 = \omega C$.

Again, for a more involved circuit such as that of Figure 1-12, we cannot in general identify the G_0 and B_0 with any particular element. In this case

$$Y_0 = \frac{I_0}{E_0} = \frac{9.83\underline{/-51.59}}{100\underline{/0}}$$
$$= 0.0983\underline{/-51.59}$$
$$= 0.0610 - j0.0770$$

and

$$G_0 = \text{Re } Y_0 = 0.0610 \text{ siemen}$$
$$B_0 = \text{Im } Y_0 = -0.0770 \text{ siemen}^\dagger$$

*The elements of the equivalent circuit were found for the particular frequency (60 Hz in this case). At another frequency the elements R and L of the equivalent circuit would be different. For power engineers many problems involve only one frequency, so this point is immaterial in such cases.

†It is important to note that G_0 is *not* equal to $1/R_0$ and B_0 is *not* equal to $-1/X_0$, as is the case in dealing with a single element where $G = 1/R$ and $B = -1/X$!

If in turn we set $G = 0.0610$ and $B = -0.0770$ of the simple parallel circuit of Figure 1-11, we have an *equivalent* circuit (valid at the frequency being considered). Since the susceptance B_0 is numerically negative, the equivalent circuit element would be an inductor. Again, it is only for simple parallel circuits (or single elements) that we can associate G_0 and B_0 with particular elements!

Many students will recognize that the circuit solution above could have been approached in other ways — as by series parallel network reduction, for example. These matters are treated in introductory circuit courses and reaffirmation will be reserved for the study exercises at the end of the chapter.

1.9 PHASOR DIAGRAMS

Complex numbers, as two dimensional quantities, are often represented in Cartesian coordinates. As with other plane vectors, complex numbers add by the parallelogram law, and the sum of two complex numbers can be found graphically if desired. Sketching the complex number phasors of voltages and/or currents in a circuit often adds clarity to the understanding of the circuit and the interrelations of the parts. Actual graphical computation is seldom used for elementary sums and differences in this present time of electronic calculators, but a graphic sketch of the phasors can often serve as a quick check to the reasonableness of a solution in case of blunders in computation; it can also give a clue as to the effect of any changes in the circuit.

Consider the simple example of the *R-L* circuit of Figure 1-5 for which Kirchhoff's law in phasor form reads

$$E = RI + j\omega LI$$

Suppose that the circuit has been solved and we have the following example numbers for all quantities involved:

$$E = 100\underline{/53.2°}, \qquad R = 3 \text{ ohms}, \qquad \omega L = 4 \text{ ohms}, \qquad I = 20\underline{/0°}$$

A sketch (phasor diagram) of the complex number quantities might show only the overall voltage and current as in Figure 1-13(a) or it might be that we would also wish to show the voltages across the resistor and inductor separately, and how they add by Kirchhoff's law to give the total impressed voltage, as in part (b) of the figure. The voltage across the resistor is designated as E_R and that across the inductor as E_L for ready reference.

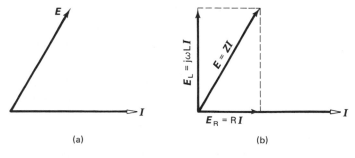

(a)   (b)

Figure 1-13   Phasor diagram for the *R-L* circuit.

For graphic convenience the quantities given above were so chosen that the current phasor had an angle of zero. Some say we are using the current as the *reference phasor*.

If we recall that the actual time functions are related to the phasors by equations such as

$$e(t) = \text{Im}[\sqrt{2}E\varepsilon^{j\omega t}]$$

the phasor diagram could be described in another way. Multiplication of a phasor such as E by the factor $\varepsilon^{j\omega t}$ would cause the phasor to rotate at a speed ω, and the projection on the imaginary axis is the actual sinusoidal time function (divided by $\sqrt{2}$ for rms phasors!) when using the sine function description. The projection on the real axis is the time function when using the cosine formulation. A phasor diagram such as Figure 1-13 is sometimes said to be a "snapshot" of the rotating phasors at time $t = 0$. Since, for steady-state analysis, the time $t = 0$ is arbitrary, the diagram as a whole could be rotated and shown at some other angle, and this would illustrate the circuit relations as well as the position shown. One must be careful, however, to observe that all the phasors must be shown in terms of the same $t = 0$; that is, the reference is arbitrary, but once chosen must be used for all phasors in a given problem!

In this book we use a common convention and show the current phasors with closed arrowheads and the voltage phasors with open arrowheads.* If the phasors are to be shown accurately for graphical purposes as contrasted with a casual sketch, then we must be careful to scale the lengths of the phasors by some convenient scale factor. The scale factor for voltage phasors does not need to be the same as the scale factor for the current phasors, since we would never add voltage and current together. We choose the two scale factors independently with a view toward spreading the diagram out as far as possible for clarity in interpretation.

At this stage let us note that the phasor voltage $E_R = RI$ has a magnitude of R times that of the current and an angle equal to that of the current. On the other hand the phasor voltage $E_L = j\omega LI$ has a magnitude of ωL (or X_L) times that of the current and an angle $90°$ greater than that of the current. The quantity j is an operator that swings the angle of a phasor ahead by $90°$ when it multiples that quantity. We could interpret j as $1\underline{/90°}$ in polar form.

The phasors E_R and E_L were added in Figure 1-13(b) by the parallelogram rule, since complex numbers add by the same rule as two-dimensional vectors.† On the other hand, the addition could be accomplished by drawing the phasors in tandem as shown in Figure 1-14(a) or (b). The sequence of drawing E_R and E_L is immaterial so far as the sum is concerned but may be of some significance in some cases as will be treated in the following.

---

*In subsequent work we will be describing magnetic flux linkages, which vary sinusoidally with time and which are represented by phasors. In such cases we use yet another distinctive arrowhead on the diagrams!

†Historically, diagrams such as the one in Figure 1-13 were called *vector diagrams,* and students may find that some texts and senior engineers refer to them as such. The possible confusion with space vector quantities such as the H vector of electromagnetic fields has led to the adoption of the term *phasor* to describe the complex number quantities representing sinusoids.

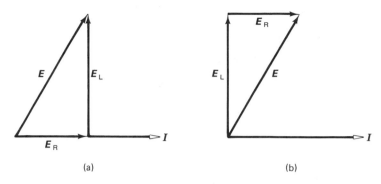

Figure 1-14   Phasor addition of component voltages.

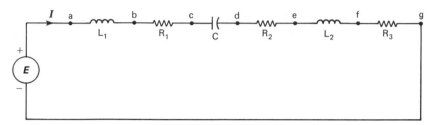

Figure 1-15   Example circuit for phasor diagram discussion.

A slightly more complicated circuit is shown in Figure 1-15, where a number of elements in series are connected to a voltage source E. Since there are a large number of voltage components that could be discussed it will be convenient to use double-subscript notation in terms of the points identified by the letters a through g on the diagram. We will assume that the actual circuit solution is no problem and the following data are available as a result of preliminary computations:

$$E = 90.67\underline{/83.03°} \qquad I = 10\underline{/0°}$$
$$X_{L_1} = 5 \text{ ohms} \qquad R_1 = 2 \text{ ohms}$$
$$X_{L_2} = 8 \text{ ohms} \qquad R_2 = 3 \text{ ohms}$$
$$X_C = -4 \text{ ohms} \qquad R_3 = 6 \text{ ohms}$$

Note that the applied voltage exists between points a and g and could be described in double subscripts as E_{ag}, and hence the plus and minus signs on the source are really redundant. A phasor diagram is sketched in Figure 1-16. The individual voltage drops across the several elements are drawn to add to the total impressed voltage, and the components are arranged in the same sequence as the circuit elements. This particular way of sketching the phasors has advantages of clarity:

1. The clutter of eight phasors emanating from the origin is eliminated by the tandem type of diagram.
2. Notational clutter is reduced by identifying the intersection of the phasors with the circuit points a through g. This is made possible by arranging the phasors

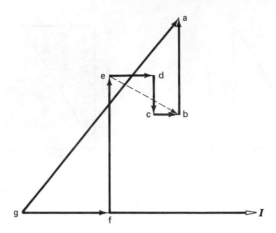

Figure 1-16   Phasor diagram for circuit of Figure 1-15.

in the same sequence as the circuit elements. The voltage component E_{ab}, for example, is easily seen as represented by the arrow between points a and b.

3. Voltages between selected points, for example b and e, are easily seen by sketching an arrow between the points as illustrated by the dotted line in the diagram. This ability to visualize the voltage components is very useful in evaluating the reasonableness of a numerical solution. (Suppose our calculator gives us a positive angle for E_{be}; a quickly sketched phasor diagram such as this tells us we have goofed!)

If in the above example we had wished to visualize E_{eb} instead of E_{be}, we note that these two phasors are merely the negatives of each other. E_{eb} is then represented by the same dotted line in the figure, but with the arrowhead on the opposite end! This sort of handling of the diagram sometimes leads to diagrams such as Figure 1-17

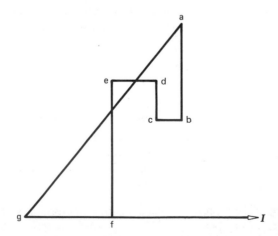

Figure 1-17   Alternate phasor diagram for the circuit of Figure 1-15.

in power problems, where no arrowheads are given on the voltage phasors. If a certain phasor is wanted, one merely draws a line between the two points involved and places an arrowhead on the end of the line at the first point in the double-subscript designation of the voltage. This method will be found especially convenient in the case of polyphase circuits to be covered later. As an example, the nameplate diagrams of large polyphase transformers with involved phase relations between the winding elements are usually presented in terms of a diagram like the one given, without arrowheads.

In the case of current phasors in parallel or series-parallel circuits, the phasors may be drawn in tandem for addition, but the double-subscript methods above are not so useful. The reason is that voltage is an "across" variable between two points and adds algebraically around closed paths, whereas current is a "through" variable adding algebraically at nodes.*

1.10 POWER: REFERENCE DIRECTION

With a voltage e across two terminals of a device or combination of devices and a current i through the device, the electric power is given by $p = ei$. Lower-case letters signify time-varying quantities for voltage and current, and therefore the power also is a time-varying quantity in the general case. In Figure 1-18(a) and (b) voltage and current symbols are shown in relation to the two terminals in question (a "black box" in classroom terminology). The two parts of the figure differ in the relative reference directions for voltage and current. In (a), positive charge is traveling from the positive terminal *inside* the box to the negative and is therefore losing potential energy; the box is *absorbing* electric energy. In part (b) of the figure, the positive charge is traveling from minus to plus *inside* the box and is increasing in potential energy; the box is therefore *delivering* electric energy. Since the voltage and/or the current may be positive, zero, or negative numerically, the product p may likewise be positive, zero, or negative numerically. In ac circuits with sinusoidal voltage and current, the power is quite often positive for part of the cycle and negative for part of the cycle. It is important to realize the significance of the reference directions in interpreting the results of a given problem. For emphasis let it be stated that:

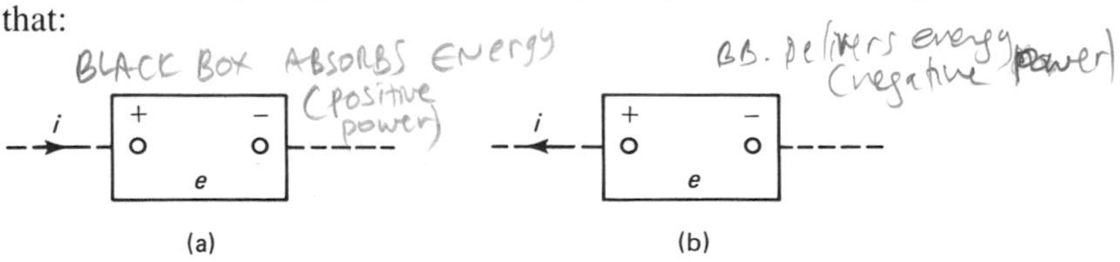

Figure 1-18   Illustration of reference direction for power, (a) into the box, and (b) out of the box.

*It may be helpful in considering the voltage diagrams to view the diagram as a sort of topographic map analogous to a map showing vector distances between points. Suppose a student is at the library and walks down to the Electrical Engineering building. A series of line segments on paper would describe the movements and a line between any two points would represent the net travel between these two points.

When the current reference *enters* the *positive* terminal of the voltage reference, a positive numerical value for power represents electrical power absorbed by that device.

When the current reference *leaves* the *positive* terminal of the voltage reference, a positive numerical value for power represents electrical power delivered by the device.

1.11 POWER IN AC CIRCUITS

When the voltage and current are in the sinusoidal steady-state, the product $p = ei$ is also a periodic function of time. For example, suppose we have

$$e = \sqrt{2}E \sin(\omega t + \theta_E)$$
$$i = \sqrt{2}I \sin(\omega t + \theta_I)$$

then

$$p = 2EI \sin(\omega t + \theta_E) \sin(\omega t + \theta_I) \qquad (1\text{-}10)$$

We may put Eq. (1-10) in various forms by means of trig identities, but prior to doing this, let us look at a sample plot (as from a computer) for some particular values of the parameters, say, $E = 100$ volts, $I = 2$ amperes, $\theta_E = 10°$, and $\theta_I = -20°$. Figure 1-19 shows a plot of power, p, and e and i versus ωt, which is given in radians.

The most striking feature of the power plot is that the power is a displaced sinusoidal function of double frequency compared to the voltage and current. It will also be noted that the power curve assumes both positive and negative numerical values (at least for the particular functions e and i). The significance of positive and negative numerical values must be interpreted in terms of the references chosen for e and i as discussed in the preceding section, that is, in terms of power *in* or power *out* of the terminals involved with e and i.

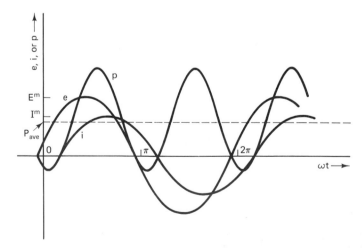

Figure 1-19   Sample plot of power variation in an ac circuit vs. ωt.

If we are interested in energy transfer with time, we integrate the expression for power

$$w = \int p \, dt$$

and we see that the particular example forms a curve such as that of Figure 1-20, starting at some arbitrary level of time and energy. The curve is characterized by small perturbations around a steadily increasing base.

Note, however, that the time interval shown is only a few cycles of the power frequency. For time intervals of usual interest, the scale of the plot would be such that the small perturbations would be invisible. As a result, we are seldom interested in the exact details of a curve such as that in Figure 1-20, but we proceed on other ways of quantifying power and energy flow.

One possible rearrangement of Eq. (1-10) becomes*

$$p = EI \cos(\theta_E - \theta_I)[1 - \cos(2\omega t + 2\theta_I)] \quad \longleftarrow \quad A$$
$$+ EI \sin(\theta_E - \theta_I) \sin(2\omega t + 2\theta_I) \quad \longleftarrow \quad B \qquad (1\text{-}11)$$

Equation (1-11) appears rather more complex than its predecessor, but it does illustrate some interesting properties of the ac power function. For discussion, the two terms of the equation are identified by the letters A and B, and these parts are plotted separately in Figure 1-21 for the example of the previous figure.

We note from the plot (or inspection of the equation) that part A is unidirectional and as time progresses, the energy passed by the component p_A, $W_A = \int p_A \, dt$, continues to grow monotonically, whereas part B of the figure shows a variation that contributes no net energy flow over a period of time.

As indicated in the discussion of energy transfer, the detailed time variation over a period of one cycle is seldom of interest. The total energy transfer over a period of at least several cycles is most easily obtained by simply multiplying the *average* power by the time interval to find energy in joules, kilowatt-hours, or whatever units are convenient. It is the *average* power that determines the median slope of the curve of Figure 1-20.

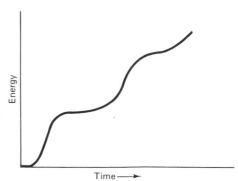

Figure 1-20   The transfer of energy with a power curve like Figure 1-19.

*The trigonometric work to obtain this form is tedious and will be left to the interested student as a study exercise in trig identities.

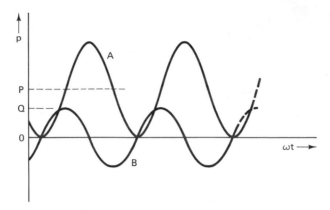

Figure 1-21   An example plot of Eq. (1-11) $p = 173.2[1 - \cos(2\omega t - 40°)] + 100 \sin(2\omega t - 40°)$.

Only part A of Eq. (1-11) contributes to the average, and we can see from inspection that the average is given by

$$P = EI \cos(\theta_E - \theta_I) \qquad (1\text{-}11a)$$

We see that the power involves the rms values of voltage and current. The product EI is known as the voltamperes (or apparent power) of the particular circuit or device and the factor $\cos(\theta_E - \theta_I)$ is known as the *power factor*.* We note that, for a given number of voltamperes, the average power is a maximum when $(\theta_E - \theta_I) = 0$ and is zero when $(\theta_E - \theta_I) = 90°$. The average power is positive when $(\theta_E - \theta_I)$ is in the first or fourth quadrant and negative when in the second or third quadrant. In some cases, we can be sure ahead of time that a certain device must absorb (or deliver) average power (consider a resistor, for example). In such a case, the voltage and current references might be chosen to give a positive number for power in the required direction. It is not necessary to do so, however, and we do well to form the habit of noting the reference direction for power from the voltage polarity and the current reference and then be prepared to accept and interpret negative numerical values for power if they occur.

Part B of Eq. (1-11) would seem to be of little significance, since the energy transferred during the positive loop flows back during the negative portion. We do find it convenient to give a name and a measure to this component, however. The maximum value of this component is named *reactive voltamperes* and is symbolized

---

*In considering electric power apparatus, we normally find that we are limited by the device to a certain maximum voltage magnitude and to a certain maximum current magnitude in order to avoid excessive heating. The rating of apparatus is therefore usually given in terms of the product EI or voltamperes. The power factor, $\cos(\theta_E - \theta_I)$, is determined by the load and does not normally have much effect on the rating of the electrical equipment used to supply the load. In power work, multiples of voltamperes are commonly encountered such as kilovoltamperes (kVA), megavoltamperes (MVA), or gigavoltamperes (GVA). To give some idea of orders of magnitude, it might be mentioned that the transformer supplying modern residences would be of the order of 25 kVA in rating, whereas a very large modern generator in a steam plant might have a rating of over one GVA.

$$ELI$$
$$\theta_E > \theta_I$$
$$\sin(\theta_E - \theta_I) > 0$$

$$Q = EI \sin(\theta_E - \theta_I) \qquad (1\text{-}11b)$$

where the product EI is again the voltamperes or apparent power and the term $\sin(\theta_E - \theta_I)$ is named *reactive factor*. We can see the physical significance of this quantity by considering the $R\text{-}L$ example circuit of Figure 1-5 once more. So long as the inductance L is non-zero, the angle $(\theta_E - \theta_I)$ is positive and the reactive voltamperes Q are non-zero. In fact, it may be shown that the integral of the component B in the figure for the duration of the positive loop is exactly the energy stored in the inductor at the maximum value of the current

$$L(I^m)^2/2 = LI^2$$

and conversely the integral during the negative loop corresponds to the withdrawal of that energy from the magnetic field of the inductor. Since many of the devices used in power work involve a magnetic field, the power engineer is accustomed to the fact that the apparatus will require the supply of a component of power like that of B in Eq. (1-11) and measured by the quantity Q of Eq. (1-11b). The supply of reactive voltamperes (or *reactive*, as called for short by some power engineers) becomes a requirement of almost equal importance to that of supplying power P of Eq. (1-11a).

If we evaluate Eq. (1-11b) for a capacitive circuit, then $(\theta_E - \theta_I)$ and Q are found to be numerically negative. A capacitor also requires a surge of energy in and out to charge the electric field, and this is reflected in the presence of a term such as that of B in the power equation (1-11). The time during the cycle of charging energy flow is different for a capacitor and an inductor, and results in reversed numerical sign for the reactive voltamperes Q of Eq. (1-11b).

If we pursue further the power engineer's viewpoint that Q is a quantity that must be supplied to the apparatus, then we sometimes take a different view of the role of the capacitor. Consider Figure 1-22. In part (a) of the figure, we indicate voltage and current reference directions in the sense of the usual choice for a passive device, and the reference direction for power is seen to be into the device. Since the current leads the voltage by $90°$ for part (a), the angle $(\theta_E - \theta_I)$ is $-90°$, P of Eq. (1-11a) is zero and Q of Eq. (1-11b) is numerically negative. The capacitor is seen to be *absorbing negative* reactive voltamperes in (a).

If, as in part (b) of Figure 1-22, we reverse the reference direction of power flow by reversing either the voltage or current reference, then the current lags the voltage by $90°$ and the angle $(\theta_E - \theta_I)$ becomes $+90°$, P is still zero but Q is numerically positive. Thus it may be said that a capacitor *supplies positive* reactive voltamperes.

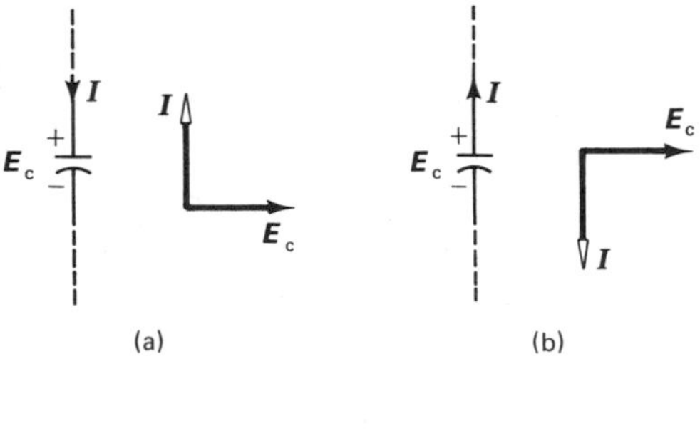

Figure 1-22 Reactive voltampere flow in a capacitor with: (a) power reference *into* capacitor, and (b) power reference *out of* capacitor.

(a)   (b)

Somewhere it must be noted that there is a certain arbitrariness in setting the trigonometric form of Eq. (1-11) and in naming the reactive voltamperes Q in Eq. (1-11b). Alternate choices might result in the angle $(\theta_I - \theta_E)$ as the argument of the cosine and sine functions of Eqs. (1-11a) and (1-11b). This choice would not change the power, P, but the reactive, Q, would be reversed in sign (reactive absorbed by an inductor would be negative and reactive absorbed by a capacitor would be positive). No harm would be done by this choice if everyone were consistent in the usage. To avoid confusion, however, the choice of $(\theta_E - \theta_I)$ in Eq. (1-11b) is the standard choice, resulting in the sign convention already given: an inductor absorbs positive reactive voltamperes, a capacitor absorbs negative reactive voltamperes (or delivers positive).

1.12 COMPLEX POWER

Since we often attach almost equal importance to power and reactive voltamperes it is convenient to use a method that carries them both through the computations together. This is done by defining a quantity called *complex power,* symbolized with the letter S and defined by the equation

$$S = P + jQ \qquad (1\text{-}12)$$

From the definition of P and Q and from Euler's theorem we have the alternate form for S

$$\begin{aligned} S &= EI \cos(\theta_E - \theta_I) + jEI \sin(\theta_E - \theta_I) \\ &= EI[\cos(\theta_E - \theta_I) + j \sin(\theta_E - \theta_I)] \\ &= EIe^{j(\theta_E - \theta_I)} \end{aligned}$$

When the voltage and current are given in terms of phasor description, it is tempting to assume that complex power must be equal to complex voltage, E, times complex current, I, but this is *not* so. Consider for example the exponential forms of E and I where the product EI gives $EIe^{j(\theta_E + \theta_I)}$ instead of $EIe^{j(\theta_E - \theta_I)}$ as shown above. Instead of EI we must use the product of the voltage phasor and the *conjugate* of the current phasor (the current phasor with the angle reversed is the conjugate) to give

$$S = EI* = EIe^{j(\theta_E - \theta_I)} \qquad (1\text{-}13)$$

If we consider the definition of complex power as in Eq. (1-12) we would note the following

$$P = \text{Re } S = \text{Re } EI* \qquad (1\text{-}14)$$

$$Q = \text{Im } S = \text{Im } EI* \qquad (1\text{-}15)$$

From the expression above we find the common usage of calling P the *real power,* but it is somewhat less common to call Q the *imaginary power. Reactive voltamperes* or *reactive power* are preferred names for this term.

Two special forms for the complex power are worthy of comment. Suppose we have the voltage across two points given as $E = ZI$ or, with Z in rectangular form, $E = (R + jX)I$ where R and X are the resistance and reactance seen at the two terminals looking into whatever circuit or device is present. The corresponding complex power is given by $S = EI^* = (R + jX)II^*$. II^* is simply I^2, the magnitude of the current squared and so $S = RI^2 + jXI^2$. In other words, the real power is given by RI^2 and the reactive voltamperes by XI^2. If, as in the case of the simple series R-L circuit of Figure 1-5, we can identify R and X with a specific resistor and inductor, respectively, then we see that the real power is consumed in the resistor and the reactive voltamperes are "consumed" in the inductor. For a more involved case, the R and X are simply the real and imaginary parts of the net impedance looking into the two terminals in question.

The second special case results from the admittance viewpoint. If we express the voltage-current relationship between two terminals in terms of admittance as $I = YE$ then complex power is developed as from

$$S = EI^* = E(YE)^* = EY^*E^* = EE^*Y^*$$
$$= E^2Y^* = E^2(G - jB) = E^2G - jE^2B$$

where E is simply the magnitude of the voltage phasor E. In this formulation the real power is given by E^2G and the reactive voltamperes by $-E^2B$.

It will be recalled that the susceptances of inductors and capacitors are given respectively by

$$B_L = -1/\omega L \; \geq \; \left| \frac{1}{z_L} \right|$$
$$B_C = \omega C \; = \; \left| \frac{1}{z_C} \right|$$

and, as a result, the expression $Q = -E^2B$ gives the correct sign for the reactive voltamperes absorbed by an inductor or capacitor.

1.13 CONSERVATION OF COMPLEX POWER

The principle of conservation of energy is surely one of the most basic postulates of science and engineering. A corollary in circuit theory requires that the power passing into a certain region in a circuit must be accounted for in the region by virtue of power being dissipated or converted, going into storage of energy, or being passed on to other regions. It is perhaps not quite so clear that reactive voltamperes must be accounted for in a similar fashion. To illustrate, consider the case of Figure 1-23,

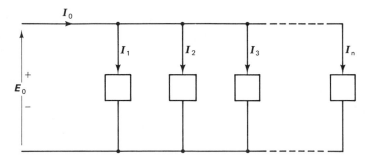

Figure 1-23 Illustration of conservation of complex power.

where a large number of "black boxes" are being fed from a common voltage E_0. The complex power being supplied at E_0 is given by $S_0 = E_0 I_0^*$. From Kirchhoff's current law we have

$$I_0 = I_1 + I_2 + I_3 + \cdots + I_n$$

but it must then follow from complex arithmetic that

$$I_0^* = I_1^* + I_2^* + I_3^* + \cdots + I_n^*$$

and complex power is then, in terms of the component currents

$$S_0 = E_0 I_1^* + E_0 I_2^* + E_0 I_3^* + \cdots + E_0 I_n^*$$
$$= S_1 + S_2 + S_3 + \cdots + S_n$$

From this last equation we see that not only real power must be accounted for in this type of situation but also reactive voltamperes, since equality of complex numbers requires that *both* the real *and* the imaginary parts must be separately equal.

By a similar method, we may treat the case of a series loop of many elements using Kirchhoff's voltage law and show the same principle at work, or indeed, by bits and pieces, conclude that any combination of elements will obey the principle of accountability of power in the complex number sense.

Let us illustrate some of the principles of complex power in terms of a specific numerical example. The circuit and element values of Figure 1-12 will serve this purpose without becoming too involved. In that case it was shown that, with a voltage $E_0 = 100\underline{/0°}$, the current $I_0 = 9.83\underline{/-51.59°}$ and therefore

$$S_0 = (100\underline{/0°})(9.83\underline{/+51.59°}) = 983\underline{/51.59°} = 611 + j770 \text{ voltamperes}$$

from which $P_0 = 611$ Watts and $Q_0 = 770$ vars.

We easily find that for this circuit $I_C = 1.51\underline{/90°}$ and $I_L = 11.05\underline{/-56.45°}$. Since $X_C = -10^6/(377 \times 40) = -66.3$ ohms and $X_L = 377 \times 0.02 = 7.54$ ohms, we have $Q_C = (1.51)^2(-66.3) = -150.8$ vars and $Q_L = (11.05)^2 7.54 = 920.7$ vars. From the idea of conservation of reactive, the total reactive into the circuit is

$$Q_0 = Q_C + Q_L = -150.8 + 920.7 = 769.9$$

as found from $Q_0 = \text{Im } E_0 I_0^*$.

It may be helpful to follow another sequence of computation to illustrate how the complex power concept gives us an alternate pathway through the mathematics in some cases. Suppose that we have computed the total S_0 as $611 + j770$ and wish to examine the internal details of the circuit. By examination of the circuit we see that there is only one place for real power to go — into the resistor — and this leads to an equation $611 = I_R{}^2 5$, from which $I_R = 11$ amperes. The same current flows through the inductor and hence $Q_L = 11^2 X_L = 11^2 2\pi f L = 921$ vars. The total vars into the circuit are only 770 and must be equal to the sum of the vars absorbed by the inductor and the capacitor; that is, $770 = 921 + Q_C$ and $Q_C = -151$ vars, and from $Q_C = I_C{}^2 X_C$ we find $I_C = 1.51$ amperes.

1.14 POWER FACTOR CORRECTION

Complex power computations are so important that it will be well to illustrate the application of the principles with yet another example. The example to follow illustrates one of the classical practical problems involved in electric power supply by alternating current; it also illustrates graphically the representation known as the *power triangle*.

Since S is a complex number, the rectangular components P and Q can be shown in Cartesian coordinates in a diagram analogous to the phasor diagrams of voltage and current. In Figure 1-24 we show a *power triangle*. If there are a large number of P and Q components we may show their combination graphically in a complex power diagram just as we show the components of voltage and current in a phasor diagram. We seldom make a completely graphic solution from the phasor diagram of voltage and current, and we also seldom make an actual scale graphic solution using the power diagram, but sketching such a diagram does help to clarify the relations and perhaps catch a blunder in the computations made with our calculators or computers.

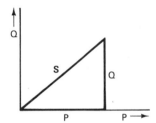

Figure 1-24   A power triangle.

From Eq. (1-11a) it is seen that a given amount of electric power can be supplied in various ways from a given voltage source E. We can supply a given number of watts with a relatively high current and low power factor, or with a smaller current and a high power factor. The minimum current at any given voltage will be drawn when the power factor is unity. Because the conductors or other apparatus supplying a load must be large enough to handle the current without overheating, a greater capital investment is required on the part of the energy supplier to supply a given number of watts at a low power factor than would be required at a higher power factor. Utility rates for large-scale energy users reflect this difference in capital investment cost by some form or other of rate penalty for drawing power at a lower power factor. It is advantageous for the user to take action to raise the power factor of loads to save on the energy bill at the end of the month.*

Figure 1-25 shows symbolically the loads on a small plant supplied from a 480-volt source. There are 50 kW of lighting load, which may be assumed to have a power factor of unity since the load is highly resistive. There are also 200 kW of

*Other advantages accrue, such as better voltage control, but such factors will not be discussed at this point.

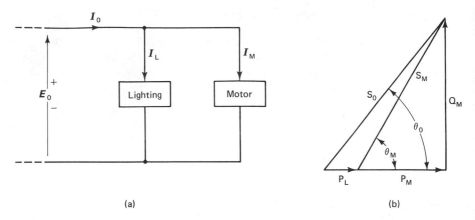

Figure 1-25   Loads on sample plant and power diagrams.

induction motor load with a power factor of 0.6, current lagging; that is, the motors draw positive vars to supply the magnetic fields within the devices. Because of a power factor penalty it is desired to improve the power factor to 0.8, current lagging.

It is convenient in problems of this nature to sketch the components of complex power in a power diagram such as (b) of the figure. The lighting load has a Q of zero and is shown by the horizontal line P_L. The motor load has a real part P_M and a reactive part Q_M, forming the smaller triangle in the figure. We know that $\cos \theta_M = 0.6$ and therefore $Q_M = 200 \tan(\cos^{-1} 0.6) = 267$ vars. The total complex power is

$$S_0 = S_L + S_M = 50 + j0 + 200 + j267 = 250 + j267 = 365\underline{/46.85°}$$

We note that, since $\cos 46.85° = 0.68$, the total load does not meet the specified power factor of 0.8.

We could conceivably improve the power factor to 0.8 by changing the motors to another type to be discussed later in this work, but this might not be economical. The more usual method of power factor improvement is to add static capacitors as shown in Figure 1-26. The capacitors draw a complex power $S_C = 0 + jQ_c$ and the

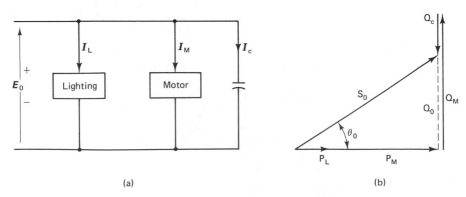

Figure 1-26   Capacitor added to loads of Figure 1-25 to improve power factor.

graphic summation of the three complex power components of the load is sketched
in part (b) of the figure. Note that Q_c is numerically negative. By adding Q_C, it is
possible to reduce the total Q_0 to a value such that the power factor is 0.8 as desired.
In particular if cos $\theta_0 = 0.8$, then

$$Q_0 = (P_L + P_M) \tan(\cos^{-1} 0.8) = 187.5 \text{ kvars}; \quad S_0 = 250 + j187.5 \text{ kVA}.$$

From the summation of the complex power at the input we have $S_0 = S_L + S_M + S_C$
or $250 + j187.5 = (50 + j0) + (200 + j267) + (0 + jQ_C)$ from which $Q_C =$
-79.5 kvars and thus the problem is solved.

In actual practice the kvars of capacitors would be matched to the nearest
commercial size available. Then we would compute the installed cost and compare
with the savings resulting from improved power factor to decide whether to proceed
with the change.

Last, it might be pointed out that we could take the alternate point of view in
this situation — that the capacitors act as a *source* of positive reactive voltamperes
and supply part of the motor magnetizing energy locally rather than buying from the
energy supplier.

1.15 MEASUREMENTS IN AC POWER CIRCUITS

In dealing with power circuits in the field or in the laboratory, we commonly wish
to measure voltage and current and other circuit quantities. It is assumed that the
student is familiar with voltmeters and ammeters used for this purpose. Perhaps the
only remarks necessary in this connection involve the fact that we usually wish to
measure the root-mean-square values of voltage and current in dealing with power
problems. Some classes of instruments inherently respond to the rms values, the
electrodynamometer element is one. That type of instrument involves two coils
arranged physically so that a torque is developed proportional to the product of the
currents in the two coils. If the same current flows in each coil, the torque is
proportional to the current squared and the mechanical inertia of the moving coil
averages the torque to produce a deflection of the indicator. The scale is calibrated
corresponding to the root-mean-square value. This type of construction is used in the
voltmeter, the ammeter, and the wattmeter (described in this section). Other types
are encountered, including the iron vane type, where torque developed is proportional to the square of the current in actuating coil.

It may be well to comment that the common multipurpose volt-ohm-
milliammeter type of instrument is often built with a dc type of movement, and
the movement is fed with rectified ac current when used as an ac voltmeter.
The response of the movement is proportional to the average of the rectified wave
form. The scale of such instruments is usually marked in terms of rms quantities
by multiplying the rectified average value by the *form factor* of the sine wave,
which is 1.11.* The instrument reads rms values correctly when the wave shape
is a sinusoid, but it may give an incorrect reading for other wave forms since it

---

*Recall that the *form factor* of a given wave is the ratio of the rms value to the rectified average.

will always read 1.11 times the rectified average, which may or may not be the true rms value.

To determine power, one could measure the voltage and the current and take the product times cos θ, the power factor, if the power factor were known. One example of a known power factor might be a purely resistive circuit, such as a water heater, where we can assume the power factor to be unity. For nonunity power factor circuits, we sometimes encounter the use of a power factor meter or a phase angle meter, based on various principles. It is more common, however, to use a *wattmeter* to measure the power directly, after which measurement (together with voltage and current) we may work backward to compute the power factor if we wish.

To measure power directly; that is, to form a *wattmeter,* requires in some fashion that we take the product of voltage and current. One method would be as illustrated in Figure 1-27 where we supply signals proportional to inputs, e and i.* Since the voltage output of the multiplier will be a displaced sinusoid similar to that of Figure 1-19, we use the simple R-C filter to derive an almost constant voltage proportional to the average power. This voltage may then be displayed by a digital voltmeter for a laboratory apparatus, or used to control a process or some function for which a reading of average power is desired.

The electrodynamometer wattmeter is an instrument built with two coils, a voltage coil and current coil, so disposed that a torque is developed on a movable element proportional to the product of the voltage and current at every instant. If the voltage and current are constant dc, the wattmeter indicator takes a position corresponding to the power inferred by the product EI. If the voltage and current are ac quantities of the same frequency, the torque developed varies with time the same as the instantaneous power. Because of the mechanical inertia of the moving element the wattmeter mechanically averages the time-varying torque and gives an indication of the average power, which is what is ordinarily wanted.

Depending on the winding directions of the coils and the connection to the circuit, the wattmeter average torque may be positive or negative. To interpret a positive or negative reading of the wattmeter, we must refer to polarity marks that

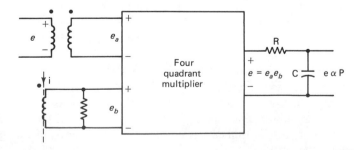

Figure 1-27   A device to measure average power.

*The input signals are shown in the figure as derived from devices known as *voltage transformers* and *current transformers,* which connect to points in the system at which the measurement is desired. Transformers will be treated in Chapter 3 and will not be discussed at this point.

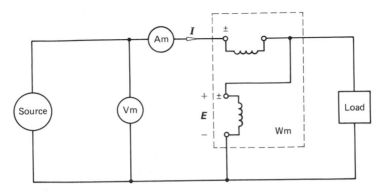

Figure 1-28   Use of instruments to measure a load.

the manufacturer places on the terminals of the voltage (potential) coil and the current coil. Figure 1-28 illustrates the connection of a wattmeter, a voltmeter, and an ammeter to measure a load supplied from a given source. Note the signs ($\pm$) on the wattmeter terminals. The manufacturer marks these terminals in such a way that the wattmeter reads the power flowing from left to right or into the load with the connection shown. If we phrase this statement more generally we would say:

> The wattmeter reads the product of the voltage across its potential coil times the current in its current coil times the cosine of the phase angle between them when the voltage and current phasors are expressed in terms of the references shown with respect to the polarity marks on the wattmeter.

The above statement may seem cumbersome, but be sure it is understood, since sometimes wattmeters are used in connection where the reading cannot be directly identified with a certain power component and we must still interpret the reading.

The watthour meter, which measures energy, is constructed somewhat similarly to the wattmeter in that it has a voltage coil and a current coil. The movable element is an aluminum disc that rotates on an axle. The voltage and current coils act together to produce a torque proportional to the product of the two. The disc is restrained by a permanent magnet so that the disc rotates at a speed proportional to the average torque (control engineers call this kind of restraint a *viscous damping*). The total angular travel of the disc is then a measure of energy or

$$w = \int p\, dt \quad \text{since} \quad \theta = \int \omega\, dt \quad \text{and} \quad \omega\, \alpha\, p$$

This total angular travel may be indicated on a dial or telemetered back to a central accounting office of the energy supplier.

ADDITIONAL READING MATERIAL

1. Brown, D., and E. P. Hamilton III, *Electromechanical Energy Conversion,* New York: MacMillan Publishing Company, 1984.
2. Chapman, S. J., *Electric Machinery Fundamentals,* New York: McGraw-Hill Book Company, 1985.

3. Chaston, A. N., *Electric Machinery*, Reston, Virginia: Reston Publications Company, Inc., 1986.

4. Elgerd, O. I., *Basic Electric Power Engineering*, Reading, Massachusetts: Addison-Wesley Publishing Company, 1977.

5. Elgerd, O. I., *Electric Energy Systems Theory: An Introduction*, New York: McGraw-Hill Book Company, 1982.

6. Gross, C. A., *Power System Analysis*, New York: John Wiley & Sons, 1979.

7. Lindsay, J. F., and M. H. Rashid, *Electromechanics and Electric Machinery*, Englewood Cliffs, New Jersey: Prentice-Hall, Inc., 1986.

8. Nasar, S. A., *Electric Energy Conversion and Transmission*, New York: MacMillan Publishing Company, 1985.

9. Shultz, R. D., and R. A. Smith, *Introduction to Electric Power Engineering*, New York: Harper & Row, Publishers, 1985.

10. Stevenson, W. D., Jr., *Elements of Power System Analysis*, New York: McGraw Hill-Book Company, 1982.

11. Weeks, W. L., *Transmission and Distribution of Electrical Energy*, New York: Harper & Row, Publishers, 1981.

STUDY EXERCISES

SECTION 1. The following exercises are basically practice problems in complex number arithmetic. It is assumed that as a minimum computational aide the student has available a calculator (perhaps programmable) with trig functions built in. More sophisticated calculators will make the problems easier, but in any case the student is well advised to master his or her particular calculator as used for this class of problem.

The exercises are framed in terms of common circuit cases such as series, parallel, or series-parallel impedances, and cases of voltage or current division. It is assumed that the student is familiar with these operations, but if the formulation of the solution as given is not clear, then some review of these methods may be in order.

The following data apply to all the problems of Section 1.

$$Z_a = 4 - j2 \qquad E_x = 150 \underline{/-20°}$$
$$Z_b = 15 \underline{/40°} \qquad E_y = -20 + j80$$
$$Y_c = 0.5 + j0.3 \qquad I_r = 20 \underline{/49°}$$
$$Y_d = 0.4 \underline{/-110°} \qquad I_s = 10 - j5$$

1.

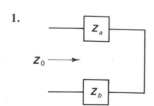

$$Z_0 = Z_a + Z_b = ?$$
$$R_0 = \text{Re } Z_0 = ?$$
$$X_0 = \text{Im } Z_0 = ?$$

2.

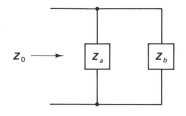

$$Z_0 = \frac{Z_a Z_b}{Z_a + Z_b} = ?$$
$$R_0 = \text{Re } Z_0 = ?$$
$$X_0 = \text{Im } Z_0 = ?$$

3.

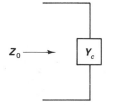

$$Z_0 = 1/Y_c = ?$$
$$R_0 = \text{Re } Z_0 = ?$$
$$X_0 = \text{Im } Z_0 = ?$$

4.

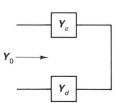

$$Y_0 = \frac{Y_c Y_d}{Y_c + Y_d} = ?$$
$$G_0 = \text{Re } Y_0 = ?$$
$$B_0 = \text{Im } Y_0 = ?$$

5.

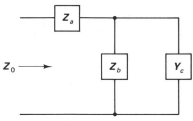

$$Z_0 = Z_a + \frac{Z_b/Y_c}{Z_b + 1/Y_c} = ?$$
$$R_0 = \text{Re } Z_0 = ?$$
$$X_0 = \text{Im } Z_0 = ?$$

6.

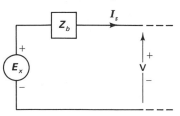

$$V = E_x - I_s Z_b = ?$$
$$S_v = VI_s^* = ?$$

7.

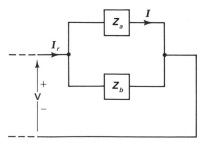

$$I = I_r Z_b/(Z_a + Z_b) = ?$$
$$V = I_r Z_a Z_b/(Z_a + Z_b) = ?$$
$$S = VI_r^* = ?$$

8.

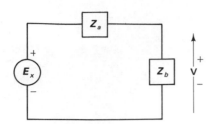

$$V = E_x \frac{Z_b}{Z_a + Z_b} = \text{?}$$

9.

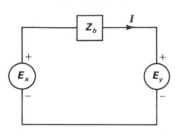

$$I = (E_x - E_y)/Z_b = \text{?}$$
$$S_x = E_x I* = \text{?}$$
$$S_y = E_y I* = \text{?}$$
$$S_x - S_y = \text{?}, \, I^2 R_b = \text{?}, \, I^2 X_b = \text{?}$$

10.

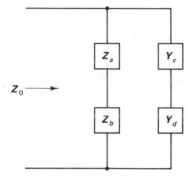

$$Z_0 = \frac{(Z_a + Z_b)(Y_c + Y_d)/Y_c Y_d}{(Z_a + Z_b) + (Y_c + Y_d)/Y_c Y_d} = \text{?}$$

SECTION 2. The following problems are slightly more involved in that the student must set up the equations for solution as well as do the arithmetic to find the numerical solution.

1. Suppose that there are three "black boxes" in the lab, each containing a battery, and each with a digital voltmeter connected to read the voltage which the battery establishes at the terminals of the box. The batteries are of random voltage and connected to the terminals of the boxes in random fashion such that the observed voltages are:

$$E_1 = -16 \text{ volts}, \qquad E_2 = 22 \text{ volts}, \qquad E_3 = -4 \text{ volts}$$

The boxes and voltmeters are sketched below, together with a dotted line indicating a series connection of the boxes to form the voltage E_{AB}.

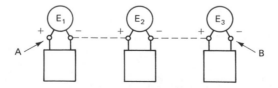

a. What is the algebraic expression for E_{AB} in terms of the component voltages, E_1, E_2, and E_3?
b. What is the numerical value for the voltage E_{AB} (magnitude *and* sign)?
c. Sketch a series connection of the boxes that will result in the maximum possible voltage for the series combination.

2. Suppose that we have been given two voltages as $E_1 = 15\underline{/30°}$ and $E_2 = 10 - j20$, as well as the information that $f = 60$ Hz.
   a. What is $e_1(t)$?
   b. What is $e_2(t)$?
   c. What is $e_1(t) + e_2(t)$ expressed as a single sinusoid?

3. An electrical engineer has arranged an ammeter and voltmeter as shown in the sketch at the right to monitor the battery in the automobile.
   a. Suppose on one nice day on the freeway I reads as 30 amperes, $V = 14$ volts. How much power is the battery *absorbing*?

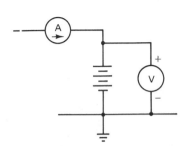

b. On a cold, rainy night $I = -15$ A, and $V = 12$ V, how much power is the battery *absorbing*?
c. By keeping records the engineer finds the readings of (a) one-fourth of the time and of (b) three-fourths of the time. What is the average power *absorbed* by the battery? Should the engineer do something about it?

4. Suppose we have a source of voltage $E = 100\underline{/0°}$ feeding the series combination of R and X, as shown in the sketch at the right. X is held at 4 ohms and R can be changed to values of -3, 0, or $+3$ ohms. Note that resistance is normally a positive number, but by special devices it is possible to form a unit that acts like a resistor of negative numerical value.
   a. What is the phasor current I for each value of R?
   b. What is the *input* power for each value of R?
   c. Sketch the phasor diagram of current and each voltage component for each value of R.

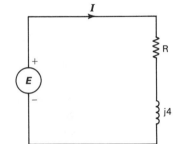

5.

For the circuit at the left it is given that $E_{ab} = 50\underline{/90°}$. Impedances of the four circuit elements are given on the diagram.
a. What are I_0, I_x, and I_y?
b. Sketch the phasor voltage, E_{ab} and the current phasors of (a), showing how the two currents add to give the total.
c. Sketch a topographic phasor diagram of the voltages like that of Figure 1-17 and from the sketch give the magnitude and phase of the voltage E_{cd}.
d. Find S_0, S_x and S_y.
e. Find the power factor in each branch and the overall power factor of the circuit.
f. Is $S_0 = S_x + S_y$? (Why or why not?)

6. On a certain 2400-volt circuit we have a motor load of 200 kW at 0.8 p.f., current lagging, also a lighting load of 50 kW at unity p.f., and 75 kvar of static capacitors.

    a. What is the total complex power being drawn from the source?

    b. What is the total current in amperes?

    c. What is the net power factor of the total load?

7. For the system of Problem 6, solve for the phasor currents in each element of the load using the voltage as the reference phasor ($E = 2400/\underline{0°}$). Sketch the phasor diagram of the voltage and the three load currents, showing how the three load currents add to give the total current.

SOLUTIONS TO STUDY EXERCISES

SECTION 1.

1. $Z_0 = 17.27/\underline{26.26°}$ $R_0 = 15.49$ $X_0 = 7.64$

2. $Z_0 = 3.88/\underline{-12.82°}$ $R_0 = 3.79$ $X_0 = -0.86$

3. $Z_0 = 1.71/\underline{-30.96°}$ $R_0 = 1.47$ $X_0 = -0.88$

4. $Y_0 = 0.63/\underline{-67.24°}$ $G_0 = 0.24$ $B_0 = -0.58$

5. $Z_0 = 6.12/\underline{-26.15°}$ $R_0 = 5.49$ $X_0 = -2.70$

6. $V = 92.95/\underline{-103.79°}$ $S_V = 229.72 - j1013.49$

7. $I = 17.37/\underline{62.74°}$ $V = 77.67/\underline{36.18°}$ $S = 1514.71 - j344.77$

8. $V = 130.26/\underline{-6.25°}$

    $= 129.48 - j14.20$

9. $I = 13.85/\underline{-79.21°}$ $S_x = 2077.18/\underline{59.21°}$ $S_y = 1141.92/\underline{183.24°}$

    $S_x - S_y = 2876.44/\underline{40.00°}$ $I^2R = 2203.48$ $I^2X = 1848.94$

10. $Z_0 = 1.49/\underline{64.00°}$

     $= 0.65 + j1.33$

SECTION 2.

1. a. $E_{AB} = E_1 + E_2 + E_3$

   b. $E_{AB} = -16 + 22 + (-4) = +2V$

   c. We wish to have $E_{AB} = +16 + 22 + 4 = +42V$

      In algebraic form, then, $E_{AB} = -E_1 + E_2 - E_3$

      This will be accomplished by the altered connection:

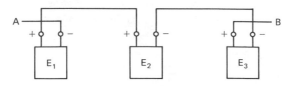

2. a. $e_1(t) = \text{Im } \sqrt{2}\,E_1\varepsilon^{j\omega t} = \text{Im } \sqrt{2} \times 15\varepsilon^{j30°}\varepsilon^{j\omega t}$

      $= \text{Im } \sqrt{2} \times 15\varepsilon^{j(\omega t + 30)}$

      $= 21.21 \sin(\omega t + 30)$ where $\omega = 2\pi 60 = 377$

b. $e_2(t) = \text{Im } \sqrt{2}\,(10 - j20)\varepsilon^{j\omega t}$
$\qquad = \text{Im } \sqrt{2} \times 22.36\varepsilon^{-j63.4}\varepsilon^{j\omega t}$
$\qquad = \text{Im } 31.6\varepsilon^{j(\omega t - 63.4°)}$
$\qquad = 31.62 \sin(\omega t - 63.4°)$
$E_1 + E_2 = 15\underline{/30°} + (10 - j20) = 26.17\underline{/-28.53°}$
$e_1(t) + e_2(t) = \text{Im } \sqrt{2} \times 26.16\varepsilon^{-j28.53}\varepsilon^{j\omega t}$
$\qquad\qquad\qquad = 37.01 \sin(\omega t - 28.53°)$

Note that some may prefer to interpret the phasors in terms of the real part of the complex expression; that is, in terms of a *cosine* function.

3. a. $P_{\text{absorbed}} = +vi = 14 \times 30 = 420$ W
   b. $P_{\text{absorbed}} = +vi = 12 \times (-15) = -180$ W
      the battery is *delivering* $+180$ W
   c. $P_{\text{average}} = (P_a t_a + P_b t_b)/(t_a + t_b)$
      $\qquad\qquad = 420 \times \frac{1}{4} + (-180)\frac{3}{4}/1 = -30$ W

   Again, this amounts to 30 watts *delivered* by the battery, so the engineer had better set up the charging rate of the car's alternator or stay home on rainy nights!

4. Case 1: $R = -3,\ Z = -3 + j4$
   $\qquad I = E/Z = (100\underline{/0})/(-3 + j4) = 20\underline{/-126.87°}$
   $\qquad P_{in} = \text{Re } EI^* = \text{Re } 100\underline{/0} \times 20\underline{/+126.87} = \text{Re}(-1200 + j1600)$
   $\qquad\quad = -1200$ W

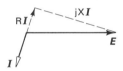

Case 2: $R = 0,\ Z = 0 + j4$
   $\qquad I = E/Z = (100\underline{/0})/(0 + j4) = 25\underline{/-90}$
   $\qquad P_{in} = \text{Re } EI^* = \text{Re } 100\underline{/0} \times 25\underline{/+90} = \text{Re}(0 + j2500)$
   $\qquad\quad = 0$

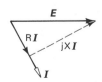

Case 3: $R = +3,\ Z = 3 + j4$
   $\qquad I = E/Z = (100\underline{/0})/(3 + j4) = 20\underline{/-53.13}$
   $\qquad P = \text{Re } EI^* = \text{Re } 100\underline{/0} \times 20\underline{/+53.13} = \text{Re}(1200 + j1600)$
   $\qquad\quad = 1200$ W

In summary, note that in Case 1 the circuit is *delivering* positive P back to the source. In Case 2 there is no average power exchange. In Case 3 the circuit is *receiving* positive average power from the source — the most common case.

5. a. $I_x = (50\underline{/90})/(100 - j100) = -0.25 + j0.25 = 0.354\underline{/135}$

   $I_y = (50\underline{/90})/(50 - j50) = -0.50 + j0.50 = 0.707\underline{/135}$

   $I_0 = I_x + I_y = -0.75 + j0.75 = 1.06\underline{/135}$

b.

c.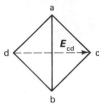

By inspection, $E_{cd} = 50\underline{/0}$

d. $S_x = E_{ab}I_x^* = 50\underline{/90} \times 0.354\underline{/-135} = 12.5 - j12.5$

   $S_y = E_{ab}I_y^* = 50\underline{/90} \times 0.707\underline{/-135} = 25.0 - j25.0$

   $S_0 = E_{ab}I_0^* = 50\underline{/90} \times 1.06\underline{/-135} = 37.5 - j37.5$

e. Power factor $= \cos(\theta_E - \theta_I) = \cos(90 - 135) = 0.707$ (same for all components)

f. $S_x + S_y = 37.5 + j37.5 = S_0$, so answer is *yes*.

6. $S_{\text{motor}} = 200 + j200 \tan(\cos^{-1}0.8)$

   $\quad\quad\quad = 200 + j150$

   $S_{\text{lghtg}} = \quad 50 + j0$

   $\underline{S_{\text{cap}} \quad = \quad\quad 0 - j75}$

   $S_{\text{tot}} \quad = 250 + j75 = 261.0\underline{/16.7}$

   $I = 261.0 \times 10^3/2400 = 108.75\ A$

   Power factor $= \cos 16.7° = 0.96$

7. $I \quad = S^*/E^*$

   $I_{\text{motor}} = (200 - j150) \times 10^3/2400\underline{/0} = 83.33 - j62.50$

   $\quad\quad\quad\quad\quad\quad\quad\quad\quad\quad\quad\quad = 104.17\underline{/-36.87}$

   $I_{\text{lghtg}} = (50 + j0) \times 10^3/2400\underline{/0} \quad = 20.83 + j0$

   $\quad\quad\quad\quad\quad\quad\quad\quad\quad\quad\quad\quad = 20.83\underline{/0}$

   $I_{\text{cap}} \quad = (0 + j75) \times 10^3/2400\underline{/0} \quad = 0 + j31.25$

   $\quad\quad\quad\quad\quad\quad\quad\quad\quad\quad\quad\quad = 31.25\underline{/90}$

   $I_0 \quad = (250 - j75) \times 10^3/2400\underline{/0} = 104.16 - j31.25$

   $\quad\quad\quad\quad\quad\quad\quad\quad\quad\quad\quad\quad = 108.75\underline{/-16.7}$

Checking: $I_0 = \Sigma I = 104.16 - j31.25 = 108.75\underline{/-16.7}$

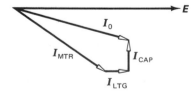

HOMEWORK PROBLEMS

1. An engineer is sent out into the field to check the phase relations of the voltages on a piece of the company's apparatus. He takes a voltmeter with him. The apparatus has six leads protruding from the case and the engineer takes voltage readings between various pairs of leads as reported in the table below.

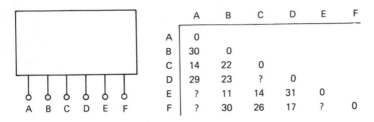

	A	B	C	D	E	F
A	0					
B	30	0				
C	14	22	0			
D	29	23	?	0		
E	?	11	14	31	0	
F	?	30	26	17	?	0

He then borrows a "scope" and finds that E_{CA} leads E_{BA}.

a. Using E_{BA} as a reference phasor ($E_{BA} = 30\underline{/0°}$), sketch the phasor diagram of all the voltages in the form of a topographic potential map as shown in the chapter.

b. What are the voltages that are missing in the table?

2.

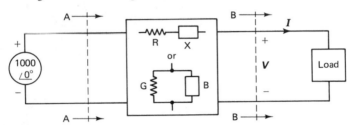

In the figure above a 1000-volt source feeds a load through a network of some kind in the box. Complex power is measured at positions A-A and B-B with the result:

$$S_{A\text{-}A} = 100 + j80 \text{ kVA}$$
$$S_{B\text{-}B} = 95 + j70 \text{ kVA}$$

a. If the elements in the box form a series impedance, as in the figure:
   i. What are the R and X in the box?
   ii. What are the voltage and current leaving the box; that is, at position B-B? (Assume that the 1000-volt source is at an angle zero—the reference phasor.)

b. If the elements in the box form a shunt admittance as in the lower part of the figure:
   i. What are the G and B in the box?
   ii. What are the voltage and current leaving the box, at position B-B?

3.

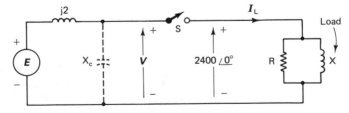

The circuit above shows a small system where a voltage E feeds a load of 150 kVA through an impedance of $j2$ ohms. The power factor of the load is 0.6, current lagging. Two situations are to be studied, with and without a capacitive reactance X_c as shown. In each case the voltage of the source is to be adjusted to result in 2400 volts across the load with the switch S closed.

a. With the switch closed (and no capacitor) and the voltage across the load as $2400\underline{/0}$, the reference:

   i. What is I_L of the load?

   ii. What are the R and X equivalents of the load (note the parallel connection)?

   iii. What must the source voltage E be?

   iv. If the switch S is opened, what will be the voltage V at the supply side of the switch?

b. It is proposed to add a capacitive reactance as shown in dotted lines on the diagram in order to draw $Q_c = -100$ kvar, and thus improve the power factor. The voltage of the source E is to be readjusted in order to set the voltage across the load at $2400\underline{/0}$.

   i. What capacitive reactance is required?

   ii. What must be the source voltage, E?

   iii. If the switch is opened, what voltage, V, appears on the source side of the switch?

4.

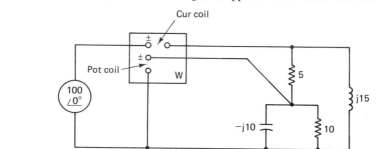

The diagram above shows a voltage source of 100 volts feeding a series-parallel combination of resistance and reactances. A technician, in trying to read the total power, has incorrectly connected the wattmeter W, as shown.

a. What will the wattmeter read in watts?

b. What is the actual total power delivered by the source?

5.

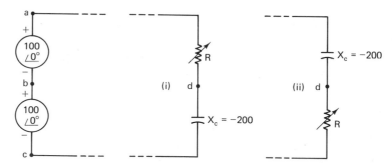

The circuit above is sometimes used to develop a voltage of variable phase angle for control of power electronic devices. The two 100-volt sources would usually be derived from transformer windings. Two possible arrangements of the load elements are shown as (i) and (ii) at the right. It is desired that the voltage E_{bd} lead the voltage E_{ac} by angles varying from 60° to 120°.

a. Which of the two R-C branches, (i) or (ii), should be used to accomplish the desired (leading) result?

b. What should be the range of the variable resistor R to cause the phase angle to vary from 60° to 120°?

Hint: A good phasor diagram will simplify the solution.

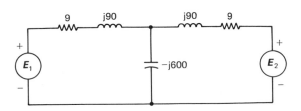

The diagram above represents one portion (phase) of a 300-mile transmission system carrying power from a generator shown as E_1 and a load shown as E_2. Circuit elements are given in ohms and form an equivalent circuit to represent the actual hardware of the system.

Suppose that we have $E_1 = 345\underline{/15°}$ and $E_2 = 340\underline{/0°}$, both in kilovolts (kV):

a. How much complex power, $S_1 = P_1 + jQ_1$ is supplied by E_1?

b. How much complex power, $S_2 = P_2 + jQ_2$ is supplied by E_2?

c. What is the efficiency (P_2/P_1) of the system?

7. a. Show by manipulation of the trig identities that Eq. (1-11) may be written in the alternate form:

$$p = EI \cos(\theta_E - \theta_I)[1 - \cos(2\omega t + 2\theta_E)] - EI \sin(\theta_E - \theta_I) \sin(2\omega t + 2\theta_E)$$

b. Show that the second term of the above equation may be interpreted as the instantaneous power surging into and out of a capacitor in a parallel G-C circuit such as that of Figure 1-11.

8. Suppose that the potential coil of a wattmeter like that of Figure 1-28 is fed through a phase-shifting network such that the voltage *lags* the applied circuit voltage by 90°.

a. Show that the wattmeter reading is proportional to Q, the reactive voltamperes.

b. Will current-lagging reactive voltamperes read positive or negative on the wattmeter scale?

Chapter 2

Three-Phase

Circuits

- A discussion of the generation of polyphase voltages.
- A study of circuit problems, solving for currents, voltages, and power primarily in balanced cases.
- Brief coverage of unbalanced cases.

2.1 GENERATION OF POLYPHASE VOLTAGES

The usual configuration of an ac generator (also called an *alternator*) involves a rotating magnetic field member called the *rotor* and a stationary magnetic ring called the *stator*. Figure 2-1 illustrates this construction in a highly symbolic elementary fashion. The rotor field is ordinarily an electromagnet fed from a dc source, although in some small machines the rotor field may be obtained from a permanent magnet. The stationary portion (or stator) of the machine is a steel ring, usually formed by means of a pile of laminations punched from steel sheets. There are slots on the inner periphery of the stator in which copper conductors are placed, connected in a multitude of coils. For simplicity, only one pair of slots and one turn of a coil is shown in the figure. As the field rotates, a voltage is generated in the coil or coils. The voltage wave is at least approximately a sinusoid by design.

In practice the entire inner periphery of the stator is filled with slots and coils. The coils are connected in groups, most often three separate groups called *phases*. Figure 2-2 shows an end view of such a machine where, again for simplicity on first exposure, each of the three groups is symbolized by a single pair of slots and one turn coil.*

In the figure, the original coil terminals are marked a and a', a second coil is marked b and b', and third coil is marked c and c'. The coil b-b' is placed 120° from the axis of a-a', and coil c-c' in turn is placed yet another 120° from the axis of coil b-b'. Each coil is threaded by the same flux from the rotor and each will have the same voltage generated per turn by the action of the rotor flux. For the counterclockwise rotation of the rotor field, the voltage in coil b-b' will be seen to lag that of a-a' by 120° and in turn that of coil c-c' will lag b-b' by another 120° or, in other words c-c' will lag a-a' by 240°. The three voltages are sketched as time functions

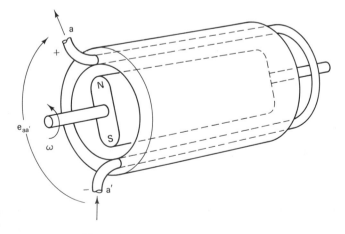

Figure 2-1   An elementary ac generator.

*Some further discussion of winding patterns — the grouping of multiturn coils — will be given in Chapter 5, where the ac generator is discussed.

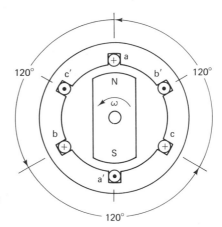

Figure 2-2 An elementary three-phase generator.

in Figure 2-3(a) and indicated symbolically by the phasor diagram of part (b) of the figure.

Each of the three coils (or coil groups) forms a voltage source, and leads from the coils are taken out of the machine for interconnection and use. One such common interconnection is shown in Figure 2-4, where the usual symbol for a voltage source is used and the three coils are connected in a Y configuration with a common point marked n (for neutral). The voltage $E_{aa'}$ thus becomes the voltage E_{an} in double-subscript notation.

It should be noted in the circuit diagram shown thus far that the voltage source symbol is oriented on the paper in the same way as the voltage phasor. It is not necessary to do this, but it is a common usage and helps to clarify our work and avoid confusion. It must be emphasized that a circuit diagram and a phasor diagram are two different things. Circuit diagram symbols can be oriented on the paper in any way that suits drafting convenience, and the location or the symbol need have no relation to the physical location of the device or a phasor description of its voltage or current. Because of this freedom of diagram orientation, however, we might as

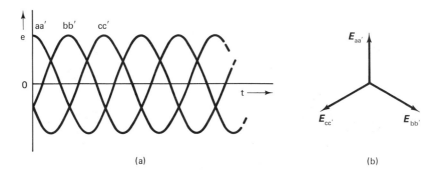

Figure 2-3 (a) Time plot of the three voltages of the machine in Figure 2-2 and, (b) a phasor diagram representation of the same voltages.

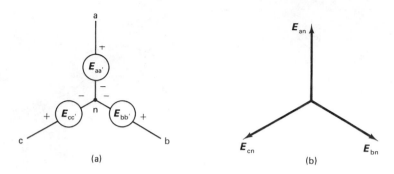

Figure 2-4 (a) Coils interconnected to form a three-phase Y-connected source. (b) Phasor diagram of the voltages.

well choose a layout that helps our understanding, and this will often be done in the work to follow.

2.2 BALANCED THREE-PHASE SYSTEMS: INTRODUCTION

A system of three-phase voltages symmetrically disposed as in Figure 2-4 is known as a *balanced* set. To be balanced, it is necessary that the three voltages be equal in magnitude and distributed evenly 120° apart around the 360° of angle space.* The orientation of the phasors as a whole is of course dependent on the arbitrary choice of $t = 0$, or as is sometimes said, "the reference phasor." Most generated voltages are balanced and only an error in winding or connecting the coils, or internal troubles in a generator, would result in unbalanced generated voltages.

The voltage magnitude of the three-phase source of Figure 2-4 could be described in two different ways. For example, a voltmeter connected from a to n would read one voltage whereas a voltmeter connected from a to b would read another. As designated above, the point n is called the *neutral* whereas the terminals a, b, and c connect to the load by means of conductors and are known as the *line* ends of the coils. In terms of this nomenclature, we note that we can speak of the *line-to-neutral* voltage *or* of the *line-to-line voltage*. The latter is sometimes called just the line voltage or the circuit voltage for short.† The relative magnitude of the two voltages is easily found from a phasor diagram as in Figure 2-5(a) where the voltages between a, b, and n are shown by a triangle with the vertices identified appropriately. The diagram is drawn according to the conventions outlined in Chap-

---

*For an n phase system to be balanced, the n voltages must also be equal in magnitude and distributed $360/n$ degrees apart; thus a six-phase system, for example, would be distributed with 60° between the phases. A two-phase system is an anomaly. Balanced two-phase voltages are 90° apart. Some people prefer to view this as half of a four-phase system!

†Sometimes one hears a three-phase system referred to as, for example, a "4160 volt system" without specification of line-to-line or line-to-neutral. In such cases it is generally understood that the speaker refers to line-to-line voltage, and the line-to-neutral voltage would therefore be 4160/$\sqrt{3}$ = 2400 volts.

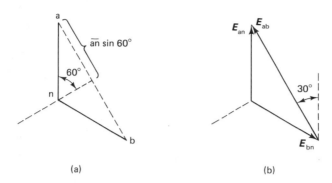

Figure 2-5   Relation between line-to-neutral and line-to-line voltage.

ter 1. By dropping a construction line from n perpendicular to $\overline{ab}$, we find the relation between the lengths $\overline{an}$ and $\overline{ab}$ as follows:

$$\overline{ab} = 2(\overline{an} \sin 60°) = \sqrt{3}\,\overline{an}$$

We now have the important relation that the line-to-line voltage of a balanced three-phase system is $\sqrt{3}$ times the line-to-neutral voltage. We also note a phase shift between any two voltages that may be compared. For example, if we wish to compare E_{an} and E_{ab}, we put arrowheads on the appropriate ends of the lines and read the fact that E_{ab} leads E_{an} by 30° from a sketch such as that of (b) above. Clearly any one of the line-to-neutral voltages could be compared to any one of the line-to-line voltages and result in the same ratio of magnitudes. The angular shift of the voltage phasors depends, however, on the specific choice of the voltages to be compared.

Now let us consider an impedance Z connected to each of the three source voltages as in Figure 2-6. So far as each impedance is concerned, the voltage is known and the current is found from the simple Ohm's law relation in each phase separately. For example, let us assume that the circuit voltage is 173.2 volts and the line-to-neutral voltage is therefore $173.2/\sqrt{3} = 100$ volts. The angle of the voltage phasors will be taken from the circuit diagram in accordance with common practice of using similar orientation for voltage symbols and voltage phasors. Thus

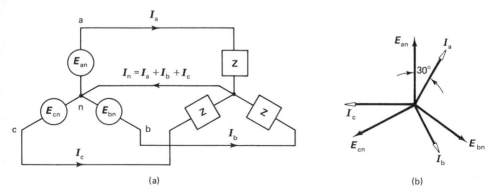

Figure 2-6   Balanced three-phase load.

$E_{an} = 100\underline{/90°}$, $E_{bn} = 100\underline{/-30°}$, and $E_{cn} = 100\underline{/210°}$. Suppose $Z = 5\underline{/30°}$ then the several currents are

$$I_a = E_{an}/Z = (100\underline{/90°})/(5\underline{/30°}) = 20\underline{/60°}$$
$$I_b = E_{bn}/Z = (100\underline{/-30°})/(5\underline{/30°}) = 20\underline{/-60°}$$
$$I_c = E_{cn}/Z = (100\underline{/210°})/(5\underline{/30°}) = 20\underline{/180°}$$

The voltages and currents are shown in a phasor diagram in part (b) of the figure. Since the three impedances are assumed equal in this example the currents are all equal in magnitude and lag their respective voltages by the same angle in the illustration of Figure 2-6(b). This angle is the negative of the angle of Z. The currents are thus seen to form a balanced set of phasors; that is, they are all equal in magnitude and spaced 120° from each other.

It might seem that this situation of exactly equal impedances in each of the three phases would be special, but it is actually the usual case. The opposite case of unbalance, where unequal impedances are placed in each of the phases, is unsatisfactory in the effect on rotating machines (generators, motors, etc.) and upon system economy. As a result, great care is taken by design engineers to distribute loads between the phases so that balanced currents are drawn from the sources. Small loads electrically far from the sources may nevertheless be unbalanced, and large unbalanced currents may flow when an overhead transmission line is struck by lightning, so we cannot ignore unbalanced cases altogether. We will, however, defer discussion of the unbalanced case for the time being in favor of discussion of the usual (balanced) case.

It is interesting to consider the current in the neutral wire, I_n. From Kirchhoff's current law at the neutral junction of the load impedances, $I_n = I_a + I_b + I_c$, but this sum may be seen to be zero either by examination of the symmetry of the phasor diagram or by mathematical addition of the phasor currents. Since the current is zero there need be no conductor to carry it! We therefore often find three-phase circuits with only three wires, the three "line" wires. In electrically examining each of the loads, we would not know whether a neutral wire was present — each of the phase currents finds its return path through the other two impedances.*

2.3 BALANCED THREE-PHASE SYSTEMS: PER-PHASE REPRESENTATION

The circuit of Figure 2-6 was indeed simple in that an Ohm's law equation sufficed for solution for the currents. A more practical situation might involve many impedances and even more than one three-phase set of source voltages. Such a case is shown in Figure 2-7, where a number of impedance elements and two separate three-phase sources are arranged in a network. It is still assumed that we have

*It should be emphasized that the conclusion that the neutral wire was not needed was based on the assumption of perfectly balanced loads, and implicitly, that we have perfect sinusoidal currents. If these conditions are not met, we need to consider the effect of the neutral wire, if present. Again this consideration is deferred.

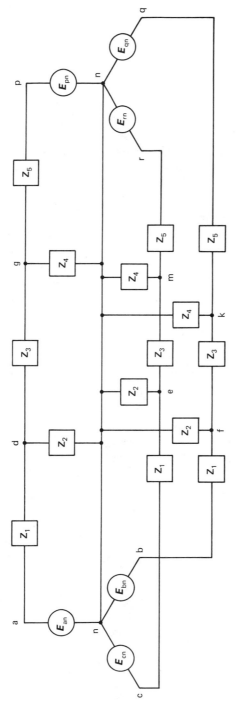

Figure 2-7 A more involved three-phase circuit.

symmetry; the voltage sources are balanced and impedances in corresponding positions in each of the phases are equal. We now have a situation where a simple Ohm's law method does not suffice, we must choose our unknowns by some convenient method such as loop currents or nodal voltages, and then write and solve simultaneous equations.

It is assumed that the source voltages and the impedances are known and it remains to find the currents everywhere or alternately the voltages of all the nodes, which in turn could lead to the currents. If we use nodal voltages as our unknowns with n as the reference node* then we would have equations such as

$$(V_{dn} - E_{an})/Z_1 + V_{dn}/Z_2 + (V_{dn} - V_{gn})/Z_3 = 0$$

which is written at node d. Similar equations may be written at each of the other five unknown nodes. We will spare the reader these details.

We note that the circuit and resulting equations, no matter how we treat them, are very cluttered even for this relatively simple three-phase circuit. We also note that treating the circuit in this manner results in unnecessary work. If, for example, we know I_{dg}, then from the symmetry of the balanced case we know that I_{fk} is the same, except that it lags I_{dg} by 120°. In turn, I_{em} is the same as I_{dg}, except that it lags I_{dg} by 240°. In short, if we know the voltages and/or currents of one of the three phases and have a balanced circuit then the other phase quantities may be found by a simple angular rotation if they are desired.

Based on the above considerations, we seldom bother to show or solve for all three phases separately in the balanced case. The circuit diagram for Figure 2-7 could be simplified as in Figure 2-8, and then this circuit solved by two nodal voltage equations (or three loop current equations if preferred). Clearly the labor and clutter are greatly reduced; even if a circuit solution program is available with a computer it is advantageous to reduce the amount of data read-in and output print-out. Treating a balanced three-phase circuit this way is said to be on a *per-phase* basis and is almost universally used for such computations. We could show any *one* of the phases, but phase a is the usual choice.

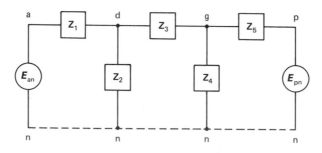

Figure 2-8   A per-phase representation of the circuit of Figure 2-6.

*It should be noted that the neutral conductor forms a giant node of equal potential through the length of the circuit. Since from symmetry we have balance and no neutral wire current, the potential of the various neutral points will not change if we remove the wire, and we can still use the concept of V_{dn}, V_{gn}, V_{kn}, etc., as though the neutral were there!

2.4 BALANCED THREE-PHASE SYSTEMS: DELTA-CONNECTED APPARATUS

If we revert for a moment's consideration to the three source voltages of the generator of Figure 2-2, we might note that there is another way of connecting the three coils symmetrically. The three coils could be connected sequentially in series with each other to form a closed loop called a *delta*.* This is illustrated in Figure 2-9. The generator coil, which was originally marked as aa', is now connected between points a and c, and this same voltage could then be described as E_{ac}. In a similar manner, $E_{bb'}$ becomes E_{ba}, and $E_{cc'}$ becomes E_{cb}. Note again that the orientation of the circuit symbols is the same as that of the source voltage phasors for ease in interpretation. The corners of the delta then become the line terminals and go on to the load.

The three equal impedances used as a load in Figure 2-6 can also be connected symmetrically in a delta configuration and thus form a balanced load. This is illustrated in Figure 2-10(a). We now find it necessary to distinguish between the line currents and the currents inside the separate legs of the delta. The phasor diagram of (b) of the figure should clarify this. Note that *if* we are dealing with the same components as before, the impedances have the same voltages across them and therefore the same currents through them. Kirchhoff's current law applies at the corners of the delta and therefore we can write

$$I_a = I_{ac} - I_{ba}, \quad I_b = I_{ba} - I_{cb}, \quad \text{and} \quad I_c = I_{cb} - I_{ac}$$

The line current I_a is then $\sqrt{3}$ times the magnitude of the leg current in the individual elements of the delta and is shifted in phase with respect to the leg currents, as may be seen from a phasor diagram like Figure 2-10(b). The geometry of the phasor difference is, of course, the same as that of Figure 2-5 and will not be repeated. In fact, we might say that the delta and Y connections are duals of each other. The line current's being greater than the leg current is a consequence of the fact that the line-to-line voltage for this particular combination of apparatus dis-

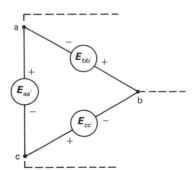

Figure 2-9   A delta-connected source.

*The reader might do well to contemplate the dangerous consequences of making a mistake of polarity and reversing one of the source coils in closing the loop! Note that, if connected correctly, a voltmeter check across an open corner of the delta should give zero because the three voltages add to zero in a phasor sense and also therefore at every instant of time (for pure sinusoidal voltages).

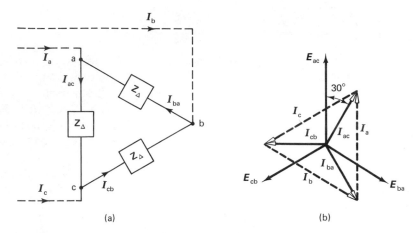

Figure 2-10   A delta-connected load.

cussed is less by a factor of $\sqrt{3}$ for the delta-connected source than for the same source coils connected in Y. In a practical sense, for understanding, it might be noted that each impedance has the original voltage across it and so consumes the same power, hence if the voltage is less by a factor $\sqrt{3}$, the current must be greater by the same factor! Although the introduction to delta connection was presented in terms of a reconnection of items from the Y connection first discussed to a delta connection for both source voltages and impedance loads, we may have all sorts of combinations of delta and Y apparatus. We may have a Y-connected source and a delta load or vice versa. We may have a combination of Y *and* delta loads. We may have a complex interconnection of many impedance elements, but if we have the symmetry necessary for a balanced load any such three terminal load network can easily be replaced with an equivalent Y or delta.

    To review the transformation from Y to delta or from delta to Y, refer to Figure 2-11. It may be shown that the two networks are equivalent if the elements bear the following relations:

$$Z_a = \frac{Z_{ab}Z_{ca}}{Z_{ab} + Z_{bc} + Z_{ca}} \qquad (2\text{-}1)$$

or

$$Z_{bc} = \frac{Z_aZ_b + Z_bZ_c + Z_cZ_a}{Z_a} \qquad (2\text{-}2)$$

as given for sample elements of the two forms. The other elements are obtained from similar expressions by merely permuting the subscripts.

    The use of Eqs. (2-1) and (2-2) is very laborious unless computer assistance is readily at hand (or at least a programmable calculator), but for the balanced case the expressions become so simple that their use can be recommended for frequent application in treating three-phase networks. Suppose that each element of the Y connection has an impedance Z_Y and each element of the delta has an impedance Z_Δ. Eqs. (2-1) and (2-2) reduce quickly to the following

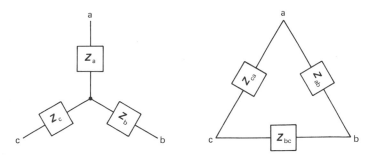

Figure 2-11   Equivalent Y and delta networks.

$$Z_Y = Z_\Delta/3 \qquad\qquad (2\text{-}3)$$
$$Z_\Delta = 3Z_Y \qquad\qquad (2\text{-}4)$$

The simple scale factor of 3 is easily remembered and quickly applied.

We can now return to the solution of three-phase circuits involving a combination of delta- and Y-connected apparatus. Let us approach this in terms of a specific example. Suppose we have three voltage sources of a three-phase generator with a voltage of 300 volts each, and the sources are connected in delta. For a load, let us consider a Y-connected group of resistors of 45 ohms each and a delta-connected group of impedances of $60\underline{/30°}$ ohms. The loads are connected to the source as shown in the circuit diagram of Figure 2-12. To solve the circuit we could of course set up a loop currents or nodal voltage unknowns and write simultaneous equations, but we can better exploit the symmetry of a balanced circuit by proceeding as follows. We convert the delta group of impedances to Y by simply dividing by three to obtain $Z_Y' = 20\underline{/30°}$. The source voltage gives us 300 volts line-to-line since it is delta connected. A Y-connected group of sources of $300/\sqrt{3} = 173.2$ volts would give us the same line voltage, so we can draw the

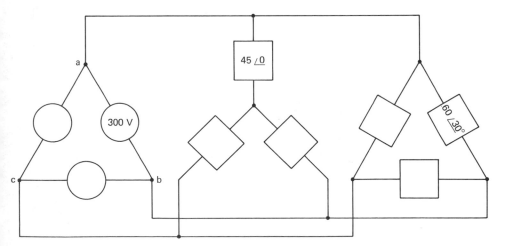

Figure 2-12   Three-phase network example.

equivalent network of Figure 2-13. Last, we draw the per-phase equivalent circuit of Figure 2-14 in preparation for solution.* If we take the voltage phasor to be oriented the same as the circuit diagram symbol we find for I_a, for example

$$I_a = \frac{173.2\underline{/90°}}{\left[\dfrac{Z_Y Z'_Y}{Z_Y + Z'_Y}\right]}$$

$$= \frac{173.2\underline{/90°}}{14.26\underline{/20.88}} = 12.15\underline{/69.12}$$

We now generalize from the lesson of this particular example and in summary make a recommendation for handling of balanced three-phase circuit problems.

1. Consider the source delta or Y connection to find the line-to-line and line-to-neutral voltage from a knowledge of the voltage of each coil of the source generator. Indicate the source in a per-phase diagram as a leg of an equivalent Y-connected generator. Usually phase a is chosen as the sample or "reference" phase.

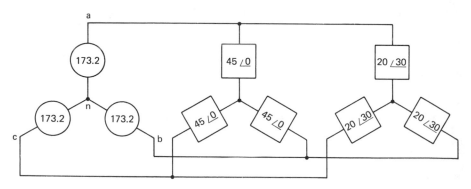

Figure 2-13   Y-connected network equivalent to Figure 2-12.

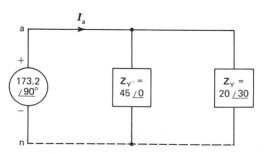

Figure 2-14   Per-phase equivalent circuit for Figure 2-13.

*The reader is advised to think through carefully just why it is that we can show a neutral wire in the per-phase equivalent circuit when no physical neutral wire was actually present.

2. Replace delta-connected impedance groups with Y equivalents by simply dividing the magnitude of the impedance by 3 (the angle remains the same).[†]

3. Sketch the complete per-phase diagram and solve by whatever method is suitable.

4. *If* the variables of the other phases, *b* and *c*, are desired (which they seldom are), simply rotate the phase *a* variables by 120° or 240° as may be appropriate.

2.5 BALANCED THREE-PHASE SYSTEMS: POWER

When a balanced circuit has been solved on a per-phase basis for voltage and current (as, for example, voltage across and current into one leg of a Y-connected impedance load), it is no great extension to compute the power and reactive volt-amperes as

$$P/\phi = E_{l-n}I_l \cos(\theta_E - \theta_I) \tag{2-5}$$

$$Q/\phi = E_{l-n}I_l \sin(\theta_E - \theta_I) \tag{2-6}$$

The somewhat unwieldy notation has been used to emphasize various things. The Greek letter phi (ϕ) is often used as a shorthand for the word *phase* and P/ϕ and Q/ϕ are therefore given as per-phase quantities. The symbol E_{l-n} stands for line-to-neutral voltage magnitude and the symbol I_l stands for the line current magnitude.[‡] The angle ($\theta_E - \theta_I$) is, of course, the difference between the angle of the voltage phasor and the current phasor. In the case of an impedance element, it is also the angle of the complex number impedance. Complex power may also be expressed succinctly as

$$S/\phi = E_{l-n}I_l^* \tag{2-7}$$

complex voltamperes per-phase just as in the single-phase case of Chapter 1.

Total power for all three phases is simply three times that per-phase or

$$P_T = 3E_{l-n}I_l \cos(\theta_E - \theta_I) \tag{2-8}$$

also,

$$Q_T = 3E_{l-n}I_l \sin(\theta_E - \theta_I) \tag{2-9}$$

or,

$$S_T = 3E_{l-n}I_l^* \tag{2-10}$$

where the phasors of Eq. (2-10) are those of any one phase, perhaps phase *a*.

Because we most often think of the voltage of a three-phase circuit in terms of E_{l-l} (line-to-line) voltage magnitude we more often use an alternate form for total power as compared with Eqs. (2-8) and (2-9) above. By substituting $E_{l-l}/\sqrt{3}$ for E_{l-n} in Eqs. (2-8) and (2-9) we have

---

[†]What if the elements are given in terms of admittance — how would you convert a delta group of admittances to a Y?

[‡]The subscript *l* for line may seem superfluous when thinking in terms of our Y-connected load example, but is included because sometimes we wish to think of a delta load and must distinguish between the line current entering the corner of the delta and the current in one branch inside the delta. Remember again that a symbol such as E_{l-n} stands for the magnitude of the voltage and is *not* a complex number!

$$P_T = \sqrt{3}\,E_{l-l}I_l\,\cos(\theta_E - \theta_I) \tag{2-11}$$
$$Q_T = \sqrt{3}\,E_{l-l}I_l\,\sin(\theta_E - \theta_I) \tag{2-12}$$
$$S_T = \sqrt{3}\,E_{l-l}I_l[\cos(\theta_E - \theta_I) + j\,\sin(\theta_E - \theta_I)] \tag{2-13}$$

A word of caution is in order in reference to the above forms for total power. The symbols E_{l-l} and I_l are magnitudes only and *not* complex number phasors. The angle $(\theta_E - \theta_I)$ is still the angle between the *line-to-neutral* voltage and the line current. For an impedance-type load it is probably best to think of the angle as the impedance angle.

For a balanced load the quantity $\sqrt{3}\,E_{l-l}I_l$ is the total voltamperes of the circuit. An example of the use of this expression would be in connection with the rating of apparatus. Suppose a certain three-phase hydro generator is rated 125 MVA at 14 kV. The rated line current can be obtained by solving the equation

$$125 \times 10^6 = \sqrt{3}\,(14 \times 10^3)I_l$$

or

$$I_l = 5155 \text{ amperes}$$

If by chance the coils of the generator are connected in delta, the current in each coil at rated voltampere load is

$$I_{\text{coil}} = 5155/\sqrt{3} = 2976 \text{ amperes}$$

Although most of this work has minimized the importance of the instantaneous time function description of ac quantities in favor of the phasor description, three-phase balanced circuits result in an interesting situation with regard to the instantaneous time function variation of total power. Suppose we have in line-to-neutral per-phase quantities $E_{an} = E_{l-n}\underline{/\theta_E}$ and $I_a = I_l\underline{/\theta_I}$, the total instantaneous power is given by the expression

$$\begin{aligned}
p(t) = {}& 2E_{l-n}I_l\,\sin(\omega t + \theta_E)\,\sin(\omega t + \theta_I) \\
& + 2E_{l-n}I_l\,\sin(\omega t + \theta_E - 120°)\,\sin(\omega t + \theta_I - 120°) \\
& + 2E_{l-n}I_l\,\sin(\omega t + \theta_E - 240°)\,\sin(\omega t + \theta_I - 240°)
\end{aligned} \tag{2-14}$$

which appears to be a formidable expression indeed! Note that the phasor descriptions of voltage and current are assumed to be rms values, whereas the coefficient of the sine function is $\sqrt{2}$ times rms or maximum value, and hence the net factor of $\sqrt{2} \cdot \sqrt{2} = 2$ in front of each term. After much tedium with trig identities, the expression (2-14) turns out to be simply

$$p(t) = 3E_{l-n}I_l\,\cos(\theta_E - \theta_I) \tag{2-15}$$

which is merely a constant equal to three times the average power of each phase.

That the total power is a constant with time seems at first a contradiction when one considers that each individual phase power is made up of a pulsating time function, as is illustrated in Figure 1-19. Each phase power is then, in general, made up of real power and reactive voltamperes. The total real power and the total reactive voltamperes are those values required to supply the load, and the combination, the magnitude of S_T, forms the required rating of the machine in voltamperes. The mechanical power input to the shaft is where the constant power of Eq. (2-15) is felt. If a generator were lossless, the mechanical power input to the shaft would balance the output. It is desirable for the mechanical power input to be a constant in minimizing noise and vibration of the machine, and this is one of the advantages of balanced three-phase operation.

2.6 UNBALANCED THREE-PHASE SYSTEMS: INTRODUCTION

Most large electric loads are inherently balanced. Large motors, electric furnaces in industry, electrochemical processing plants, and other large-scale uses of electric energy are constructed in such a way as to draw balanced currents from balanced sources. Even the smaller loads like our own homes, which are single-phase in nature, are carefully distributed so that an equal load is placed on each phase of the supplier's system.

Despite the foregoing, some loads *are* unbalanced and the analysis of the system is thereby complicated. The approach to analysis might be divided into two cases:

1. If the characteristics of rotating machines form a large factor in system performance, the analysis is very difficult, since the differential equations of rotating machines involve time-varying coefficients instead of constants, and the methods of system analysis such as phasor methods for steady-state ac and Laplace transform methods for transients are not directly applicable. In such cases, a change of variable method yields a solution. The classical change of variable used for electric power problems is known as *symmetrical components*. That method is almost universally used to solve problems in unbalance such as short circuits on high-voltage transmission lines. We will not attempt to cover this method here but reserve such treatment for more specialized study by engineers primarily interested in this area.

2. *If* we can model a system in terms of system elements such as voltage sources and constant impedances, we can solve unbalanced three-phase problems by usual circuit methods. This case typically involves things like an industrial plant with power fed from a utility bus with such excellent voltage control on the bus that a voltage source model is very accurate. This is the case covered in the following paragraphs.

2.7 UNBALANCED THREE-PHASE SYSTEMS: SOME EXAMPLE PROBLEMS

Once we have developed a model for a system problem in terms of a circuit diagram comprised of the usual circuit elements, it might be tempting to conclude that we have just another circuit solution problem and pass on to discussion of other matters. There are some special viewpoints and some unexpected happenings in dealing with three-phase unbalanced cases, however, and we will proceed to discuss some typical situations, however briefly.

Consider again the circuit of Figure 2-6 with a Y-connected load connected to a Y-connected source with the same voltages as before. Instead of equal impedances in each of the three phases, let us consider the following:

$$Z_a = 5 + j0; \qquad Z_b = 0 - j5; \qquad Z_c = 3 + j4 \qquad \text{ohms}$$

Note that the impedances are unequal in a complex number sense even though their magnitudes are all equal to 5 ohms. Figure 2-15 shows the circuit again for ready reference. We now find the currents by application of Ohm's law as

$$I_a = 100\underline{/90°}/(5 + j0) = 20\underline{/90°} = 0 + j20.0$$
$$I_b = 100\underline{/-30°}/(0 - j5) = 20\underline{/60°} = 10 + j17.3$$
$$I_c = 100\underline{/210°}/(3 + j4) = 20\underline{/156.9°} = -18.4 + j7.9$$

A phasor diagram in part (b) of the figure shows that these phasors do indeed form an *un*balanced set. Furthermore, the neutral current is no longer zero, but is given by $I_n = I_a + I_b + I_c = -8.4 + j45.2 = 45.9\underline{/100.5°}$, and it is shown as the dotted current phasor in the diagram.

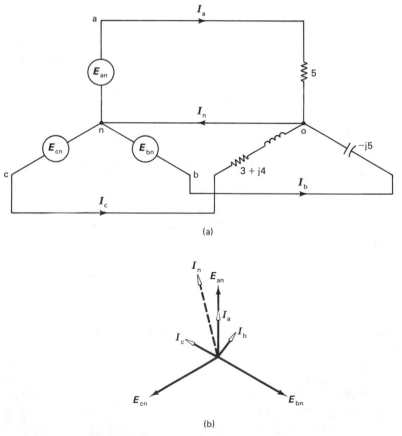

(a)

(b)

Figure 2-15   An unbalanced circuit example.

Since the neutral current in the example of Figure 2-15 is not zero, we are no longer free to say that it does not matter if the neutral is present or not. If we remove the neutral, we interrupt current and the whole situation changes, as will be seen in the following computations. Clearly, in the unbalanced case, we need to know if we have a three- or four-wire circuit, since the two types behave differently.

With the neutral wire removed from the circuit as shown in the next figure (Figure 2-16) we no longer know the voltage across each of the impedance elements, and the circuit is harder to solve. We now identify the neutral of the Y-connected load as point o and note that the points o and n have a potential difference between them.

We *could* solve the new circuit by means of two loop current equations, but it is easier to use the nodal voltages and have just the one equation summing currents at the load neutral in terms of the unknown voltage E_{on}. Kirchhoff's current law leads to the equation:

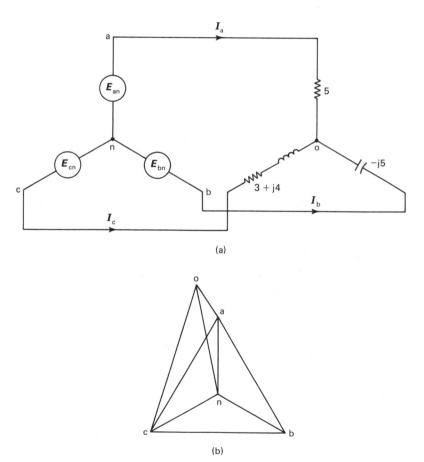

(a)

(b)

Figure 2-16   Circuit of Figure 2-15 with the neutral wire removed.

$$I_a + I_b + I_c = 0$$

$$\frac{E_{an} - E_{on}}{5 + j0} + \frac{E_{bn} - E_{on}}{0 - j5} + \frac{E_{cn} - E_{on}}{3 + j4} = 0$$

If we substitute the specified values for the source voltages and solve for E_{on} we find $E_{on} = 142.5\underline{/93.4°}$. This is the potential difference between the neutral of the load and the neutral of the source. The several voltage components are best displayed in the topographic form of diagram introduced in Chapter 1 and shown in part (b) of the figure. We say that there has been a *neutral shift* because of the non-zero voltage E_{on}.

Since we can find the voltages across each of the impedance elements in terms of the source voltages and the voltage E_{on}, we can immediately find the three line currents:*

$$E_{ao} = E_{an} - E_{on} = 100\underline{/90°} - 142.5\underline{/93.4°}$$
$$= 8.45 - j42.23 = 43.07\underline{/-78.69}$$
$$I_a = E_{ao}/Z_a = (8.45 - j42.23)/5$$
$$= 1.69 - j8.45 = 8.61\underline{/-78.69°}$$
$$E_{bo} = E_{bn} - E_{on} = 100\underline{/-30°} - 142.5\underline{/93.4°}$$
$$= 95.05 - j192.23 = 214.45\underline{/-63.69°}$$
$$I_b = E_{bo}/Z_b = (95.05 - j192.23)/5\underline{/-90°}$$
$$= 38.45 + j19.01 = 42.89\underline{/26.31°}$$
$$E_{co} = E_{cn} - E_{on} = 100\underline{/210°} - 142.5\underline{/93.4°}$$
$$= -78.16 - j192.23 = 207.51\underline{/-112.13°}$$
$$I_c = E_{co}/Z_c = (-78.15 - j192.23)/(3 + j4)$$
$$= -40.14 - j10.56 = 41.5\underline{/-165.26}$$

After the foregoing laborious arithmetic, a check is in order and we see by addition that $I_a + I_b + I_c = 0$ as it must, or phrasing it another way, we see that the neutral point of the load must shift to the point *o* of Figure 2-16(b) in order that the currents shall add to zero!

Other examples of unbalanced three-phase circuits will be encountered in the exercises at the end of this chapter. For now, let us merely comment that unbalanced cases are usually far more laborious to solve (if indeed they *can* be solved by elementary circuit methods). Perhaps it might be said somewhat casually that the number of unknowns for an unbalanced network is approximately three times that of a similar balanced network and the number of simultaneous equations and the labor of solution expands greatly.

If we consider total power for an unbalanced network, there is little we can say except that the total power (or reactive voltamperes) for an unbalanced circuit must be found by separately finding the power of each phase and adding, as contrasted

---

*Remember again that in using this form of diagram we read any voltage we like by considering the line between two desired points with an arrowhead on the end of the first point in double-subscript notation.

with the balanced case, where the symmetry permits simple expressions for total power.

In particular for the foregoing example problem, the total complex power is found from summing the power of each phase as follows:

$$S_a = E_{ao}I_a^* = (43.07\underline{/-78.69})\,(8.61\underline{/78.69})$$
$$= 370.83\underline{/0} = 370.83 + j0$$
$$S_b = E_{bo}I_b^* = (214.45\underline{/-63.69})\,(42.89\underline{/-26.31})$$
$$= 9198\underline{/-90} = 0 - j9198$$
$$S_c = E_{co}I_c^* = (207.51\underline{/-112.13})\,(41.50\underline{/165.26})$$
$$= 8611\underline{/53.13} = 5167 + j6889$$

from which

$$S_0 = S_a + S_b + S_c$$
$$= 5537 - j2309$$

The reader may find it instructive to check this value for S_0 vs. $\Sigma I^2 R + j\Sigma I^2 X$.

Although totals of real and imaginary portions of complex power do indeed add to the total as given above, the value of the total must be used with some discretion. Each phase of the supplying device must be capable of supplying the voltamperes of that phase, so the total rating of a machine supplying the above load would not be merely the magnitude of the total complex power. As an extreme example, consider the possibility that one phase supplies a pure inductive reactor load and another a pure capacitive reactor load such that the two reactive voltampere quantities add to zero. We might incorrectly conclude from the total complex power expression that the machine does not have to have any voltampere capability at all!

2.8 UNBALANCED THREE-PHASE SYSTEMS: THE ROLE OF PHASE SEQUENCE

Three-phase systems have a property that has been ignored thus far in this treatment and that is *phase sequence*. By phase sequence we mean the way the phases are ordered in time sequence in arriving at some particular point on the wave, say the positive maximum. Figure 2-17 illustrates the time function plot of two balanced sets of three-phase variables. The only difference between the two is that in (a) the

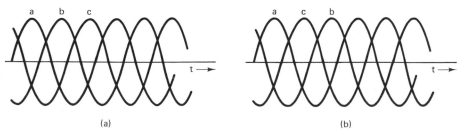

(a)   (b)

Figure 2-17   Illustration of phase sequence.

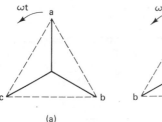
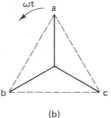

Figure 2-18 Phasor diagrams for waves of Figure 2-17.

variables come to the positive maximum point in the sequence *abc*, and in (b) they come up in the sequence *acb*. The phasor diagrams describing the two cases might be given as in Figure 2-18. The difference between the cases may appear to be trivial on first acquaintance with the term *phase sequence*; after all, it seems to be largely a question of a name. If we were free to name the phases at will we would probably choose to letter the phases in normal sequence of the alphabet.* If we take the opposite view that we have three unmarked terminals of a voltage source and we mark them *a*, *b*, and *c* at random, the actual sequence of the phases may turn out to be either *abc* or *acb*. By reversing the name tags of any two leads, or by reversing any two leads brought to the marked terminals, we would reverse the phase sequence.

In the case of balanced impedance-type loads, a change of phase sequence really makes little difference, except for the names of the variables. In the case of polyphase ac motors, the reversal of phase sequence makes a big difference because the direction of motor rotation reverses with phase sequence. In the case of unbalanced impedance-type loads, the reversal of phase sequence also makes a significant difference. For the unbalanced case, not only are the names of the variables changed but their actual magnitudes may be different, depending upon the phase sequence of the source connected to the terminals.

An illustration of the effect of phase sequence on an unbalanced impedance load is given by the simple circuit in Figure 2-19. Let it be supposed that we have three terminals of a balanced three-phase voltage source of 100 volts and someone ʜas marked the terminals *a*, *b*, and *c* by arbitrarily placed tags on the terminals. We now connect a series *R-C* branch from *a* to *c*, as shown. There are two possibilities: we may have *abc* or *acb* phase sequence. Phasor diagrams of the topographic type are shown to illustrate the two cases.† Note that I_c leads the voltage drop E_{ca} by 45°. The voltage drop E_{da} is in phase with I_c and the voltage drop E_{cd} lags the current I_c by 90°. Because of the equality between the magnitude of the resistance and the capacitive reactance, the diagram may be quickly drawn to scale without extensive computations. The difference between the two cases is rather marked. Point *d* is

*In British books, the phases are often named for colors such as the red, yellow, and blue phases, and in such a case we would have no preconceived idea of the "normal" sequences of phases.

†The student should be sure that he or she understands the nature of such a diagram by mentally placing arrowheads on the line segments to represent specific voltage drops in the circuit.

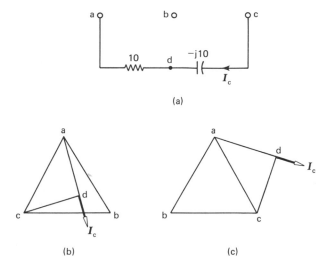

Figure 2-19   Effects of phase sequence on unbalanced load.

inside the voltage triangle for sequence *abc* and outside the triangle for sequence *acb*. If any student doubts that there is a real physical difference between the two cases, such doubt should be erased by placing a voltmeter between points *b* and *d*. For phase sequence *abc*, the voltmeter should read 36.6 volts, whereas with phase sequence *acb* the voltmeter should read 136.6, as may be seen by scaling the diagram or by making a few computations. The large difference in magnitude has been used as a means of building an instrument to indicate phase sequence for guidance to electricians when making system connections.

Other examples could be given of more involved systems, showing the effect of phase sequence as not only a change in name or arbitrary phase angle reference of the variables but an actual change in the magnitudes of currents or voltages. This will be further brought out in the exercises following this chapter.

2.9 MEASUREMENT OF THREE-PHASE POWER

If we wish to measure the power in a three-phase circuit, as for example the Y-connected group of impedances in Figure 2-20, a straightforward approach would be to use one wattmeter to measure the power in each phase. In such a case the total power for the three-phase circuit would be equal to the sum of the three wattmeter readings. It is not always necessary or desirable to use three wattmeters, however, and that is the subject of the material in this section.

In the figure, the connection from point *o*, the wattmeter neutral of the potential coils, to point *n*, the neutral of the load, is shown dotted because of an interesting situation that occurs if the connection *o-n* is broken.

If the load is balanced, and if the wattmeters are identical, such that their Y-connected potential circuits form a balanced load, then disconnection of point *o* from point *n* will not change any voltages or any wattmeter readings. If, on the other

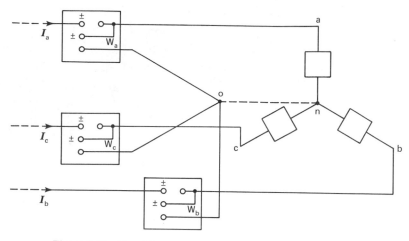

Figure 2-20   Use of three wattmeters to measure three-phase power.

hand, either the load or the wattmeter potential coils (or both) form an unbalanced load then there will be a voltage E_{no}, and the wattmeter reading will change from their readings with n and o connected. With the neutral connection broken, the sum of the three wattmeter readings will be:

$$P_{TOT} = \text{Re } E_{ao}I_a^* + \text{Re } E_{bo}I_b^* + \text{Re } E_{co}I_c^*$$
$$= \text{Re}[(E_{an} + E_{no})I_a^* + (E_{bn} + E_{no})I_b^* + (E_{cn} + E_{no})I_c^*]$$

We can rearrange this last equation as

$$P_{TOT} = \text{Re}[E_{an}I_a^* + E_{bn}I_b^* + E_{cn}I_c^* + (I_a^* + I_b^* + I_c^*)E_{no}] \qquad (2\text{-}16)$$

The term $(I_a^* + I_b^* + I_c^*)$ is actually zero because, from Kirchhoff's law, $I_a + I_b + I_c = 0$, and hence the sum of the conjugates of these three complex numbers is also zero. As a consequence, we have

$$P_{TOT} = \text{Re}[E_{an}I_a^* + E_{bn}I_b^* + E_{cn}I_c^*] \qquad (2\text{-}17)$$

but this is also the sum of the three wattmeter readings *with* the neutrals o and n connected! In other words, even with a difference in potential of the neutrals, the *sum* of the three wattmeter readings is still equal to the total power, regardless of the magnitude of the voltage E_{no}. Note, however, that the individual wattmeter readings may change when the neutral connection is broken; it is their sum that remains fixed.

   If we connect point o to any one of the three lines, a, b, or c, that particular wattmeter will read zero (actually, we would not bother to put it into the circuit), and the sum of the remaining two wattmeters still gives the total power. Figure 2-21 illustrates such a connection where o has been connected to line c and wattmeter c eliminated. The voltage E_{no} then becomes simply E_{nc}, and since it doesn't matter what value this voltage has (short of infinity), the sum of the wattmeters is still the total power!

   We should review some of the things that are said or inferred in the above development for emphasis — and to avoid misunderstandings.

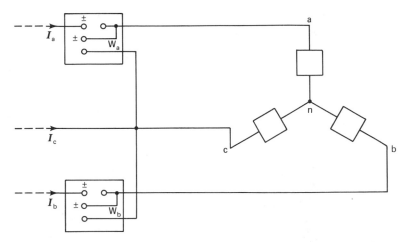

Figure 2-21   Connection of two wattmeters to measure total power in a three-phase, three-wire circuit.

1. There was no power system neutral connection to the Y-connected load and that fact was the basis for the equation $I_a + I_b + I_c = 0$. If there were a neutral with current (unbalanced load) then the "two-wattmeter" method would not give the true value for total power. The development above could be generalized to cover circuits of n wires, where n is any integer. It can be shown that to measure total power requires (n-1) wattmeters.

2. It was convenient to think of the load as a Y-connected group of impedances, but a little thought will reveal that such is not necessary; a delta group could be used as well, or some more general combination of elements, *provided* that there are only three wires feeding the load.

3. It was convenient to present the development in terms of phasors; that is, in terms of a sinusoidal waveform, but the same proof could be modified to show the same results of two wattmeters for a three-wire circuit in terms of any periodic waveform. In other words, the method is still valid if the waveform is distorted, as discussed in the section to follow.

4. Even if the net power flow is positive from left to right in a case like Figure 2-21, it is not to be inferred that the individual wattmeter readings will be positive. Commercial wattmeters have a polarity reversing switch in case the indication is negative.

Although two wattmeters would be used as shown in Figure 2-21 in laboratory work, in actual practice it is not uncommon to find polyphase wattmeters, where one unit measures the total power in a polyphase circuit. For a three-wire circuit, a polyphase wattmeter would contain two elements with a common shaft, so arranged that their torques are additive mechanically and their scale indication is calibrated in terms of total power for the circuit.

2.10 EFFECT OF NONSINUSOIDAL WAVEFORMS IN THREE-PHASE CIRCUITS

It is assumed in the work of Chapters 1 and 2 that the voltage or current waveforms are sinusoidal and this is usually a very good approximation. If the waveforms are not sinusoidal they are at least periodic, so they can be represented by a Fourier series and, with linear circuits, superposition can be used to treat each frequency component separately.

In representing a waveform by a Fourier series, the wave is broken down into a fundamental frequency sinusoid (often 50 Hz or 60 Hz for power work) and higher frequency components that are integer multiples of the basic frequency. These higher frequencies are called *harmonics*. In power work it usually happens that the negative half cycles of the wave are exactly the same as the positive half cycles, except "upside down." In such cases, Fourier theory tells us that only odd-numbered multiples of the basic frequency are present. It happens then that the most significant distortion of waveform is usually that due to the addition of a triple frequency term (third harmonic) and this also happens to be a case of special significance in three-phase work so it will be used as an example.

Consider Figure 2-22, which shows how a fundamental sine wave and a third harmonic add to give a distorted waveform.

A general mathematical description of a distorted wave by use of a Fourier series is given by an infinite series as

$$e(t) = \sum_{n=1}^{\infty} \sqrt{2} E_n \sin(n\omega t + \theta_n) \qquad (2\text{-}18)$$

using a voltage wave as an example and with an index of summation given as n. Although in the general case the summation extends over values of n from zero to infinity, in the example of Figure 2-22 it is assumed that values of n equal to 1 and 3 are the only significant values.

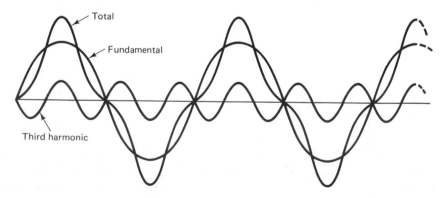

Figure 2-22   Sine wave with a third harmonic distortion.

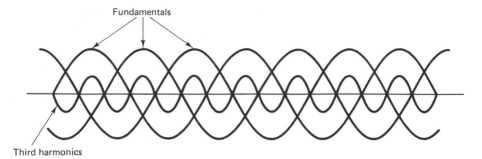

Figure 2-23   Balanced three-phase variables with third harmonic components.

If we consider the waveform of Eq. (2-18), delayed by τ (tau) seconds, the expression becomes

$$e(t - \tau) = \sum_{n=1}^{\infty} \sqrt{2}\,E \, \sin(n\omega t - n\omega\tau + \theta_n) \qquad (2\text{-}19)$$

It will be noted that each component suffers a phase shift of $(n\omega\tau)$. From Eq. (2-19) we see that, if a waveform is delayed by 120° on a fundamental frequency scale, the third harmonic ($n = 3$ component) will be delayed by 360°. The significance of this is shown in Figure 2-23, where a three-phase set of variables is sketched, each fundamental frequency component being 120° in phase from each of the others. The third harmonics, however, all appear at the same angle; that is, they all lie on top of each other!

Now let us reconsider the balanced Y-connected circuit of Figure 2-6, where it was concluded that the neutral current was zero. This will not be true if there are third harmonic components present; instead the neutral current will be equal to three times the third harmonic component of any one phase. If there is no neutral conductor, the third harmonic currents cannot flow—but probably the line-to-neutral voltage will be distorted as a consequence. Further discussion of this had best wait for specific examples. A first approximation solution often ignores waveform distortion, and fortunately it is only rarely that we find waveform distortion in power networks approaching that of Figure 2-22.

2.11 THREE-PHASE CIRCUITS: SUMMARY

This chapter has attempted to build a background in the understanding of the definitions and workings of three-phase circuits. Primary emphasis has been given to the balanced case, since it is the usual case. An unbalanced network of voltage sources and impedances (or admittances) can be solved by the usual methods of network theory, but there are certain viewpoints and properties peculiar to three-phase systems, and these have been discussed briefly.

The methods of this chapter will be used in subsequent chapters dealing with power equipment.

ADDITIONAL READING MATERIAL

1. Brown, D., and E. P. Hamilton III, *Electromechanical Energy Conversion*, New York: MacMillan Publishing Company, 1984.

2. Chapman, S. J., *Electric Machinery Fundamentals*, New York: McGraw-Hill Book Company, 1985.

3. Chaston, A. N., *Electric Machinery*, Reston, Virginia: Reston Publications Company, Inc., 1986.

4. Del Toro, V., *Electric Machines and Power Systems*, Englewood Cliffs, New Jersey: Prentice-Hall, Inc., 1985.

5. Elgerd, O. I., *Basic Electric Power Engineering*, Reading, Massachusetts: Addison-Wesley Publishing Company, 1977.

6. Elgerd, O. I., *Electric Energy Systems Theory: An Introduction*, New York: McGraw-Hill Book Company, 1982.

7. Fitzgerald, A. E., C. Kingsley, Jr., and S. D. Umans, *Electric Machinery*, New York: McGraw-Hill Book Company, 1983.

8. Gross, C. A., *Power System Analysis*, New York: John Wiley & Sons, 1979.

9. Lindsay, J. F., and M. H. Rashid, *Electromechanics and Electric Machinery*, Englewood Cliffs, New Jersey: Prentice-Hall, Inc., 1986.

10. Nasar, S. A., *Electric Energy Conversion and Transmission*, New York: MacMillan Publishing Company, 1985.

11. Shultz, R. D., and R. A. Smith, *Introduction to Electric Power Engineering*, New York: Harper & Row, Publishers, 1985.

12. Stevenson, W. D. Jr., *Elements of Power System Analysis*, New York: McGraw-Hill Book Company, 1982.

STUDY EXERCISES

1. Suppose that we have eight voltage sources, each equal to 100 volts and spaced in phase angles to form a balanced set of eight phases. If the sources are connected into a symmetrical star:
   a. What is the voltage between adjacent terminals of the star?
   b. What is the maximum possible voltage between any two terminals?

2. Suppose that we have a balanced Y-connected group of impedances of $Z_Y = 10\underline{/30°}$ each and also a balanced delta-connected group of impedances of $Z_\Delta = 15\underline{/45°}$ each, and both groups are connected to the same 300-volt three-phase source.
   a. What is the net equivalent impedance per-phase (line-to-neutral)?
   b. How much current flows in the line connecting to the source?
   c. What is the total power drawn from the source?

3. The network shown at the right is made up of nine 18-ohm resistors that are connected in a symmetrical pattern. If connected to a 173.2 volt (100-volt *l-n*) source, how many amperes will flow in each line from the source?

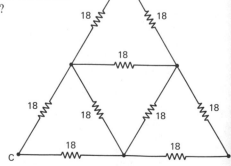

4. We have a balanced three-phase load of 10 MW (total), which is being supplied from a 115 kV circuit at 0.8 power factor, current lagging.
   a. What is the line current?
   b. If we want to improve the power factor to 0.9, current lagging, what must be the reactance of each leg of a Y-connected capacitor bank used for power factor correction?

5. Given the balanced load shown at the right connected to a 4000-volt three-phase source. Voltage *source* phasors are to be assumed aligned in the same way as the circuit symbols between points A, B, and C. For example, V_{BC} is taken as angle zero.
   a. What are I_A, I_B, and I_C?
   b. What are V_{DE}, V_{EF}, and V_{FD}?
   c. What is the total power?

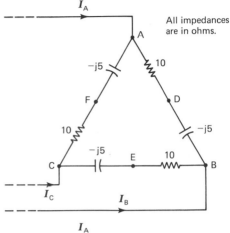

6. The *unbalanced* load at the right is connected to the same 4000-volt three-phase source as in the problem above. Again the voltage *source* phasors are assumed to be aligned the same as the circuit symbols between A, B, and C. What are:
   a. I_A, I_B, and I_C?
   b. V_{DE}?
   c. The total power?

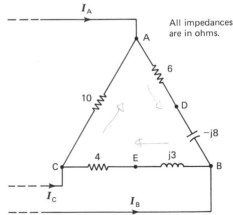

7. A certain large induction motor is fed from a 2400-volt three-phase circuit. Power is measured by two wattmeters with current coils in lines a and c and the free end of the potential coils connected to line b. A three-phase induction motor draws balanced current-lagging power and under light load these wattmeters read $P_a = 45$ kW and $P_c = -5$ kW.
   a. What is the total power drawn by the motor?
   b. What is the phase sequence of the applied voltage? *Hint:* Sketch a phasor diagram for both sequences and see which is consistent with the wattmeter readings.

8. A certain three-phase alternator generates identical (but distorted) waveforms in each of the three phases. The phase sequence of the fundamental frequency components is *abc*; the zero crossings of phases b and c are 120° and 240° respectively after that of phase a. The waves may be described by means of Fourier series, with the significant harmonics as the odd numbers from the third to the eleventh, $(3, 5, 7, 9, 11)$. For each of the harmonics:
   a. What are the phase angle differences (on the harmonic angle scale) between phases a, b, and c?
   b. What is the phase sequence of each set of harmonics (zero, *abc*, or *acb*)?

SOLUTIONS TO STUDY EXERCISES

1. $\theta = 360/n = 360/8 = 45°$
$E_{l-l} = 2 \times 100 \sin(45°/2)$
$\qquad = 76.5$ V
Maximum voltage is diametrical; that is,
$2 \times 100 = 200$ V

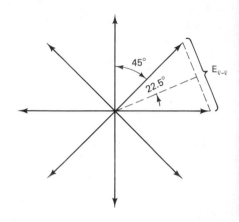

2.

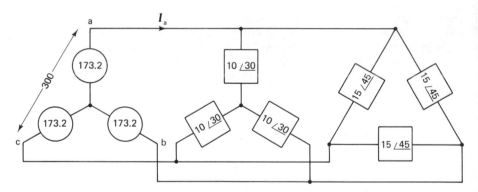

First, convert the right-hand delta to a Y.
$Z_Y = (1/3)Z$
$\qquad = (15\underline{/45°})/3 = 5\underline{/45°}$
Next combine the two Y's in parallel:

$$Z = \frac{10\underline{/30°}\ 5\underline{/45°}}{10\underline{/30°} + 5\underline{/45°}} = 3.36\underline{/40°} \quad \longleftarrow \quad a$$

$$I_a = \frac{(300/\sqrt{3})\underline{/90°}}{3.36\underline{/40}} = 51.57\underline{/50°} \quad \longleftarrow \quad b$$

$$P = \sqrt{3} \times 300 \times 51.57 \cos(90° - 50°) = \underline{20522\ W} \quad \longleftarrow \quad c$$

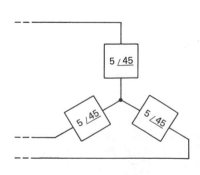

3. We *might* solve by loop equations like those designated on the graph at the right. This approach will involve six equations in six unknowns—formidable! A simpler approach is to make a series of delta-Y transformations, as below:

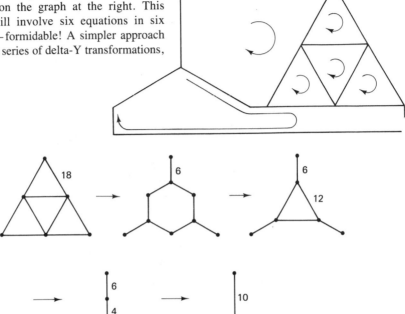

From the last equivalent we see that any one of the line currents is simply obtained as
$I = 100/10 = \underline{10\ A}$ ———

4. a. $P = \sqrt{3}\,E_{l-l}I\,\cos(\theta_E - \theta_I)$
$10 \times 10^6 = \sqrt{3} \times 115 \times 10^3 \times I \times 0.8$
$I = \underline{62.76\ A}$ ——— a

b. *Method 1.* Compute the required current increment.
Take any sample voltage to neutral as
$E = (115 \times 10^3)/\sqrt{3}\ \underline{/0°}$
$I = 62.76\underline{/-36.86°} = 50.20 - j37.65$

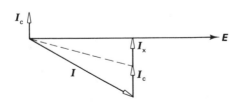

$I_x = 50.20\ \tan(\cos^{-1}0.9) = 24.32$
$I_c = 37.65 - 24.32 = 13.34$
$X_c = -(115 \times 10^3/3)/13.34 = \underline{-4978\ ohms}$ ⟵ b

Method 2. Compute in terms of complex power.

$$S = 10 \times 10^6 + j10 \times 10^6 \tan(\cos^{-1}0.8)$$
$$= 10 \times 10^6 + j7.5 \times 10^6$$
$$Q = 7.5 \times 10^6 \text{ var}$$

We want: $S = 10 \times 10^6 + j10 \times 10^6 \tan(\cos^{-1}0.9)$
$$= 10 \times 10^6 + j4.84 \times 10^6$$
$$Q = 4.84 \times 10^6 \text{ var, therefore}$$
$$7.5 \times 10^6 + Q_c = 4.84 \times 10^6$$
$$Q_c = -2.657 \times 10^6$$
$$Q_c/\text{phase} = -2.657 \times 10^6/3 = -0.8856 \times 10^6$$
$$X_c = (115 \times 10^3/\sqrt{3})^2/(-0.8856 \times 10^6) = \underline{-4978 \text{ ohms}} \quad \longleftarrow$$

5. a. If we convert to an equivalent Y:

$$Z_Y = Z_\Delta/3 = (10 - j5)/3$$
$$E_{AN} = (4000/\sqrt{3})\underline{/90°}$$
$$I_A = \frac{(4000/\sqrt{3})\underline{/90°}}{10/3 - j5/3}$$
$$I_A = \underline{619.7\underline{/116.57°}}$$

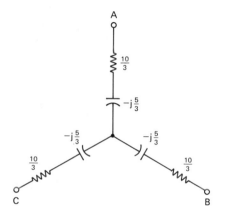

We *can* find I_B and I_C by similar means, but why not exploit the symmetry of a balanced system and merely rotate I_A by $-120°$ and $-240°$, thus:

$$I_B = \underline{619.7\underline{/-3.43°}} \quad \text{and} \quad I_C = \underline{619.7\underline{/-123.43°}} \quad \longleftarrow \quad \text{a}$$

b. $V_{DE} = V_{DB} + V_{BE}$

The components of V_{DE} may easily be found by voltage divider action. (See for example the method of Study Exercise 1-8 of Chapter 1.)

$$V_{DE} = V_{AB}\frac{-j5}{10 - j5} + V_{BC}\frac{10}{10 - j5}$$

$$= \frac{1}{10 - j5}[(300\underline{/120°})(-j5) + (300\underline{/0°})(10)]$$

$$= \underline{5204\underline{/36.45°}} \quad \longleftarrow$$

Again we *can* solve for V_{EF} and V_{FD} by similar means but why not just rotate by $-120°$ and $-240°$, thus:

$$V_{EF} = \underline{5204\underline{/-83.54°}} \quad \text{and} \quad V_{FD} = \underline{5204\underline{/-203.54°}} \quad \longleftarrow$$

$$P_{tot} = 3 \times 619.2^2 \times (10/3) \times 10^{-6} = \underline{3.84 \text{ MW}} \quad \longleftarrow \quad \text{c}$$

$$\left.\begin{array}{c} \\ \\ \\ \\ \\ \\ \end{array}\right\} \text{b}$$

6. a. To convert to a wye in this case is laborious and would still not result in a circuit that would be simple to solve. Instead we may apply Ohm's law directly to each leg of the delta.

$$I_{AB} = (4000\underline{/120°})/(6 - j8) = -397.1 + j47.84$$
$$I_{BC} = (4000\underline{/0°})/(4 + j3) = 640 - j480$$
$$I_{CA} = (4000\underline{/-120°})/(10 + j0) = -200 - j346.4$$

$$I_A = I_{AB} - I_{CA} = 440.8\underline{/116.57°}$$
$$I_B = I_{BC} - I_{AB} = 1163.7\underline{/-26.97°} \quad \}\longleftarrow \quad a$$
$$I_C = I_{CA} - I_{BC} = 850.6\underline{/170.96°}$$

b. $V_{DE} = V_{DB} + V_{BE} = I_{AB}(-j8) + I_{BC}(j3)$
$$= 541.3\underline{/70.3°} \quad\longleftarrow\quad b$$

c. $P_{\text{tot}} = (|I_{AB}|^2 6 + |I_{BC}|^2 4 + |I_{CA}|^2 10) \times 10^{-6}$
$$= \underline{5.12 \text{ MW}} \quad\longleftarrow\quad c$$

7. a. $P_{\text{tot}} = P_a + P_c$
$$= 45 - 5 = 40 \text{ kW}$$

b. Note that the line current *lags* the *line-to-neutral* voltage by an angle less than 90°. The two possible diagrams are shown below:

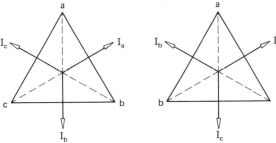

It is given that wattmeter *a* uses E_{ab} and wattmeter *c* uses E_{cb}. We see from the diagram at the left that wattmeter *a* would read negative and wattmeter *c* positive. Conversely from the diagram at the right wattmeter *a* would read positive and wattmeter *c* negative. Therefore the phase sequence is *acb*.

8. If the fundamental frequency wave is, for example:
$$e_{a(1)} = E_1^m \sin(\omega t + \theta_1)$$
then phase b is retarded by a time delay, τ, such that $\omega\tau = 120°$ and
$$e_{b(1)} = E_1^m \sin(\omega t + \theta_1 - \omega\tau)$$
likewise phase c is retarded by 2τ such that $2\omega\tau = 240°$ and
$$e_{c(1)} = E_1^m \sin(\omega t + \theta_1 - 2\omega\tau)$$
For the *n*th harmonic then we would have $\omega_n = n\omega$ and
$$E_{a(n)} = E_n^m \sin(n\omega t + \theta_n)$$
$$E_{b(n)} = E_n^m \sin(n\omega t + \theta_n - n\omega\tau)$$
$$E_{c(n)} = E_n^m \sin(n\omega t + \theta_n - 2n\omega\tau)$$
We now tabulate the angles of $e_{a(n)}$, $e_{b(n)}$, and $e_{c(n)}$ for each of the specified harmonics. The phase sequence follows from observation of the angles.

n	$\underline{/e_{a(n)}}$	$\underline{/e_{b(n)}}$	$\underline{/e_{c(n)}}$	Phase Sequence
1	θ_1	$\theta_1 - 120$	$\theta_1 - 240$	*abc*
3	θ_3	$\theta_3 - 0$	$\theta_3 - 0$	zero
5	θ_5	$\theta_5 - 240$	$\theta_5 - 120$	*acb*
7	θ_7	$\theta_7 - 120$	$\theta_7 - 240$	*abc*
9	θ_9	$\theta_9 - 0$	$\theta_9 - 0$	zero
11	θ_{11}	$\theta_{11} - 240$	$\theta_{11} - 120$	*acb*

HOMEWORK PROBLEMS

1. A 1000-kVA, 4160-volt, three-phase, 60-Hz generator delivers full rated kVA at 0.5 power factor, current lagging, and at rated voltage at its terminals. Part of the load is 200 kW of balanced lighting (pure resistance) load.
   a. What would be the complex number impedance per phase of a Y-connected load that would represent the remainder of the total load?
   b. What is the total complex power S, which is absorbed by the Y-connected impedance load?

2.

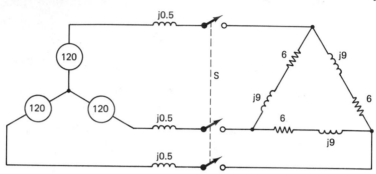

The figure above illustrates a 120/208Y volt source (balanced three-phase) which has an "internal" impedance of $j0.5$ ohm per phase. The generator is to be connected to a balanced, delta-connected impedance load by closing a switch, S. Prior to switch closure the line-to-line voltage is 208 volts at the switch.
   a. By what percentage of the no-load voltage does the terminal voltage at the switch drop when the load is connected?
   b. What is the total complex power S, delivered to the delta-connected load?

3.

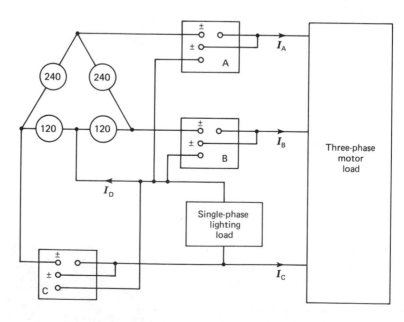

Frequently in supplying a small commercial load such as a machine shop we wish to supply power at three-phase 240 volts for the motors and single-phase power at 120 volts for lighting and miscellaneous small devices. The diagram shows one way of doing this, where as usual the circuit symbols for the voltage *sources* are oriented the same way as the phasors. Suppose that we have the following loads:

Lighting	Motors
2 kW at 120 volts (single phase)	30 kW at 240 volts (three phase)
(Unity pF)	(0.8 pF, current lagging)

a. What are the phasor currents I_A, I_B, I_C, and I_D?
b. If three wattmeters are connected as shown, what is the reading of each wattmeter?
c. Is the sum of the three wattmeter readings the true total power? (Check by addition.)

4. An accident (regrettably not uncommon) occurs when a farmer contacts an overhead line with a piece of irrigation pipe. Many such lines are connected in Y at the source with the neutral grounded, and line-to-neutral voltage is therefore present between line and ground as a hazard to life. It has been suggested that a delta-connected source would avoid this hazard, since there would be no return to ground. If we ignore load apparatus (which might be connected in Y-grounded), we still have capacitance to ground as shown on the circuit diagram below where the source is delta connected and the line-to-line and line-to-ground capacitances are given for each of the phase wires. All capacitors are estimated from line parameters to have an admittance of $j0.1 \times 10^{-4}$ S at 60 Hz. We estimate the resistance of a human body as 1000 ohms under typical conditions on a damp field without rubber boots. What is the current in the victim shown touching the line with a length of metal pipe?

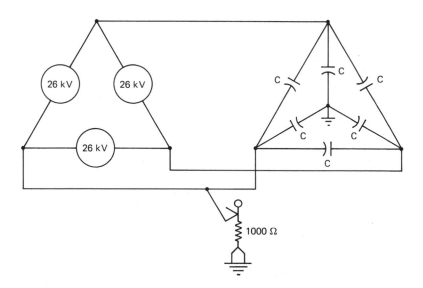

5. The following circuit illustrates an old electrician's trick to find the phase sequence of a three-phase source. Two lamp bulbs are connected in a Y circuit with a static capacitor and ungrounded neutral. If we assume that the bulbs can be represented by a constant resistance, then which bulb burns the brighter for a phase sequence *abc*? What is the voltage across the brighter bulb?

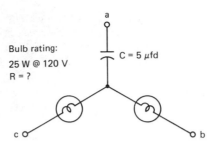

Bulb rating:
25 W @ 120 V
R = ?

C = 5 μfd

Three-phase circuit voltage = 125 volts
f = 60 Hz

6.

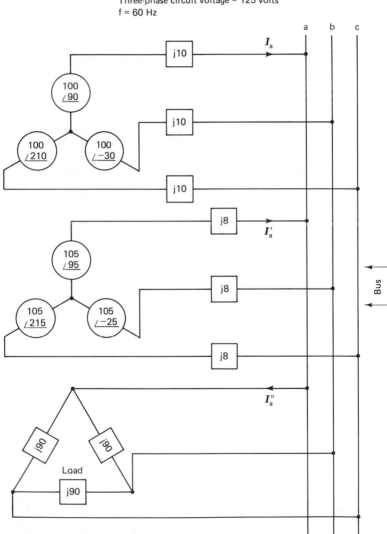

The circuit diagram on the preceding page represents two generators feeding a common load through the vertical connections, which are called a *bus*. Note that the voltage phasors vary slightly in angle from each other and from the orientation of the circuit symbols.

a. What are the phasor currents I_a, I'_a, and I''_a?
b. What is the total complex power delivered to the load?
c. What is the voltage (line-to-line) across the load?
d. What is the complex power supplied by each generator *at the point of connection to the bus?*

7. Suppose that the circuit elements of the circuit of Figure 2-21 are those of the unbalanced example of Figure 2-16, for which the solution is given.
   a. What will each wattmeter, W_a and W_b read?
   b. Check by comparing the sum of the two wattmeter readings with the total power computed from the sum of the I^2R values.

8. We have a three-phase, balanced, Y-connected inductive reactor at one end of a 345 kV (line-to-line) transmission line. The reactor is tested by applying 345 kV. We measure a current of 17 amperes in each phase. Wattmeters are connected in lines b and a, with the standard two-wattmeter connection, with the free end of the potential coils connected to line c. Wattmeter readings are as follows:

$$W_b = 3770 \text{ kW}$$
$$W_a = -2006 \text{ kW}$$

   a. What must the phase sequence of the applied voltage have been in order to result in these readings?
   b. What is the total (all three phases) complex power, $S = P + jQ$, drawn by the inductive reactor at the test voltage?
   c. What is the complex number impedance of the reactor in ohms per phase, $Z = R + jX$?

Chapter 3

The Transformer

- The behavior of an "ideal" transformer.
- The departures from ideal transformer theory.
- A circuit model for an actual transformer.
- Model and variations used in typical problems encountered in power engineering.

3.1 MAGNETIC FIELD DEVICES — AN INTRODUCTION

Many of the devices of concern to the power engineer operate on principles involving the magnetic field. Students using this book have ordinarily studied the magnetic field in electrophysics courses and are familiar with field relations, including the application of Maxwell's equations and the material relationships. The geometry of power apparatus is quite involved and makes the direct application of Maxwell's equations rather difficult. Fortunately, however, the low frequencies involved in power apparatus permit the use of quasi-static field relationships and the presence of steel cores (while introducing the complication of nonlinearity) simplifies the problems by confining the major field effects to rather well-defined paths or regions.

non-ferrous cores give constant permeance

As an illustration in explanation of the last statement, consider the device shown in Figure 3-1, where a coil is shown wound around a steel core of toroidal shape. If a current i is passed through the N turns of the coil, a magnetic field is set up in the toroid. From symmetry considerations, we assume the flow lines of the H field to be concentric circles within the toroid. If we make a line integral of the H vector around the toroid, Ampère's rule says

$$\oint \vec{H} \cdot d\vec{l} = Ni$$

An integration around one of the concentric circles that form a flow line at a radius r from the center of the toroid gives

$$H 2\pi r = Ni$$

$$H = \frac{Ni}{2\pi r}$$

For a material of magnetic permeability* μ we have for the B vector

$$B = \mu H$$

$$= \frac{\mu Ni}{2\pi r}$$

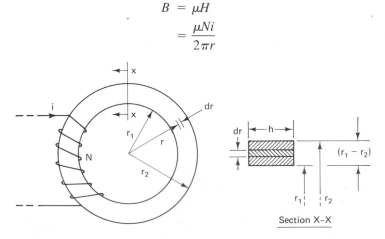

Figure 3-1   Illustration of a magnetic field device.

*In these introductory pages, we will assume linearity of the material; that is, a constant permeability. Some discussion of the nonlinear effects will be given in the following pages and chapters.

The flux of the B vector, ϕ, is given by the surface integral

$$\phi = \int_s \vec{B} \cdot d\vec{s}$$

where ds is an element of area.

A suitable element of area, normal to the B vector flow lines, is the infinitesimal rectangle of area $h\,dr$, as shown in the figure. On this basis we may indicate the flux, ϕ, by

$$\phi = \int_{r_1}^{r_2} B(r)h\,dr$$

$$= \int_{r_1}^{r_2} \frac{\mu Nih}{2\pi} \frac{dr}{r}$$

and integrating, we have

$$\phi = \frac{\mu Nih}{2\pi} \ln \frac{r_2}{r_1}$$

A common approximation for this expression may be derived from a power series expansion for the natural logarithm,

$$\ln \frac{(n+1)}{(n-1)} = 2 \left[\frac{1}{n} + \frac{1}{3n^3} + \frac{1}{5n^5} + \cdots \right]$$

The term $\ln(r_2/r_1)$ may be placed in this form by defining

$$(r_2 + r_1)/2 = r_{ave}, \quad \text{and} \quad \Delta r = r_2 - r_1$$

$$\text{and therefore } r_2/r_1 = \frac{2r_{ave}/\Delta r + 1}{2r_{ave}/\Delta r - 1}$$

if we approximate $\ln r_2/r_1$ by the first term of the power series above based on the usual situation where $r_{ave}/\Delta r$ (and hence n in the power series) is large and the series converges rapidly, we have

$$\phi \simeq 2 \frac{\mu Nih}{2\pi} \left[\frac{\Delta r}{2r_{ave}} \right] = \frac{\mu Nih\,\Delta r}{2\pi r_{ave}}$$

where we note that the quantity $h\,\Delta r$ is simply the cross-sectional area of the core A and we have

$$\phi = \frac{\mu A}{2\pi r_{ave}} Ni \tag{3-1}$$

The approximation involved is seen to be equivalent to assuming that the B vector is of constant magnitude determined by the average radius r_{ave}. Unless the inner diameter of the toroid is very small compared with the outer diameter, this is a reasonably good approximation.

We see from Eq. (3-1) that the flux, ϕ, is proportional to the term Ni, which is often called the *magnetomotive force,* and symbolized $\mathscr{F}$. The proportionality constant is called *permeance* and symbolized $\mathscr{P}$. Thus we have

$$\phi = \mathscr{P}\mathscr{F} \tag{3-2}$$

where

$$\mathcal{P} = \frac{\mu A}{2\pi r_{\text{ave}}} \tag{3-3}$$

Sometimes we see the reciprocal of permeance used, this is called *reluctance* and symbolized $\mathcal{R}$, where $\mathcal{R} = 1/\mathcal{P}$.

It should be emphasized that Eq. (3-2) is a consequence of the linear assumptions of the preceding work. For cases involving steel, the permeability is not a constant and hence the permeance is not a constant. We do find it convenient, however, to use the permeance idea even in such cases as a first approximation and to gain an understanding of the magnetic field behavior.

The total flux of the B vector through the N turns of the coil can be closely approximated by multiplying the flux, ϕ, in the core cross section by the number of turns to get $\phi_{\text{TOT}} = N\phi$. This procedure is justified by the fact that the permeability of the steel is very much greater than that of the air or nonmagnetic materials outside the core and so the B vector outside the core is very small. The total flux of the B vector linking the coil is usually called flux linkages and symbolized by the greek letter lambda, λ, rather than ϕ or ϕ_{TOT}, hence

$$\lambda = N\phi$$
$$= N^2 \mathcal{P} i \tag{3-4}$$

The coefficient of i in Eq. (3-4) is known as the coefficient of self-induction, or *inductance* for short and symbolized L. Thus we have

$$\lambda = Li \tag{3-5}$$

where

$$L = N^2 \mathcal{P} \tag{3-6}$$

The expression for permeance in Eq. (3-3) above is relatively simple for the example structure. For more complex geometries it would be more difficult to evaluate $\mathcal{P}$. We will extrapolate the observations of the sample case to conclude that there *is* a permeance constant for other structures (subject to the linearity assumption) but it may be necessary to use elaborate computer programs to evaluate the quantity. Designers of power apparatus will need to become familiar with such approaches but, for our present purposes we will content ourselves with assuming that a permeance can be found for cases of interest.*

If we have two coils, with turns N_1 and N_2, involved in a problem then we can find inductance coefficients for the two from the use of Eq. (3-6) as

$$L_1 = N_1^2 \mathcal{P}_1$$
$$L_2 = N_2^2 \mathcal{P}_2$$

In addition to the self-inductance terms, we may have flux in one coil set up by current in another as

$$\phi_{12} = \mathcal{P}_{12} N_2 i_2$$

---

*We will investigate *somewhat* more complex structures as we need to do so, particularly in the next chapter, where we must include the effects of air gaps in the magnetic path.

and

$$\lambda_{12} = N_1 \phi_{12}$$
$$= N_1 N_2 \mathcal{P}_{12} i_2 \tag{3-7}$$

The coefficient of i_2 in Eq. (3-7) is called the coefficient of mutual induction or *mutual inductance* for short, symbolized M, as in

$$M = N_1 N_2 \mathcal{P}_{12} \tag{3-8}$$

It may be shown that, for a linear case the permeance coefficient is the same in either direction; that is, $\mathcal{P}_{12} = \mathcal{P}_{21}$ and the mutual inductance is the same from coil 1 to coil 2 as it is from coil 2 to coil 1.

For the theoretical case where $\mathcal{P}_1 = \mathcal{P}_2 = \mathcal{P}_{12}$, we can show the interesting relationship

$$M = \sqrt{L_1 L_2} \tag{3-9}$$

To have the three permeance coefficients equal would require that the two coils be wound in exactly the same space, and this cannot happen, so the actual relation between the inductance coefficients is given by

$$M = k\sqrt{L_1 L_2} \tag{3-10}$$

where k is known as the *coupling coefficient* and is a quantity less than unity for actual apparatus.

From Faraday's law

$$e = d\lambda/dt$$

we have the expressions for voltages of self- and mutual induction when the currents vary as

$$e_1 = L_1 \frac{di_1}{dt}$$

or, for the component of voltage e_1 in coil 1 owing to a varying current i_2 in coil 2.

$$e_1 = M \frac{di_2}{dt}$$

Since inductance coefficients are functions of the geometry, and physical relations may change with time, the above relations may include another term to account for the variation in inductance with time. This will be discussed in the next chapter under electromechanical energy conversion, but our current objective does not require the inclusion of such terms.

It is important in dealing with the parameter, mutual inductance, to know whether two currents aid or oppose each other in the direction of setting up magnetic flux. To determine this, we need to know how the devices are physically oriented with respect to each other. Consider, as an example, two coils as in Figure 3-2, which shows several quasi-physical pictures of two coils with mutual magnetic flux coupling the two together. It is common to mark the terminals with a distinguishing mark such as a dot to show which terminals share a common directional relation to the mutual flux. In part (a) of the figure it is easily seen that the left-hand terminals of the two coils bear the same relation to mutual flux, so they are both marked with

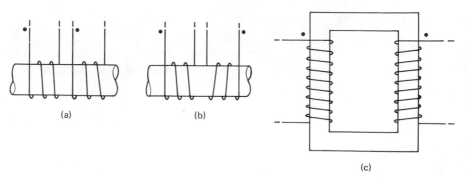

Figure 3-2   An illustration of marks to show polarity of coupled coils.

dots. We could just as well have marked the two right-hand terminals, since they too have a corresponding relation to mutual flux. If we are marking terminals, the choice of the first one to dot is arbitrary; the second then must follow from the physical relation. In part (b) of the figure it is only slightly more difficult to tell from the picture how the leads should be marked. In part (c) it may be necessary to use the right-hand rule or some such aid to determine which leads have the corresponding relation to the mutual flux.

To use the right-hand rule in part (c) of the figure, we imagine a test current into one lead of one coil. We then wrap the fingers around the coil in the direction of the test current and observe the thumb to give the corresponding direction of the magnetic flux. A trial will indicate the terminal of the second coil into which test current must be introduced to cause flux in the same direction. Then we may place dots on the corresponding ends of the two coils. It is important to note that we are talking about an imaginary test current in this procedure. The actual reference direction for currents in the coils is not constrained by the choice we make to determine the relative polarities of the coil ends.

When two coils are coupled magnetically, it is common to represent them by symbols such as those shown in Figure 3-3, where the physical orientation and/or winding direction information is given by the dots on the coil symbols rather than

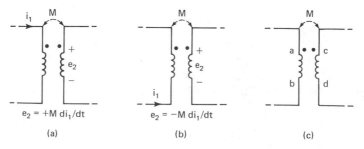

Figure 3-3   Mutual inductance effects in a circuit situation.

by a quasi-physical picture. Here a voltage, e_2, is induced in the coil 2 by the current i_1. The equation giving this voltage comes from

$$e_2 = d\,\lambda_{21}/dt = d(Mi_1)/dt = M\,di_1/dt \qquad (3\text{-}11)$$

It must be noted that, once you have chosen a current reference direction and a voltage reference polarity, the voltage $M\,di/dt$ must be written with a plus or minus sign, depending upon the relation of these references to the physical coil as given by the dots on the circuit symbols.

The sign of the $M\,di/dt$ terms in Figure 3-3 deserves comment. When we write Faraday's law as $e = +d\lambda/dt$, we infer that the positive direction of flux is related to a current direction by the right-hand rule, and further that the voltage e is a drop in the current direction; the current enters the plus sign of the voltage reference. This is as it must be from the viewpoint of conservation of energy: if λ is increasing in a positive sense, stored energy in the field is increasing and this energy is being absorbed from the circuit. Remember that when the current enters the plus sign of the voltage then $p = ei$ is an expression for the power *absorbed*. The two-coil case is a simple extension of this reasoning: If both reference current directions enter the dots, the voltage terms $L\,di/dt$ and $M\,di/dt$ will both be positive in terms of drops in the current direction. If, as in part (b) of the figure the current enters the undotted terminal of one winding the voltage $+M\,di/dt$ is positive in terms of a drop (plus to minus) from the undotted terminal of the second winding to the dot.* We have the opposite voltage reference in the figure, and hence the term is $e = -M\,di/dt$.

Rather than go through the mental gymnastics of a long chain of argument as above, most people, having once understood the physical happenings, would prefer some kind of short memory aid. Let us try to formulate a statement to describe the sign of terms involving mutual inductance:

> When a current reference enters the marked (dotted) end of a coil, the voltage of mutual induction in another coil is $+M\,di/dt$ in terms of a reference polarity of plus $(+)$ on the marked (dotted) end of the second coil.

> Each reversal of any reference multiplies the term $M\,di/dt$ by minus one (-1).

Once more before we leave this tedious but regrettably necessary discussion of plus and minus signs, let us refer to part (c) of the figure. In terms of double subscripts, we would write equations for the voltage of mutual induction as for example

$$e_{cd} = +M\,di_{ab}/dt \quad \text{or} \quad e_{dc} = -M\,di_{ab}/dt$$

As with the first two chapters of this book, we quickly leave the equations in terms of instantaneous time functions in favor of the phasor representation of

---

*In this treatment, we are assuming that M is inherently a positive number. Sometimes in the multicoil case, the dots are assigned arbitrarily without regard to the physical relation of the coil ends. The mutual inductance coefficients L_{ij} are then assigned positive or negative numerical values to conform to the physics of the situation.

sinusoids, which are our primary interest in the present work. If $i(t) = \text{Im } I\varepsilon^{j\omega t}$, $di(t)/dt = \text{Im } j\omega \, I\varepsilon^{j\omega t}$ where I is the phasor representation of a sinusoidal current. The voltage $e(t) = \pm M \, di(t)/dt$ is then given by $e(t) = \pm \text{Im } j\omega MI\varepsilon^{j\omega t}$ and the phasor representation of this sinusoid is simply $E = \pm j\omega MI$ where the sign of the terms derives from the same polarity considerations as that of the instantaneous form. The quantity ωM is simply called the *mutual reactance*.

3.2 A COUPLED-COIL EXAMPLE

An application of the foregoing ideas to a typical situation is illustrated in Figure 3-4, and consideration of this circuit leads us ultimately into our principal topic for this chapter, which is the *transformer*.

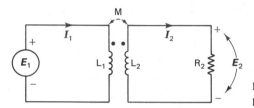

Figure 3-4 Example of magnetically coupled circuits.

If we write the Kirchhoff's law equations around the two loops shown, we have the equations

$$E_1 = j\omega L_1 I_1 - j\omega M I_2 \tag{3-12}$$

$$j\omega M I_1 = j\omega L_2 I_2 + R_2 I_2 \tag{3-13}$$

It will be noted that each loop has two inductive voltage terms, one term is due to the current in the loop itself, and the other term is due to the current in the other loop and its mutual magnetic effects. The student is well advised to think through the reasons for the signs on the various terms in these equations.

If we solve Eq. (3-13) for I_2 we have

$$I_2 = \frac{j\omega M}{R_2 + j\omega L_2} I_1 \tag{3-14}$$

If M is not equal to zero there will be a current I_2 whenever there is a current I_1, and, since electrical energy is thereby consumed in the resistance, R_2, this energy must have been transferred through the magnetic coupling from the source on the left.

If we now insert the value for I_2 of Eq. (3-14) into Eq. (3-12) we have an indication of how this transfer is reflected into loop 1.

$$E_1 = \left[\frac{\omega^2 M^2 - \omega^2 L_1 L_2 + j\omega L_1 R_2}{R_2 + j\omega L_2} \right] I_1 \tag{3-15}$$

The coefficient of I_1 is simply the complex impedance "seen" by the source E_1 and involves R_2 modified by a somewhat complex relationship owing to the circuit.

Because the relation of Eq. (3-15) is complex, it is common to consider some idealized relations and their consequences. If we wish to maximize the energy transfer into R_2 for a given I_1 we see from Eq. (3-14) that M should be large. We make M large for a given structure by winding the coils in close proximity. Carried to the theoretical limit, we would wind the coils in the same physical space which would cause the permeance coefficients to all be equal as $\mathcal{P}_1 = \mathcal{P}_2 = \mathcal{P}_{12}$. With this assumption we have $M = \sqrt{L_1 L_2}$ (unity coupling) and Eq. (3-15) simplifies to

$$E_1 = \frac{j\omega L_1}{R_2 + j\omega L_2} R_2 I_1 \tag{3-16}$$

If the permeance coefficients are made very high by using a highly permeable medium for the core then $\omega L_2 \gg R_2$ and Eq. (3-16) simplifies to the forms

$$E_1 = \frac{j\omega L_1}{j\omega L_2} R_2 I_1$$

$$= \frac{\mathcal{P}_1 N_1^2}{\mathcal{P}_2 N_2^2} R_2 I_1$$

$$= \frac{N_1^2}{N_2^2} R_2 I_1 \tag{3-17}$$

The coefficient of I_1 in Eq. (3-17) is seen to be the resistance R_2 modified by the square of the turns ratio under the idealized conditions outlined above. We say that the resistance R_2 has been *transformed* by the device.

  With the same idealized assumptions (high and equal permeances for all terms), we can find similar relations between the voltages and currents on either side of the coupled coils.

  From Eq. (3-14)

$$I_2 = \frac{M}{L_2} I_1$$

$$= \frac{\mathcal{P}_{12} N_1 N_2}{\mathcal{P}_2 N_2^2} I_1$$

$$= \frac{N_1}{N_2} I_1 \tag{3-18}$$

Also, since $E_2 = R_2 I_2$, combining Eqs. (3-18) and (3-17) we have

$$E_2 = \frac{N_2}{N_1} E_1 \tag{3-19}$$

The three preceding equations are those of a postulated device known as an *ideal transformer*. The perfect unity coupling and the infinitely high permeance required to realize these relations exactly is not actually attainable. It *is* possible to approach these conditions so closely, however, that the idealized relations of Eqs. (3-17) through (3-19) are approximately correct for many applications. At power frequencies (say 16-2/3 Hz to 400 Hz) these effects are accomplished by winding the coils of the device on a highly permeable steel core.

3.3 THE IDEAL TRANSFORMER: CLASSICAL THEORY

Modern power transformers are highly effective and energy-efficient devices. Very large transformers may run over 99 percent energy efficiency! For analysis of transformers, it is convenient to start by using the *ideal transformer* relations rather than treating the device by mutual inductance relations *per se*. Many problems of the practical world are solved by assuming that the actual transformer is very closely approximated by the ideal transformer. Regrettably, not all transformer problems can be solved using this approximation, so we must ultimately study the departures of the actual transformer from the idealized assumptions. This, then, will be our approach in the following pages: study the characteristics of an ideal device, illustrate the application to practical problems, study the imperfections, and put the resulting model to work solving problems.

Suppose we consider two coupled coils on a steel core of high magnetic permeability. Such coils are symbolized in Figure 3-5 where (a) shows a quasi-physical sketch and (b) shows the circuit symbol representation. The polarity dots on (b) eliminate the need for the quasi-physical sketch, since we can arrive at our voltage and current relations without actually having to see the relative winding direction.

The considerations of self- and mutual inductances of two coupled coils in Section 3.1 provide a link with the circuit theory courses the student is presumed to have completed. When it comes to the design and use of power and distribution transformers, the idealizing assumptions are often stated in different ways and lead to certain viewpoints and vocabulary useful in design and application. Let us again approach the equations of the ideal transformer (3-17) through (3-19) by means of the classical power engineer's viewpoint.

If we postulate that the coils are wound on a steel core of very high permeability then we can make the assumption that the magnetic field in regions outside the core is negligible. If this is the case, each turn of either winding is threaded by the same flux. This assumption disallows flux, which links only one winding or a portion of the one winding and not the other. Such flux is called *leakage flux,* which will be discussed later in these pages. This viewpoint is equivalent to the assumption that the permeances $\mathcal{P}_1$, $\mathcal{P}_2$, and $\mathcal{P}_{12}$ were equal in the preceding section.

If the flux varies sinusoidally, there will be a sinusoidal voltage generated in each turn of each coil according to Faraday. This quantity is called *volts-per-turn*

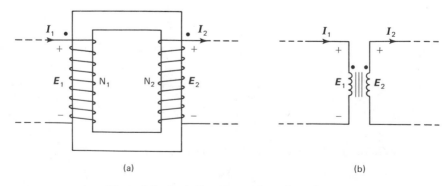

(a)   (b)

Figure 3-5   Symbolic representations of transformers.

and occupies an important role in transformer design. Let us represent this voltage per turn by the phasor E/turn. It follows that if winding 1 has N_1 turns, the total voltage E_1 will be N_1 times the voltage per turn, and similarly, if winding 2 has N_2 turns, E_2 is N_2 times the voltage per turn; that is, $E_1 = N_1 E$ and $E_2 = N_2 E$ and hence

$$\frac{E_1}{E_2} = \frac{N_1}{N_2} \tag{3-20}$$

This is one of the fundamental relations seen before describing the properties of the hypothetical ideal transformer. We say that the voltage ratio is equal to the turns ratio.*

Note that, in order for Eq. (3-20) to carry a positive sign, the polarity references of the voltage must be similarly oriented with respect to the reference dot (after all the same flux is causing both voltages). With a positive sign, the two voltage phasors are in phase (have the same angle).

Another of the ideal transformer relations follows from the idea that no energy can be stored or lost in such a device. Recall that the energy density in a magnetic field is given by

$$w_m = \int \vec{H} \cdot d\vec{B} \qquad \text{joules per } m^3 \quad \text{inside the material core}$$

and since $\vec{B} = \mu \vec{H}$, an infinite permeability infers zero $\vec{H}$, no matter how big the finite $\vec{B}$ has to be to provide the flux to induce the requisite voltage. Since $\vec{H}$ is zero, there can be no energy absorbed by the core and, if the windings have zero resistance, whatever complex power enters one winding must leave the other. In terms of the references of Figure 3-5 we have

$$E_1 I_1^* = E_2 I_2^* \tag{3-21}$$

by working with Eqs. (3-20) and (3-21) we have

$$\frac{I_1^*}{I_2^*} = \frac{N_2}{N_1}$$

and in turn

$$\frac{I_1}{I_2} = \frac{N_2}{N_1} \tag{3-22}$$

which is another of the ideal transformer relations. We say the currents are inversely proportional to the turns ratio. Again, the student would be well advised to study the interrelations of reference directions for power flow, voltage, and current that result in positive numbers in the foregoing equations. The two currents are in phase with each other with the choices used.

If we rearrange Eq. (3-22) we have $N_1 I_1 = N_2 I_2$, which says that the ampere turns of the two windings are equal. Since I_1 and I_2 are oppositely directed with respect to the dots, this means that there is a net of zero ampere turns applied to the

---

*This is close enough to the truth for actual power transformers that American standards call for the nameplate voltage ratio to be given as the turns ratio, even though the actual voltage will differ somewhat under load conditions.

core. This also follows from Ampère's rule, which states that

$$\oint \vec{H} \cdot \vec{dl} = \sum Ni$$

This conclusion (that the net ampere turns are zero) is consistent with the ideal transformer assumption of infinite permeability and zero magnetic field intensity, H.

Sometimes there are three separate windings on the same core, as indicated in Figure 3-6. If we extend the ideal transformer postulates to the three-winding case we find similar relations to those of the two winding-case.

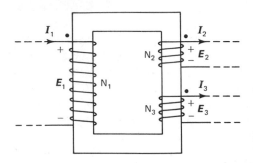

Figure 3-6   An ideal transformer with three windings.

The assumption that every turn links the same flux and the corollary that the volts per turn are equal for all windings leads to

$$\frac{E_1}{N_1} = \frac{E_2}{N_2} = \frac{E_3}{N_3} \tag{3-23}$$

The assumption of infinite permeability of the core leads to the equation

$$N_1 I_1 = N_2 I_2 + N_3 I_3 \tag{3-24}$$

which says that the net ampere turns must be zero. Again it is well to note and understand the reference directions that have led to the particular signs of the terms in Eqs. (3-23) and (3-24).

Extension of the properties to transformers with even larger numbers of windings follows the same pattern and will not be covered here.

In Figure 3-7 we show an ideal transformer connected between an impedance load, Z_2, and a source voltage, E_1. The various components are assumed to obey

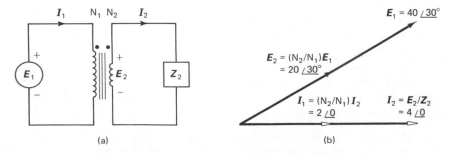

Figure 3-7   Illustration of transformer coupling.

Ohm's law and the ideal transformer relations, Eqs. (3-20) and (3-22); a phasor diagram illustrates these relations in part (b) of the figure based on an assumed impedance of $5\underline{/30°}$ and a voltage source, E_1 of $40\underline{/30°}$. The turns ratio is taken as $N_1/N_2 = 2/1$. The complex number arithmetic is given directly on the diagram since it is simple. If we take the values given for E_1 and I_1 we see that the source "thinks it sees" an impedance $E_1/I_1 = 20\underline{/30°}$ ohms. This is a specific illustration of the general case. From Eqs. (3-20) and (3-22) we may derive the expression

$$\frac{Z_1}{Z_2} = \left(\frac{N_1}{N_2}\right)^2 \tag{3-25}$$

which says that a transformer transforms impedances as the square of the turns ratio.* Note that, since the turns ratio is a real number (scalar), only the magnitude and not the angle of the impedance is changed. Lastly it should be pointed out that even though the apparent value of the impedance is changed, the power is invariant ($S_1 = 80\underline{/30°} = S_2 = 80\underline{/30°}$ in the example) as it must be from the basic postulates of the ideal transformer.

3.4 TRANSFORMER RATINGS

A further example of the use of the ideal transformer relations is found in the ratings of transformers. Although in theory an ideal transformer might be capable of handling any voltage or any current, no matter how large, there are practical limits to both voltage and current. The reasons for these limits will be more clear after the later sections dealing with imperfections of the transformers. Since both voltage and current have upper limits, a transformer's capability is given by the product of the two, in *voltamperes*. From the ideal transformer relations we see that the voltamperes *into* one winding of a two-winding transformer must equal the voltamperes *out* of a second winding. Consider the example of the preceding section where $S_1 = S_2$. The voltampere rating of a transformer is then given as the voltampere rating of either winding since the two are equal.† (Not the sum of the two!) In large power transformers the nameplate gives a voltamperes (or kVA or MVA) rating for the device as well as the voltage ratings of the two windings. The current ratings then follow from these data since $S = EI$. Small transformers, for example, those used in electronic power supplies, are often rated by giving the voltage and current ratings of each winding, from which the voltampere rating would follow if desired.

No mention was made of *power* in the above statements. In a practical transformer the relative phase angle of voltage and current has almost no effect on the voltage and current capabilities of the windings, and hence the *magnitude* of S is the important factor and how S is divided into P and Q is immaterial to the rating.

---

*The transformation may work in either direction, to make the impedance look larger or smaller. It is like a pair of binoculars, which may make objects look larger, or, if inverted, make them look smaller. It is worth remembering that the largest impedance is "seen" looking into the largest number of turns.

†When a transformer has three or more windings, it is necessary to define the rating of each winding separately and then to be careful that the loading does not exceed the capabilities of any one winding.

As an example of some of the things said above, suppose that we read from the nameplate of a large transformer at a hydroelectric generating station the following rating data:

$$\boxed{100 \text{ MVA}, \ 138\text{-}13.8 \text{ kV}}$$

These data now tell us other things by using the ideal transformer relations, for example:

$$N_1/N_2 = 138/13.8 = 10$$

since the voltage ratio and the turns ratio are the same under rating standards of large transformers. Also the rated current of the high voltage winding, which we call I_1 is given by

$$I_1 = 100 \times 10^6/138 \times 10^3$$
$$= 724.6 \text{ amperes rated current}$$

and

$$I_2 = 100 \times 10^6/13.8 \times 10^3$$
$$= 7246 \text{ amperes rated current for the low voltage winding}$$

It will be noted that this latter figure for I could also have been obtained by using the ideal transformer relation $I_1/I_2 = N_2/N_1$ if more convenient.

3.5 TRANSFORMER CONNECTIONS

In many cases transformer windings may be connected in other than the simple source-load arrangement of Figure 3-7, and then may also be used in combination with other transformers to perform some given function. In dealing with various connections it is necessary to compute the voltage and current ratings required of the several windings or winding parts. The ideal transformer relations are ordinarily used in such cases as an adequate approximation to the actual transformer values. Following are a few examples to indicate the pattern of applying the ideal transformer equations to sample situations.

As a first example, consider a case where it is desired to boost a voltage slightly when feeding a load from a 2400-volt source. Rather than purchase a transformer wound for 2400 volts on the one winding, and the desired voltage on the other, make use of a small 25 kVA, 2400-240–volts transformer by means of the connection shown in Figure 3-8(a). The analysis may perhaps be made more clear by drawing the connection as in part (b) of the figure. The source voltage, E_{in} is connected to winding 1 of the transformer to become E_1. Because of the turns ratio (10:1) the voltage E_2 is 240 volts. Because of the orientation of the dots on the windings vis-a-vis the connection scheme the voltage $E_{\text{load}} = E_1 + E_2$. Note that E_1 and E_2 are in phase by the ideal transformer relations and therefore $E_{\text{load}} = 2400 + 240 = 2640$ volts.

The current I_{load} is determined by the load and is seen to flow directly in winding 2 as I_2. Since the rated current of winding 2 is $25,000/240 = 104.2$ am-

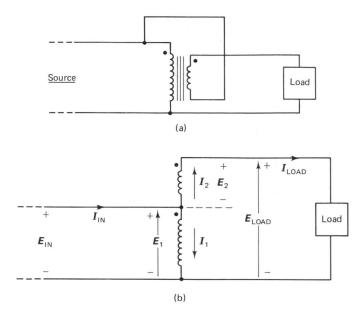

Figure 3-8 Illustration of a transformer connection scheme, an autotransformer.

peres, this will be the maximum allowable I_{load}. The current I_1 is given from the turns ratio relation $I_1 = (N_2/N_1)I_2 = (1/10)104 = 10.4$ amperes, the rated current of winding 1. The current I_{in} from the source is found from Kirchoff's current law to be $I_{in} = I_1 + I_2 = 114.4$ amperes.

It is interesting to note that the voltamperes into the load are $2640 \times 104 \times 10^{-3} = 275$ kVA and are equal to the volt-amperes from the source $2400 \times 114.4 \times 10^{-3} = 275$ kVA as must be from the ideal transformer postulate that all complex power going into one winding must come out of the other.

It is also interesting to note that the voltampere capability of the device in this connection is over ten times as large as in the simple source-load arrangement of Figure 3-7(a). This property of this type of connection is known as an *autotransformer* relation.

A second example of the use of the ideal transformer relations is shown in Figure 3-9, which illustrates a common connection of three transformers for three-phase transformation. Such a group is called a *bank*. Since there are three transformers and six windings, some simplification of notation is desirable. The conventions involved in this diagram are commonly used. First, let it be understood that two coil symbols drawn parallel are located on the same transformer core. Second, the orientation of the coil symbols is the same as the voltage phasors using the double subscript methods of earlier discussion. This implies that, if the top end of the winding from A to N bears a polarity dot, then the top end of the winding from a to c is the dotted end, thus $E_{AN} = (N_1/N_2)E_{ac}$.

By use of these conventions, the connection diagram serves double duty as a voltage phasor diagram also. We might, for example, read E_{ab} as a phasor of

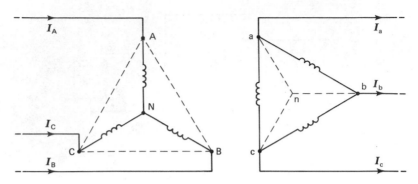

Figure 3-9 A three-phase Y-delta transformer bank.

$26\underline{/150°}$ by placing an arrowhead on the first subscript end and visualizing the position of the voltage phasor.

As a specific numerical example, suppose that we wish to transform three-phase power of 6000 kVA from 69 kV to 26 kV by means of a Y-delta bank. The Y-connected windings must accept the line-to-neutral voltage of $69/\sqrt{3} = 39.8$ kV and have a current rating of $6000/\sqrt{3}\,69 = 50.2$ amperes. Each transformer of the three-phase bank must therefore be rated at $39.8 \times 50.2 = 2000$ kVA as might have been expected. The low-voltage winding must be equal to the line-to-line voltage of 26 kV, and therefore the turns ratio is $39.8/26$ or $1.53:1$. The current rating of the low-voltage winding must then be that of the high-voltage winding times the turns ratio or $50.2 \times 1.53 = 76.92$ amperes. The current of 76.92 amperes flows inside the delta and it will be remembered that, for a balanced three-phase set the line current is $\sqrt{3}$ times this value, or 133.23 amperes.

It should be noted that, in addition to the change in magnitudes of voltages and currents, there is a phase shift between the two sides of the transformer bank. The triangle of line voltage phasors on the high side is *ABC*, while that on the low side is *abc*. From symmetry of the figure we see that the low side triangle is shifted 30° ahead of the high side. The same effect is noted in dealing with the line-to-neutral voltages *AN* and *an*, respectively. The delta winding does not have a neutral *per se* but point *n* is the geometrical neutral of the voltage triangle and may exist physically also because of Y-connected apparatus elsewhere on this side of the transformer. (At the very least there is capacitance to ground forming a Y-connected "load.")

Just as in the case of the voltages, there is a 30° phase shift in the currents between the two sides of the transformer bank. This is shown in Figure 3-10 by drawing the current phasors for a sample balanced case. Note that current I_{ca} is in phase with I_A and I_{ab} is in phase with I_B by ideal transformer theory. By Kirchhoff's law, $I_a = I_{ca} - I_{ab}$ and this is seen on the diagram to be a phasor $\sqrt{3}$ times as long as the coil currents and advanced 30° in phase from the current I_A on the high-voltage side. It should be pointed out that it is necessary that the voltage phasors and the current phasors be shifted by the same angle from one side of the bank to the other so that complex power is invariant to the transformation.

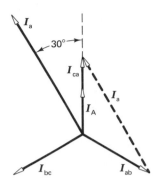

Figure 3-10   Relation between coil currents
and line currents in the delta.

Another example of the use of ideal transformer relations is given in terms of the Y-Y connection of Figure 3-11. Again, the connection diagram serves the dual purpose of a voltage phasor diagram by agreeing on the same conventions of notation already covered. The Y-Y connection seems much simpler. There is no phase shift of voltage or current and the corresponding quantities on each side simply bear the turns ratio relation in the proper direction.

In spite of the apparent simplicity of the Y-Y connection, it is not commonly used as shown for practical reasons. One reason may be appreciated by considering a single-phase load to neutral as shown on the sketch. (This might also be considered a small unbalanced component of an otherwise balanced three-phase load.) If we postulate a current I_a in the load, we must have a current $(N_2/N_1)I_a$ in the other winding for phase A. I_b and I_c are zero, however, so I_B and I_C must be zero and I_A cannot be other than zero from Kirchhoff's law at the neutral. We conclude then that the connection as shown cannot support a line-to-neutral unbalanced load. With zero current in the impedance, the voltage E_{an} must be zero; that is, the neutral collapses and point n potential shifts to point a.

The difficulties outlined above may be eliminated if a fourth wire is connected to the neutral on the high side, but this has attendant disadvantages of economy and performance and may not be desirable. As a result, a Y-Y connection as shown in Figure 3-11 is seldom used.

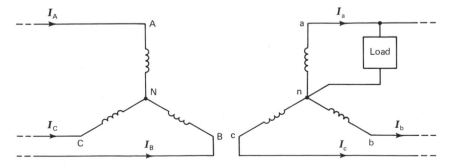

Figure 3-11   A Y-Y three-phase transformer bank.

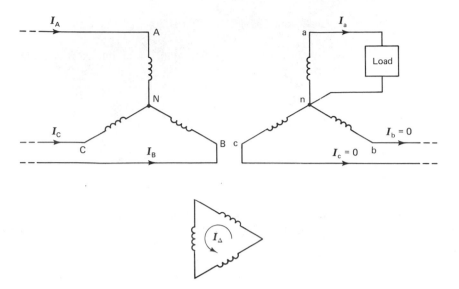

Figure 3-12 A Y-Y-Delta transformer bank using three winding transformers.

If for some reason it is desired to use the Y-Y connection in spite of the attendant difficulties, another way of curing the troubles is shown in Figure 3-12, illustrating the use of third windings on each transformer connected in delta as shown. Now if we consider the previously troublesome case of a line-to-neutral load on the low side we find rather different performance by the transformer bank. First, in regard to voltage, we see that the three-coil voltages around the delta must add to zero, therefore the three line-to-neutral voltages of the Y-connected windings must add to zero and *there can be no neutral shift.* Under these conditions there will be a current circulating in the delta. The delta current plus the currents in the Y-connected coils must satisfy the ampere turn balance of the ideal transformer and also Kirchhoff's current law, and we result in the following relations. For simplicity let us assume that the turns ratio between all the coils is $1:1$.

$$I_A + I_\Delta = I_a \qquad \text{(ampere turn balance)}$$
$$I_A + I_B + I_C = 0 \qquad \text{(Kirchhoff)}$$
$$I_B = I_C = -I_\Delta \qquad \text{(ampere turn balance)}$$

By solving the above equations we find:

$$I_\Delta = I_a/3$$

$$I_A = \frac{2}{3} I_a$$

$$I_B = I_C = -\frac{1}{3} I_a$$

The required ratings of the several windings in the transformer bank of this type are thus obtained from the ideal transformer relations.

The number of different combinations of transformer windings into connection schemes is very large, and space will not permit discussion of any more examples, but the examples thus far shown will serve to give insight into how the ideal transformer relations are applied to find voltages and currents in various connection patterns.

3.6 TRANSFORMER IMPERFECTIONS: THE CORE

The ideal transformer relations give very good answers to many transformer problems, as in the examples preceding this section. For some problems, however, we must take account of the departures from perfection to get an adequate answer to a transformer problem. The first imperfection we will discuss is that of the core. The core is *not* infinitely permeable, it does require ampere turns to establish the flux, and in addition, there are internal energy losses in the core when the flux varies with time.

To illustrate the influence of the core properties, suppose we have a transformer with N_1 turns on coil 1 and a core with flux varying according to the equation $\phi = \phi^m \sin \omega t$. This flux variation is sketched in Figure 3-13. The induced voltage is then given by

$$e_1 = N_1 \frac{d\phi}{dt} = N_1 \omega \phi^m \cos \omega t .$$

We see that the induced voltage is a sinusoid having a peak value of $N_1 \omega \phi^m$ volts and rms value of*

$$E_1 = \frac{N_1 \omega \phi^m}{\sqrt{2}} \tag{3-26}$$

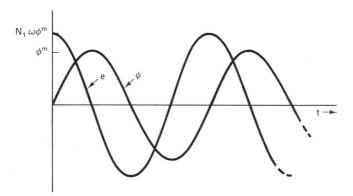

Figure 3-13   Relation between flux and induced voltage.

*We might note that for the practically interesting case of $f = 60$ Hz ($\omega = 377$), a peak flux of 0.00375 weber will induce one volt per turn in a transformer winding, a constant useful in design.

If we choose to treat this situation in complex numbers, we have

$$\phi = \text{Im } \boldsymbol{\phi}^m \varepsilon^{j\omega t}$$

$$e = N_1 \frac{d\phi}{dt} = \text{Im } j\omega N_1 \boldsymbol{\phi}^m \varepsilon^{j\omega t}$$

and thus we see that if the flux is represented by a phasor $\boldsymbol{\phi}^m$ then the voltage would be represented by the phasor $j\omega N_1 \boldsymbol{\phi}^m$ or, as is more usual, by an rms phasor

$$j \frac{\omega N_1 \boldsymbol{\phi}^m}{\sqrt{2}}$$

By whatever means we use (phasor or time plot), we see that the voltage and flux are related by a constant and that the flux lags the voltage by 90°, or conversely, the voltage leads the flux by 90°. This last sentence is of course based on the references chosen for voltage and flux.*

The postulate of a certain $\phi(t)$, above, implies the presence of a B vector field in the material where the magnitude of the B vector is given approximately by $B(t) = \phi(t)/A_{core}$. The presence of a B field also requires an H field. If the flux, and hence the B, vary sinusoidally at any given frequency, the B vs. H plot will form a loop called the *hysteresis* loop after several cycles have passed and a steady-state is reached. The arrows on the loops indicate the direction of traverse around the loop as B goes through a sequence of values according to the sinusoidal variation required. The word *hysteresis* relates to the fact that the value of B lags behind the values of H, for example, when H is reduced to zero the B magnitude remains non-zero. For any specific maximum value of B, noted as B^m for convenience, there will be a perfectly definite loop. For other values of B^m, there will be other loops, with larger values of B^m causing larger loops. Part (a) of Figure 3-14 illustrates these loops for two values of B^m. A laboratory test will give us a whole sequence of these loops for various values of B^m. By connecting the tips of this family of loops we form a curve like that of part (b) of the figure, and this curve is known as the *normal magnetization curve*. Since the curves are nonlinear (and [a] is even multivalued), they are difficult to use mathematically. We often approximate by various means and work with curves like (a) or (b) of Figure 3-14, depending on the problem. In any case the curves are characterized by high values of H for high values of B. We say that the steel *saturates* and practical designs use the steel in the region near or below the bend or *knee* of the normal magnetization curve.

Figure 3-15 shows the sinusoidal time variation of the B vector and relates this to corresponding points on the B-H loop. In particular, point P is identified and the direction of traverse around the loop is shown by arrowheads. Sample point P is in the rising portion of B vs. t, and is so shown on the loop in (b). It will be noted that the maximum value of B corresponds to the maximum value of H. The current

---

*A suitable reference relation is the positive direction of flux to a current reference by the right-hand rule, where the current enters the $+$ mark of the voltage reference. Note that this choice means energy is being absorbed as the magnetic field increases.

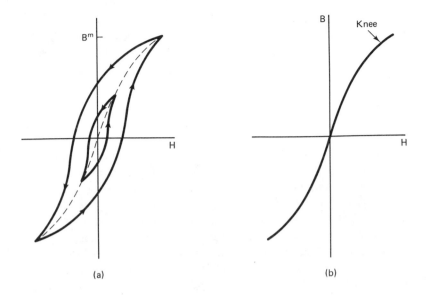

Figure 3-14   (a) Typical Hysteresis loops of steel, and (b) the normal magnetization curve.

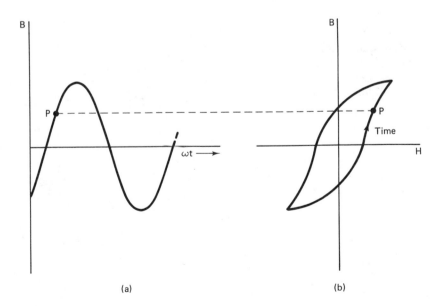

Figure 3-15   (a) Variation of B with time, and (b) corresponding points traveling around the B-H loop.

and the H vector are related by Ampère's circuital law, $Ni = \oint \vec{H} \cdot \vec{dl}$. For our present purposes, we might approximate by $Ni = Hl$ which is valid if the core has a constant cross-sectional area and then the current required to establish a given value of H is given by

$$i = \frac{Hl}{N}$$

where l is the path length in the direction of the H vector and N is the number of turns on the coil supplying the magnetomotive force, Ni.

A succession of points in time will give a succession of current values which, when plotted, ordinarily look much like the plot of Figure 3-16 where the sample point P is again identified. The curve is very much like that of Figure 2-22, in Chapter 2. This current is called the *exciting* current.

It will be noted that the maximum value of the exciting current corresponds to the maximum value of B, B^m, and hence the maximum of H. The significance of the normal magnetization curve of Figure 3-14 is thus revealed. In design of a transformer we must not use B^m values too far out on the B-H curve, lest the exciting current peak up too high (and losses described below be excessive). Again let it be noted that the distorted waveform like Figure 3-16 may be described by a Fourier series, since the waveform is periodic. The relative magnitude of the harmonics is also affected by the value of B^m, as may be imagined by considering the two B-H loops of Figure 3-14(a). In practice it is sometimes necessary for the manufacturer to guarantee the harmonic composition of the exciting current waveform, and a Fourier series analysis on a discrete sampling basis of the waveform may be made. We will not pursue this aspect further, however.

Before going on with this general topic, it may be helpful to digress with a short design illustration relating to some of the foregoing material. Suppose that we need a nonstandard voltage source of 38 volts to supply a one-ampere load and wish to build a transformer to feed this load from a 120-volts, 60-Hz source. As a first trial, we find a transformer core is available that has a cross-sectional area of 2.5 × 3 cm², which seems from comparison with other units to be of about the right size to supply the 38 voltamperes of the load. Our problem is to find how many turns of what size wire to wind on each of the two coils.

As a preliminary, we might wind a trial coil on the core, and go to the laboratory and find the normal magnetization curve—and on this basis choose a

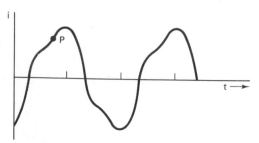

Figure 3-16 Typical exciting current waveform.

suitable B^m. On the other hand, manufacturer's data or previous experience may lead us to skip this step and simply select a value of $B^m = 1.25$ teslas (Webers per square meter), which we do as follows: The maximum value of ϕ is then found from the equation $\phi = BA$ or $\phi^m = 1.25 \times 2.5 \times 3 \times 10^{-4} \times 0.95 = 8.91 \times 10^{-4}$ weber. The factor 0.95 in the foregoing comes from our realization that the core is not solid steel but is built up from flat sheets or *laminations* and we estimate that only 95 percent of the gross area is actually steel. The factor 95 percent is called the *stacking factor*. This value of ϕ^m will generate $8.91 \times 10^{-4}/0.00375 = 0.24$ volt per turn at 60 Hz per the fact previously cited that 0.00375 weber will induce one volt per turn at 60 Hz. The 120-volt winding will then require $120/0.24 = 500$ turns and the 38-volt winding will require $38/0.24 = 160$ turns.

Reference to copper wire tables tells us that No. 22 wire would be a suitable choice to carry the 1-ampere load of the 38-volt winding. Since the current in the 120-volt winding is $1 \times (38/120) = 0.32$ ampere, we find that No. 26 wire size would be a good choice for this winding.

A sketch of one possible core configuration (called the *shell* type) is given in Figure 3-17.* On a core such as this the coils are wound on the center leg (and it is this leg that has the cross section described above). The coils are shown in cross section in the figure where it will be noted that they pass through holes in the core structure called the *window*.

Our wire table tells us that we can place 1000 turns of No. 22 wire in one square inch of cross section and the corresponding figure for No. 26 wire is 2300 turns per square inch. If we allow about 30 percent of the window area for insulation, we require a window area of $(500/2300 + 160/1000) \times 1.3 = 0.49$ in^2 or 3.16 cm^2. If our core has a window area of at least this size, we can proceed. If the window area is not big enough, we cannot get the requisite number of turns of the wire size chosen through the area. We might then sharpen our pencil and operate with a higher B^m (maybe 1.5 w/m^2) or a smaller wire size, but either recourse may lead to excessive heating of the transformer. A larger core would then be the probable answer.

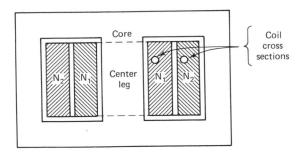

Figure 3-17   A shell-type transformer.

*See also Figure 3-24 for a more detailed view of the shell type of construction. The type of construction of Figure 3-5(a), by contrast, is known as a *core* type.

Following the hypothetical design example, we pursue the details of the core imperfection in greater detail with a view toward developing an equivalent circuit model that will account for the fact that the ideal transformer relations do not *exactly* describe the behavior of an actual transformer.

For every value of B in curves like Fig. 3-15 there is a corresponding value of flux, $\emptyset$, and for every value of $\emptyset$ there is a corresponding value of flux linkages, λ. For a simple core structure the flux linkages, λ, might be approximated by $\lambda = NBA$. Likewise, for every value of H there is a corresponding value of current, and for the simplified core relations treated earlier, this is given by $i = Hl/N$. For some purposes of analysis we may wish to study a plot of λ vs. i and we see that such a plot will be very similar in form to the loops of Fig. 3-14(a) as in the example of Figure 3-18.

As a loop like that of Fig. 3-18 is traversed, there is an energy interchange that may be analyzed as follows. By Faraday we have $d\lambda = e\,dt$, and, if we multiply both sides by the current, we have $i\,d\lambda = ei\,dt$, which is an incremental amount of energy in watt-seconds or joules; $dW = i\,d\lambda$ is the incremental amount of energy *absorbed* by the coil over the time increment dt. This is shown graphically by the small shaded area in the figure.

The significance of the loop can be understood by considering the sequence of flux linkages, λ, and current as time passes and the state of the core passes sequentially from point 1 to 2 to 3 to 4 and back to 1. As the current passes from zero at point 1 to the maximum positive value at point 2, energy is absorbed proportional to the area to the left of the branch of the trace from 1 to 2. As the state passes from 2 to 3 the energy absorbed is numerically negative; that is, energy is returned to the source. The area to the left of this branch is less than that of the left of the ascending branch, so not all the energy is returned. The story is the same for the second half of the loop from 3 to 4, and to 1 and the total energy absorbed is thus proportional to the loop area. We might quantify this as

$$W/\text{cycle} = k_1 k_2 A \qquad \text{joules}$$

where k_1 is the scale constant of the vertical axis in webers per cm, k_2 is that of the horizontal axis in amperes per cm, and A is the area of the loop in square centimeters. If we traverse the loop at the rate of f cycles per second or Hertz then the *rate* of energy absorption is $Wf = k_1 k_2 Af$ joules per second or *watts* of power.

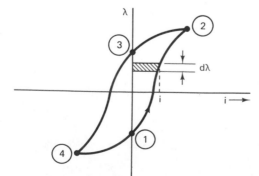

Figure 3-18 Typical λ vs. i plot of open-circuited transformer.

The power absorbed by the transformer under the above conditions is lost to the circuit and goes into heat in the core. It is called *core loss*. The reasons for the loss are twofold: *hysteresis* and *eddy currents*.

The hysteresis loss occurs as an inherent property of the magnetic material. The internal structure of a ferromagnetic material is organized into domains and these domains are reoriented as the *B* vector goes through a cyclic change in magnitude or direction. An internal energy loss appearing is the result. The energy loss may be minimized by suitable alloying and heat treatment of the metal. The treatment processes may also affect the mechanical properties, however, so compromises must be made.

If the frequency of the applied voltage were reduced but the range of *B* in the core maintained (by applying lower voltage), a similar loop would be observed but with smaller area than that originally observed. The reason for the larger area with higher frequency is the effect of eddy currents in the steel. The steel is a conductor and, as the flux in the steel varies, voltages are induced within the closed contours in the material. Currents flow as a result of the voltage and an I^2R loss occurs known as eddy current loss. The loss is reduced by building the core from sheets of steel called laminations and by increasing the resistivity of the material by alloying.

It is instructive to consider analogous relations to the λ vs. i plot with linear circuit elements rather than the nonlinear transformer.

Since voltage is given by $e = d\lambda/dt$, λ is given by the integral $\lambda = \int edt$. If we have a voltage $e = E^m \sin \omega t$ across a resistor R, the current in the resistor is $i = E^m/R \sin \omega t$. If we now consider $e' = \int edt = -E^m/\omega \cos \omega t$, the integral is analogous to the quantity λ and plotting e' vs. i results in a plot like Fig. 3-19(a) since e' and i form the equations of an ellipse in parametric form. It is reserved as a problem exercise for the students to show that the area of the ellipse is proportional to the energy dissipated in the resistor in joules per cycle.

If we consider a linear (constant) inductor where $e = L(di/dt)$, the parametric equations with $e = E^m \sin \omega t$

$$i = -\frac{E^m}{\omega L} \cos \omega t$$

$$e' = \int edt = -\frac{E^m}{\omega} \cos \omega t$$

which are the equations of the straight line of Fig. 3-19(b).

Last, if we use the resistor and inductor in parallel we combine the two shapes. At any given instant the inductor and resistor have the same voltage across them and their currents add. The result is that the ellipse is tilted as shown in Figure 3-20. The similarity of this figure to that of the actual transformer shape in Figure 3-15(b) is very evident. It appears that, with a proper choice of R_0 and L_0, we could approximate the actual characteristics very closely. The circuit of Figure 3-20(a) is commonly used, together with the ideal transformer symbol, to give a model that accounts for the actual transformer's departure from the ideal because of the lack of infinite permeability of the transformer core. Such a representation is shown in Figure 3-21.

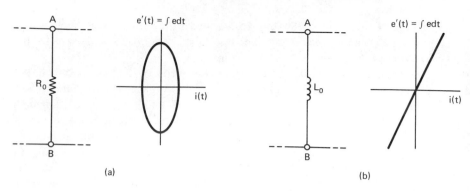

Figure 3-19   Relation of $e'(t)$ and $i(t)$ for linear elements.

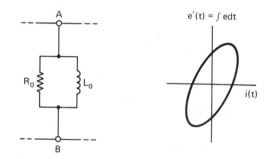

Figure 3-20   A linear model for transformer exciting characteristics.

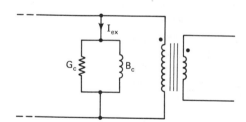

Figure 3-21   The use of a fictitious circuit added to the ideal transformer to account for exciting current.

The loops on the *B-H* or λ-i plane may be easily shown in the laboratory on a cathode-ray oscilloscope screen using the circuits of Figure 3-22. In the circuit the voltage developed across the capacitor C is proportional to the integral of the applied voltage e (provided that the product RC is great with respect to the period of the applied voltage). The voltage across the resistor r is, of course, proportional to the current i. The result is a plot of λ vs. i on the scope screen.

The equivalent circuit of Figure 3-21 is, of course, an approximation, but it is valid for most purposes.* The advantages of using a linear model far outweigh the slight error introduced by a linear model. In any case the exciting current of a

---

*The circuit diagram shows the resistor and inductor labeled in terms of admittance components. This is a common choice because of the parallel connection.

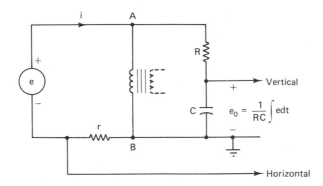

Figure 3-22   Laboratory set up to study core characteristics.

modern transformer of any size at all is only a very small percentage of the full load current, so if there is a slight error in the linear representation, it amounts to very little in terms of the total current passed by the transformer. The model to use for a certain study is of course a matter of experience and engineering judgment. For heavy load or short circuit studies the exciting current branch is normally omitted entirely. For light load or no load the exciting current branch may by included. For studies involving the wave form distortion of the exciting current the linear model is completely inadequate.

Evaluation of G_c and B_c of Figure 3-21 involves an approximation to best model the actual transformer in some sense of a most useful model. The usual method of evaluating the parameters is to choose G_c and B_c so that the exciting current has the same rms value as the actual exciting current and the power loss in G_c is the same as the actual core loss. For example, suppose that a certain transformer is tested by applying rated voltage to a 2400-volt winding with no load on the other winding, and it is observed that a current of 12 amperes flows and a power of 5500 watts is drawn. We solve for G_c and B_c as follows:

$$P = G_c E^2$$
$$5500 = G_c(2400)^2$$
$$G_c = 0.000955 \text{ siemen}$$
$$I_{ex} = EY_c$$
$$12 = 2400Y_c$$
$$Y_c = 0.005 \text{ siemen}$$
$$Y_c^2 = G_c^2 + B_c^2$$
$$B_c = \sqrt{Y_c^2 - G_c^2} = -0.00491 \text{ siemen}$$

3.7 TRANSFORMER IMPERFECTIONS: SERIES IMPEDANCE ELEMENTS

In addition to the core loss and the requirement for exciting current, there are other ways in which an actual transformer differs from the ideal transformer model. For one thing, some of the applied voltage is absorbed in *IR* drop in the winding resistance. Modifying the ideal transformer model to account for winding resistance is not too difficult, as is shown in Figure 3-23. Each winding has a resistance which, while not zero, is kept low in order to minimize losses and increase efficiency. A more significant factor in transformer performance is a series reactance, which will be explained below.

Let us consider again the geometry of the transformer core and windings as was shown for a shell type core in Figure 3-17. A similar sketch is shown in Figure 3-24. In part (a) of the figure we are looking down on the core along the axis of the center leg and see the two coils, of turns N_1 and N_2, which are wound around

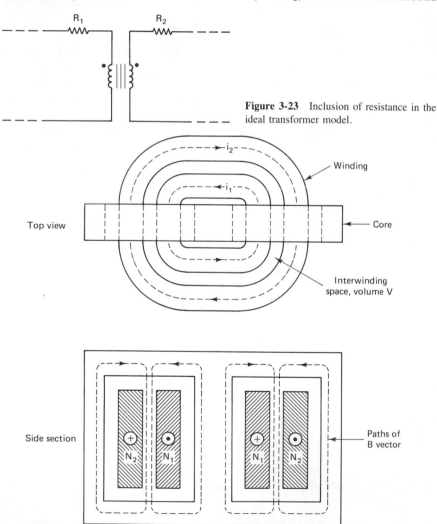

Figure 3-23   Inclusion of resistance in the ideal transformer model.

Figure 3-24   Illustration of leakage flux.

the center leg. In part (b) of the figure a cross section of the two coils is shown as they look going through the window of the core. Since the two windings must be insulated from each other, there will be a space between them as seen by the sketch. This space is filled with nonmagnetic material having a magnetic permeability equal to μ_0, that of free space. If we evaluate a line integral of the H vector around either winding 1 or winding 2 as shown, Ampère's law says that we must sum to the total current enclosed. If we disregard the small difference owing to exciting current, the ampere turns are the same around either of the dotted paths shown and we can write the equations

$$\oint \vec{H}_1 \cdot d\vec{l} = N_1 i_1 = N_2 i_2 = \oint \vec{H}_2 \cdot d\vec{l}$$

The permeability of the steel, while not *really* infinite, is very high compared to that of the space between the windings, and almost the entire contribution to the integrals above occurs in the portion of the path in that region. If we designate that distance as d, we can write the approximate equations

$$H_1 = H_2 = \frac{N_1 i_1}{d} = \frac{N_2 i_2}{d}$$

Since the H is the same around either path, we can use either H as our bookkeeping quantity; we will use H_1.

Field theory tells us that we can assign an energy density to each point in an H field, $w_m = (\mu_0 H^2)/2$ joules per meter^3, and the total energy stored in a magnetic field can be obtained by a volume integration of the energy density over the entire region. In our case, if we assume that H is everywhere the same in the region of interest, the integration amounts to the simple form of multiplying the energy density by the total volume, V.*

$$W_m = \int w_m dV = \frac{\mu_0 H^2}{2} V$$

Since H varies with i, W_m varies with i and the peak or maximum value of W_m occurs when $i = \sqrt{2} I_{rms}$. In terms of the geometry above, the peak magnetic energy stored is

$$W_m = \frac{\mu_0 N_1^2 I_1^2}{d^2} V \tag{3-27}$$

This energy passes into and out of the interwinding space as the current varies from zero to the maximum value and back to zero again.[†] In terms of power this *energy transfer is* represented by a curve such as B of Figure 1-21. The peak value of curve

---

*If we consider paths that pass right *through* the copper windings, we find that H is smaller (but not zero) inside the windings themselves, since such paths encircle only part of the ampere turns. This effect can be treated by a more complex integration, but an old designer's rule of thumb (which can be given a rational basis) simply increases the interwinding volume by extending one-third of the way into each copper winding.

[†]Note that if H in the core is almost zero, owing to the high permeability of the core, there is no energy stored in the steel core—it is all in the interwinding spaces.

B was defined as Q, the reactive voltamperes. In terms of Q, the peak energy of Eq. (3-27) is the integral of the curve B over one-quarter of a period, $T/4$. Since the average of a sine loop is $2/\pi$ times the max value, energy transfer into the field during a positive loop (and out during a negative loop) is $(Q2T)/4\pi = Q/(2\pi f) = Q/\omega$. This leads to the equation

$$\frac{Q}{\omega} = \frac{\mu_0 N_1^2 I_1^2}{d^2} V$$

$$Q = I_1^2 \left[\frac{\omega \mu_0 V N_1^2}{d^2} \right] \qquad (3\text{-}28)$$

To summarize the meaning of this somewhat lengthy development, we point out that for the ideal transformer we assumed that no complex power (real power or reactive voltamperes) was consumed in the transformer. In the actual transformer with load current flowing, both real power and reactive voltamperes are "consumed." The real power goes into the $I^2 R$ of the winding resistance and the reactive voltamperes go into the charging of the magnetic field between the windings. By placing resistance in series with the ideal transformer model, we can account for real power loss, and similarly, since $Q = I^2 X$, we can account for the reactive voltamperes by placing a reactance X in series with the ideal transformer windings, where*

$$X = \frac{\omega \mu_0 V N_1^2}{d^2} \qquad (3\text{-}29)$$

The magnetic flux in the spaces between the windings is called *leakage flux*, since it does not link both windings in the mutual flux path, and the reactance associated with this flux is called the *leakage reactance* of the transfomer.

In practice, it is customary to associate half of the magnetic energy stored in the leakage space with each of the two windings. It is very difficult to establish any more rational allocation in view of the small but significant role of the nonlinear core. In any case, any argument about the assignment of the stored energy between the two coils is moot in view of the approximations introduced in the next section that are almost universally used in transformer analysis.

If we assign half the stored magnetic energy to each coil, we visualize a leakage reactance in series with each winding where

$$X_{l1} = \frac{w \mu_0 V N_1^2}{2d^2}$$

and

$$X_{l2} = \frac{w \mu_0 V N_2^2}{2d^2}$$

Note that the two are not equal in number of ohms but vary as the square of the number of turns, as we might expect from Eq. (3-25).

---

*Note that the quantity $\mu_0 V N_1^2/d^2$ is dimensionally an inductance, in henries.

3.8 THE COMPLETE MODEL OF AN ACTUAL TRANSFORMER

We are now in a position to present a model for an actual transformer, accounting for the imperfections from the ideal transformer, as discussed in the preceding sections.* By "hanging on" the elements that account for the actual transformer characteristics, we arrive at the model of Figure 3-25. This model represents the transformer quite well at ordinary power frequencies. With this model embedded in a circuit diagram for some system we may wish to analyze, we can write the necessary simultaneous equations and solve for the variables of interest.

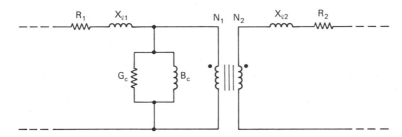

Figure 3-25   The model of an actual transformer at power frequencies.

In actuality, the model shown is seldom used because simpler models are usually adequate and lead to simpler solutions and easier understanding. We have already indicated that for problems of transformer ratings and interconnections we often use only the ideal transformer portion of the model. Likewise, for problems involving load currents or short circuit currents where I_{ex} is negligibly small in comparison to the other currents, we often ignore the shunt branches G_c and B_c. For open circuit conditions, or extremely light loads, we sometimes ignore the series resistance and reactance elements. Even if we include the effects of both the series and shunt elements, we usually rearrange the elements as in the following discussion for greater convenience.

If we refer back to Eq. (3-25) concerning the impedance transforming properties of an ideal transformer, we can replace the model of Figure 3-25 with the equivalent circuit of Figure 3-26. In this figure the imperfection elements are lumped together and the ideal transformer is separated. The equivalence is of course external to the replaced elements; we cannot necessarily expect to put a voltmeter on the points inside the equivalent and get a reading! As a further simplification, we note that the voltage drop owing to I_{ex} flowing through the series elements is very small compared with that due to the load current. In large transformers, the I_{ex} may be less

---

*It is probably safe to say that no model of a physical device is ever "complete." In this case, an engineer interested in high frequencies would go on to investigate capacitive effects in and between the windings. Fortunately, the power engineer is rarely concerned with these except in considering the high rates of change associated with lightning strokes and switching surges. We will not go into these things in this work.

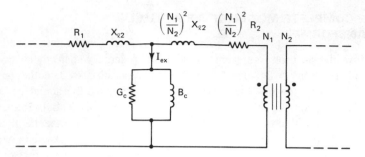

Figure 3-26   Alternate equivalent circuit for the transformer.

than 1 percent of the full load current. As a consequence, it makes little difference if we move the exciting current branches, G_c and B_c to one side of the series impedance and combine the series impedances into single resistance and reactance elements as in Figure 3-27. The side to which we move the elements is chosen to suit the convenience of the problem at hand. The net series resistance and reactance are known simply as the *impedance* of the transformer. This is an item of data that is available from the manufacturer of the transformer or that, on the other hand, may be specified by the purchaser (within limits), if the purchaser so wishes. We designate the net resistance and reactance by the symbols R_{eq} and X_{eq} in the figure where:

$$R_{eq} = R_1 + (N_1/N_2)^2 R_2$$

and

$$X_{eq} = X_{l1} + (N_1/N_2)^2 X_{l2}$$

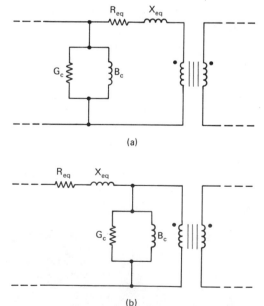

(a)

(b)

Figure 3-27   Simpler models for the transformer.

3.9 OPEN- AND SHORT-CIRCUIT TESTS

The parameters of the equivalent circuits of Figure 3-27 may be determined by the designer from the physical dimensions and material properties of the transformer. On the other hand, an actual transformer may be tested electrically to determine these values. On the test floor we may apply rated voltage to the left side of the transformer of Figure 3-27(a) with the right side open-circuited. Since the output current is zero, the current through R_{eq} and X_{eq} is zero from properties of an ideal transformer. We thus "see" only the shunt branch and determine G_c and B_c from the instrument readings. This procedure was illustrated in Section 3.6. To determine the series impedances, we short-circuit one side, and it is convenient to use the model of Figure 3-27(b) in this case. With a short circuit on N_2, the voltage is zero on this winding and also is zero across N_1 according to the properties of an ideal transformer. As a result we can ignore the shunt exciting current branch and we "see" only R_{eq} and X_{eq}.

As a specific example, suppose that we have a transformer rated 2400 volts on the N_1 side and 500 kVA. We short the N_2 side and apply a reduced voltage to the N_1 side just sufficient to cause rated current of $500 \times 10^3/2400 = 208$ amperes to flow.* Suppose that this requires an impressed voltage of 120 volts, and we record an input power on a wattmeter of 10 kW. The impedance is then $120/208 = 0.58$ ohm. The 10 kW of input power evidently was taken by the $I^2 R_{eq}$ of the transformer and so we solve

$$10,000 = (208)^2 R_{eq}$$
$$R_{eq} = 0.23 \text{ ohm}$$

Lastly

$$X_{eq} = \sqrt{Z_{eq}^2 - R_{eq}^2}$$
$$= \sqrt{0.58^2 - 0.23^2}$$
$$= 0.53 \text{ ohm}$$

We have thus found the elements of the equivalent circuit of the transformer.

3.10 CIRCUIT PROBLEMS INVOLVING TRANSFORMERS

If we have a model available for the transformer involving an ideal transformer and circuit elements, the use of a transformer in a systems problem simply involves the usual methods of circuit solution, but with the added complication of the presence of the ideal transformer. Consider the example of Figure 3-28, where an ideal transformer is associated with other circuit elements. The impedances Z_1 and Z_2 might be regarded as the transformer resistance and leakage reactance equivalent, or perhaps as the result of combining these impedances with external circuit impedances from some other apparatus. The exciting current branch is neglected for simplification.

*In practice it is usually most convenient to short the low-voltage winding and apply voltage to the high-voltage side in order to work within the range of available sources and instruments most effectively.

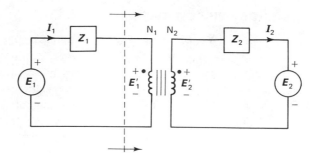

Figure 3-28 A transformer in a circuit problem.

The following equations may be written from Kirchhoff's voltage law and the properties of an ideal transformer.

$$E_1 = Z_1 I_1 + E_1'$$
$$E_2' = Z_2 I_2 + E_2$$
$$E_1' = (N_1/N_2)E_2'$$
$$I_1 = (N_2/N_1)I_2$$

Assuming that the source voltages have been given as well as the circuit elements, there are four equations and four unknowns, and solution can proceed by algebraic means.

Although the algebraic approach outlined above is rather basic and straight-forward, transformer problems are rarely solved in this way. A common alternate approach might be explained in terms of a Thevenin's theorem substitution for the portion of the network to the right of the dotted line in Figure 3-28. "Looking through" the ideal transformer, the side 2 quantities appear changed in scale by the transformer turns ratio in the proper amount and direction. Figure 3-29 shows an equivalent circuit, with the turns ratio applied to the quantities involved. A diagram such as this is said to refer side 2 quantities to side 1. The problem now involves a single loop and if we recognize that $(N_2/N_1)I_2$ is simply equal to I_1 we arrive at the equation

$$E_1 = Z_1 I_1 + (N_1/N_2)^2 Z_2 I_1 + (N_1/N_2)E_2$$

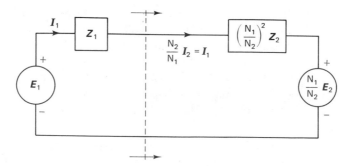

Figure 3-29 Figure 3-28 altered to give side 2 quantities referred to side 1.

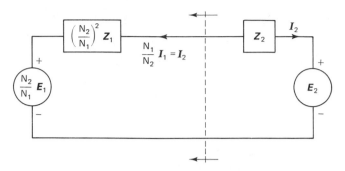

Figure 3-30   Side 1 quantities referred to side 2.

and this equation may be quickly solved for the unknown current I_1. If I_2 is desired, we merely multiply by (N_1/N_2). Of course, we might arrive at this equation by simply rearranging the algebra of the first equations; after all this is what the network theorems amount to!

As another alternate, we might choose to refer to side 1 quantities to side 2 as is shown in Figure 3-30. The equation for this equivalent circuit is then

$$(N_2/N_1)E_1 = (N_2/N_1)^2 Z_1 I_2 + Z_2 I_2 + E_2$$

This equation in turn may be regarded as a simple algebraic variation on the preceding equation. The choice of reference is a matter of convenience or personal preference.

The technique illustrated above may be extended to large-scale system problems, where there may be many different voltage levels on the system separated by transformers. For example, an industrial plant might purchase power at 4160 volts from the utility and transform the power down to 480 volts for distribution around the plant and usage in medium-sized motors. The 480 volts, in turn, may be transformed down to 120 for lighting in various sections of the plant. A study of short-circuit currents in the plant, preliminary to selection of fuse sizes, will often be performed by first drawing the circuit diagram with all levels referred to some portion (perhaps the 480 volt level) and solving. Actual fuse sizes at the other levels are determined by applying the proper turns ratio to the currents computed from the circuit solution. A very simple example of this sort of treatment of a multivoltage level system is given in Figure 3-31, where part (a) shows the original system and part (b) shows the equivalent in terms where everything is referred to the central region voltage level. Where very large-scale systems are involved, or where preferred by the engineer, another approach may be used to eliminate the transformer turns ratio from the computations. This approach is known as the *per-unit* approach and is discussed in the next section.

3.11 NORMALIZED VARIABLES: PER-UNIT

It is common in engineering and scientific work to *normalize* the variables in a problem. Many advantages accrue, among the advantages being the possibility of

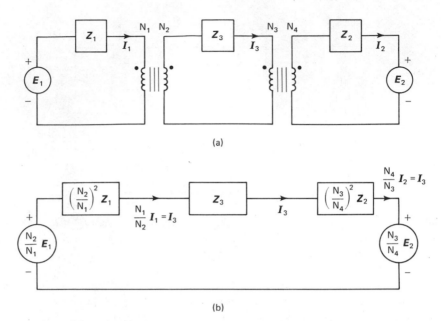

Figure 3-31 An equivalent circuit where there are three different voltage levels.

eliminating nuisance constants such as the transformer turns ratio in transformer problems.

Mathematically, the process of normalizing simply amounts to dividing both sides of an equation by the same number. For example, consider the following equation (Ohm's law)

$$E = ZI$$

If we divide both sides by the same number, E_b, called the *base,* then we have

$$E/E_b = ZI/E_b$$

where the voltage is said to be expressed in per-unit of the base quantity. We will indicate a quantity that has been normalized to some selected base by a bar over the symbol, hence $\overline{E} = E/E_b$ in the example. If we wish, we may factor the base quantity as into

$$E_b = Z_b I_b \qquad (3\text{-}30)$$

and in this case the normalized equation becomes

$$\overline{E} = \frac{E}{E_b} = \frac{ZI}{E_b} = \frac{Z}{Z_b} \times \frac{I}{I_b} = \overline{ZI}$$

or

$$\overline{E} = \overline{ZI}$$

The choice of the normalizing base is arbitrary, but in engineering work it is often related to the rated magnitude of the quantities being normalized, and thus we might expect per-unit quantities in the neighborhood of unity for some of the variables in the problem. It should be noted that, although the choice of base is arbitrary, the constraint of Eq. (3-30) must be followed if our equations are not to have new scale constants popping up to cause possible error.

The complex number nature of our ac circuit quantities is retained in the normalizing process. The voltage base, for example, is chosen as a real number (scalar). In dividing by the base when a voltage is in rectangular form, both the real and imaginary components of the voltage are included as in

$$\overline{E} = E/E_b = \frac{E_r}{E_b} + j\frac{E_i}{E_b} = \overline{E}_r + j\overline{E}_i$$

If the voltage is in polar form, it will be the *magnitude* only of the phasor that is affected by the division by E_b as

$$\overline{E} = \frac{E\underline{/\theta}}{E_b} = \frac{E}{E_b}\underline{/\theta} = \overline{E}\underline{/\theta}$$

In addition to per-unit voltage, current, and impedance, it is common to express power in per-unit. If we define

$$S_b = E_b I_b \qquad (3\text{-}31)$$

as the *voltampere base,* power equations may be normalized as, for example, if

$$P = EI \cos(\theta_E - \theta_E)$$

$$\frac{P}{S_b} = \frac{E}{E_b}\frac{I}{I_b} \cos(\theta_E - \theta_I)$$

$$\overline{P} = \overline{E}\overline{I} \cos(\theta_E - \theta_I)$$

The same base volt amperes are used for Q, the reactive voltamperes, and hence also for complex power S

$$S = EI^*$$

$$\frac{S}{S_b} = \frac{E}{E_b}\frac{I^*}{I_b}$$

$$\overline{S} = \overline{E}\overline{I}^*$$

Again it must be noted that, if the power equations are not to require awkward constants in order to be true, then S_b must be chosen consistent with the constraint of Eq. (3-31) above.

In partial summary of the things covered so far, note that there are four base quantities involved, E_b, I_b, Z_b and S_b. If the constraints of Eqs. (3-30) and (3-31) are followed, only two of the four quantities are independent. We could arbitrarily choose any two, then the others must follow from the constraining equations. The usual choices are the voltampere base S_b, and the voltage base E_b. If these two are chosen, it follows that

$$I_b = \frac{S_b}{E_b}; \qquad Z_b = \frac{E_b}{I_b} = \frac{E_b^2}{S_b} \qquad (3\text{-}32)$$

by applying the constraining equations.

If we are concerned with a particular piece of equipment, most often the base quantities chosen for per-unit computation are simply the rated quantities of that item of equipment. On the other hand, for system studies, many pieces of equipment are involved, and seldom do all have the same rating. It is then necessary to change the per-unit values to a common base for all segments of the system; that is, it is necessary that all terms of an equation be divided by the same normalizing constant for the equation to remain true! The most common situation like this occurs when the manufacturer gives the impedance of a piece of equipment in per-unit on the equipment's own rated quantities as base (After all, the manufacturer doesn't know what base quantities you might wish to use!). We may then wish to change to another base as follows:

$$Z(\text{ohms}) = \overline{Z}Z_b(\text{given})$$
$$\overline{Z}(\text{on new base}) = Z(\text{ohms})/Z_{b(\text{new})}$$
$$= \overline{Z}_{(\text{old})}Z_{b(\text{old})}/Z_{b(\text{new})}$$
$$= \overline{Z}_{(\text{old})}\left[\frac{E_{b(\text{old})}}{E_{b(\text{new})}}\right]^2 \frac{S_{b(\text{new})}}{S_{b(\text{old})}} \qquad (3\text{-}33)$$

In words, the per-unit impedance $\overline{Z}$ varies inversely as the square of the voltage base and directly as the voltampere base.

3.12 NORMALIZED VARIABLES: APPLICATION
TO TRANSFORMERS

Suppose that we have an ideal transformer embedded in a circuit such as that of Figure 3-32, where the transformer is shown in combination with an impedance. The impedance might be that of the transformer itself, or an external element, or a combination of both. To treat by means of per-unit we choose base quantities and divide the circuit equations by the appropriate base to normalize. We must divide both sides of an equation by the same number, but this does not require the same base quantities at all locations. To illustrate, suppose that the transformer turns ratio is given by $n = (N_1/N_2)$ and we have a base voltage on side 1 as E_{1b}. If we define the voltage base on side 2 such that $E_{1b} = aE_{2b}$, where a is a real number scale constant, then we have the following forms for the voltage relations

$$E_1 = nE_2, \quad \text{or} \quad E_1/E_{1b} = \frac{nE_2}{aE_{2b}}, \quad \text{or} \quad \overline{E}_1 = (n/a)\overline{E}_2$$

and we see that if we choose $a = n$, then $\overline{E}_1 = \overline{E}_2$; that is, the transformer ratio becomes unity between the per-unit voltages. In a similar fashion, we find that

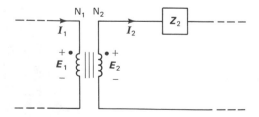

Figure 3-32 Ideal transformer and associated impedance.

$$\text{If: } I_{1b} = (1/n)I_{2b} \qquad \text{Then: } \overline{I}_1 = \overline{I}_2$$
$$Z_{1b} = n^2 Z_{2b} \qquad\qquad \overline{Z}_1 = \overline{Z}_2$$
$$S_{1b} = S_{2b} \qquad\qquad \overline{S}_1 = \overline{S}_2$$

and the relations of all quantities are those of a unity turns ratio transformer. If we wish, we may not even bother to show the unity turns ratio ideal transformer, and the writing of the circuit equations is greatly simplified.

The foregoing relations may be simply remembered by noting that if we change the base quantities the same way that the ideal transformer changes the actual quantities, the transformer ratio goes to unity and the transformer disappears from the problem.* The sequence of base choice usually proceeds as follows: choose S_b, which is the same everywhere in a problem, next choose E_b at a *specific location*, next determine E_b at every other location by applying the transformer ratios to the base, and last compute I_b and Z_b at each location if they are needed.

A numerical example is in order. Suppose that we have a 100-kVA, 2400-480 volts transformer used in the circuit of Figure 3-33, with two voltage sources and a series impedance as shown. We wish to solve for the current flowing in per-unit. Lacking any overwhelming counter-reason, we choose the base quantities as the rated quantities of the apparatus involved; we choose $S_b = 100$ kVA, $E_{1b} = 2400$ volts, and $E_{2b} = 480$ volts. Note that, since the nameplate voltage ratio and the turns ratio are the same for transformers, we have followed the rule given above of transforming the base quantities just the same as an ideal transformer would transform actual quantities. We now have

$$\overline{E}_1 = \frac{2500\underline{/30°}}{2400} = 1.042\underline{/30°}$$

$$\overline{E}_2 = \frac{470\underline{/0°}}{480} = 0.979\underline{/0°}$$

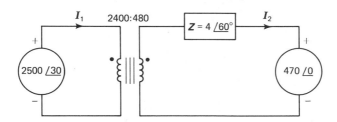

Figure 3-33   An example circuit for per-unit treatment.

*The inclusion of the scale constant a in the foregoing development is a recognition of the fact that we may not *always* wish to (or be able to) choose the voltage bases in the ratio of the transformer turns. If we do not choose the bases in the same ratio as the transformer turns, an ideal transformer of ratio a/n will still be present in our model. This situation seldom arises in the class of problems with which we deal in this work.

$$Z_{2b} = \frac{E_{2b}^2}{S_b} = \frac{480^2}{100 \times 10^3} = 2.30 \text{ ohms}$$

$$\overline{Z} = \frac{4\underline{/60°}}{2.30} = 1.74\underline{/60°}$$

A new circuit diagram with quantities labeled in per-unit (and the turns ratio eliminated by our base choice) is given in Figure 3-34.

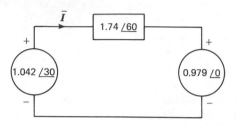

Figure 3-34 Circuit of Figure 3-32 with elements converted to per-unit.

We now proceed to solve from the diagram of Figure 3-34 as

$$\overline{I} = \frac{\overline{E}_1 - \overline{E}_2}{\overline{Z}} = \frac{1.042\underline{/30°} - 0.979\underline{/0°}}{1.74\underline{/60°}}$$

$$= 0.303\underline{/38.42°} \quad \text{per-unit}$$

Thus the problem is solved, in per-unit. If we wish to know the actual amperes of I_1 and I_2, we must find the base current at each location

$$I_{1b} = \frac{100 \times 10^3}{2400} = 41.67 \text{ amperes}$$

$$I_{2b} = \frac{100 \times 10^3}{480} = 208.33 \text{ amperes}$$

then

$$I_1 = (0.303\underline{/38.42°})41.67 = 12.6\underline{/38.42°}$$

$$I_2 = (0.303\underline{/38.42°})208.33 = 63.1\underline{/38.42°} \text{ in amperes.}$$

The complex power delivered by $\overline{E}_1$ is given by

$$\overline{S}_1 = \overline{E}_1\overline{I}* = 1.042\underline{/30°} \times 0.303\underline{/-38.42°} = 0.316\underline{/-8.42°}$$

and that received by $\overline{E}_2$ is

$$\overline{S}_2 = \overline{E}_2\overline{I}* = 0.979\underline{/0°} \times 0.303\underline{/-38.42°} = 0.2966\underline{/-38.42°}$$

In terms of actual complex voltamperes

$$S_1 = 0.316\underline{/-8.42°} \times 100 \times 10^3 = 31.6\underline{/-8.42°} \text{ (kVA)}$$

$$S_2 = 0.2966\underline{/-38.42°} \times 100 \times 10^3 = 29.66\underline{/-38.42°} \text{ (kVA)}$$

Another example of application of per-unit to transformers is taken from the practically important case of testing a transformer by means of a short-circuit test to find the series impedance of the transformer (see Section 3.9). If we short-circuit the low-voltage winding and apply a reduced voltage to the high-voltage winding sufficient to cause rated current (equal to I_b) to flow, the applied voltage is absorbed in the transformer series impedance. The magnitude of this voltage is then given by

$$E = I_b Z$$

If we divide both sides of this equation by E_b we have

$$\frac{E}{E_b} = \frac{I_b Z}{E_b} = \frac{I_b Z}{I_b Z_b} = \frac{Z}{Z_b} = \overline{Z}$$

This is the magnitude of Z in per-unit. In words we can say that the per-unit Z is the per-unit fraction of E_b, which would be absorbed by the impedance with I_b flowing.

From the example of Section 3.9 (using rated values as the base quantities) we have

$$\overline{Z} = \frac{120}{2400} = 0.05 \qquad \text{per-unit}$$

In a similar manner, if we observe a wattmeter reading of P_w on the same test with base current flowing, we know that this power is going into the series resistance of the transformer and we can write

$$P_w = I_b^2 R$$

Dividing both sides by S_b we have

$$\frac{P_w}{S_b} = \frac{I_b^2 R}{S_b} = \frac{I_b^2 R}{E_b I_b} = \frac{I_b^2 R}{I_b^2 Z_b} = \frac{R}{Z_b} = \overline{R}$$

This is the magnitude of R in per-unit, and again in words we say that the per-unit R is the power absorbed when I_b flows, expressed as a per-unit fraction of the S_b.

Back to the example figures again, if $P_w = 10$ kW, then

$$\overline{R} = 10/500 = 0.02$$

and, if we finish the example in per-unit

$$\overline{X} = \sqrt{\overline{Z}^2 - \overline{R}^2} = \sqrt{0.05^2 - 0.02^2} = 0.045$$

The student may wish to check these figures by computing Z_b from the transformer rating and dividing the Z_b into the ohmic values computed previously in Section 3.9.

It is common commercial practice for manufacturers to multiply the per-unit impedance values by 100 and thus give the impedance in *percent*. When the data is given in percent, it is advisable to first divide by 100 to convert to per-unit before performing any computations. If we fail to do this, we may end up with errors, owing to an extraneous factor of 100 that creeps into the math. For example, 100 percent voltage is *not* given by 100 percent current times 100 percent impedance!

3.13 PER-UNIT IN THREE-PHASE CIRCUITS

In dealing with balanced three-phase circuits, we customarily describe power, reactive voltamperes, and voltamperes in terms of the total value for all three phases (three times the value per phase). Likewise, in describing the voltage of a balanced three-phase circuit, we usually give the line-to-line voltage ($\sqrt{3}$ times the line-to-neutral voltage). In setting bases for three-phase per-unit computations, it is normal to consider S_b as applying to total complex power quantities, thus, if we have a per-unit power of 1 per unit it is the *total* power that is equal to S_b. Likewise, we associate E_b with line-to-line voltage, and if we have 1 per-unit of voltage it is the *line-to-line* voltage that is equal to E_b. If per-phase quantities are wanted, it is then a simple matter to find them.

Another nuisance constant in power computation is the factor $\sqrt{3}$ in the expression for total three-phase power. This can be eliminated by a suitable choice of base. Consider the expression for total three phase-power (balanced case)

$$P + jQ = \sqrt{3}E_{l-l}I_l[\cos(\theta_E - \theta_I) + j\sin(\theta_E - \theta_I)]$$

Normalizing

$$\frac{P}{S_b} + j\frac{Q}{S_b} = \frac{\sqrt{3}E_{l-l}I_l}{S_b}[\cos(\theta_E - \theta_I) + j\sin(\theta_E - \theta_I)]$$

Now if we choose to define

$$S_b = \sqrt{3}E_bI_b$$

(Note that this really defines I_b, since we normally start with S_b and E_b given) it follows that

$$\overline{P} + j\overline{Q} = \overline{E}\,\overline{I}[\cos(\theta_E - \theta_I) + j\sin(\theta_E - \theta_I)]$$

When impedances are given in ohms, they are usually expressed in so-called per-phase quantity, which might be thought of as that of one leg of an equivalent Y-connected group. We then define a Z_b for per-phase impedances such that a voltage of one per-unit ($E_b/\sqrt{3}$ volts from line-to-neutral) would cause one per-unit current (I_b) to flow in one per-unit impedance (Z_b), thus

$$Z_b = \frac{E_b/\sqrt{3}}{I_b} = \frac{E_b^2}{\sqrt{3}E_bI_b} = \frac{E_b^2}{S_b}$$

It is fortuitous that the expression for Z_b thus turns out to be the same for the three-phase case as it does for the single-phase case. We sometimes see a $Z_{b(delta)}$ defined as three-times the above, but it is probably less confusing to treat delta-connected impedances by first converting to an equivalent Y.

3.14 VOLTAGE REGULATION

Because of the "internal" impedance, $R_{eq} + jX_{eq}$, the output voltage of a transformer ordinarily changes under load from the no-load or open-circuit voltage, even though the input voltage is maintained constant. Since most of our systems are constant potential systems, it is desirable that the voltage change be small. A figure of merit used to compare the relative performance of different transformers is the *voltage regulation*. Voltage regulation, usually expressed in percent, is defined by

$$\text{Voltage Regulation} = \frac{\text{Change in voltage magnitude with load}}{\text{Rated voltage magnitude at full load}} \times 100$$

The matter may be clarified by reference to Figure 3-35, showing (a) an equivalent circuit of a transformer feeding a load from a constant potential source, and (b) a

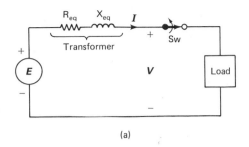

(a)

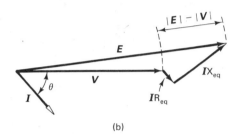

Figure 3-35 For illustration of voltage
regulation.

(b)

phasor diagram showing load current and voltage components for a typical load. It
is assumed that the circuit quantities are given in per-unit, or else have been referred
to one side or other of the transformer, and hence no ideal transformer is included
in the equivalent circuit. It will be further noted that the circuit ignores the shunt-
exciting current branch of the transformer.

In the Figure 3-35, a load is shown with a voltage V across it. The load draws
a current I, which in the example is at a lagging angle, θ, with respect to the
voltage. The voltage E is adjusted to whatever value is required to deliver rated
voltage at the specified load. Hence,

$$E = (R_{eq} + jX_{eq})I + V$$

If the load is removed, as by opening a switch, Sw, the voltage V will change to
equal the source voltage E. The change in voltage magnitude is then

$$\Delta V = |E| - |V|$$

and the voltage regulation is

$$\text{Regulation} = \frac{|E| - |V|}{|V|} \times 100 \qquad (3\text{-}34)$$

Ordinarily, the voltage regulation of a transformer is just a few percent. It will be
noted from the phasor diagram that the voltage regulation will be greatest for current
lagging load for typical values of transformer parameters where $X_{eq} > R_{eq}$.

As a numerical example, consider a transformer rated 50 kVA, 7200-240 volts
with $\overline{R}_{eq} = 0.015$ and $\overline{X}_{eq} = 0.04$, both in per-unit on the transformer's own rating
as base. The voltage regulation at full load, 0.8 power factor, current lagging, will
be computed as follows:

$$\overline{E} = (0.015 + j0.04)1.0\underline{/-36.87} + 1.0\underline{/0}$$
$$= 1.036\underline{/1.272}$$
$$\text{Regulation} = \frac{1.036 - 1.0}{1.0} \times 100 = 3.6\%$$

The convenience of working in per-unit is evident from this example. The student may wish to work the problem in terms of volts and amperes referred to a particular side of the transformer as an exercise.

3.15 LOSSES AND EFFICIENCY

It has already been indicated that for a real world (nonideal) transformer, we do not get all the power out of the device that is delivered into it. Thus the efficiency, defined as

$$\eta = \frac{\text{Output}}{\text{Input}} \tag{3-35}$$

is always less than unity.* Alternately, the expression for efficiency is given by

$$\eta = \frac{\text{Output}}{\text{Output} + \text{Losses}} \tag{3-36}$$

where the output and the losses are in terms of *real power* as contrasted with the load sometimes given in *voltamperes*.

The power losses occur in two ways: in the iron core as hysteresis and eddy currents, and in the windings as I^2R losses. The two loss components are conventionally called *iron losses* and *copper losses*. The iron and copper losses may be calculated from the physical properties of the transformer by the designer, or for an actual transformer they are measured by the open and short circuit tests as described in Section 3.9. The iron losses vary with the maximum flux density and the frequency, and hence with applied voltage. It is usual, however, to assume that the iron losses are constant at the value corresponding to rated voltage and frequency on open circuit, even though the flux density may vary slightly under load. The copper losses will vary as the square of the current, and therefore the efficiency will vary not only with the power output, but also with the power factor (and hence the current), at which a given real power output is supplied.

We might rephrase Eq. (3-36) as

$$\eta = \frac{P_{\text{out}}}{P_{\text{out}} + P_{\text{Fe}} + P_{\text{Cu}}} \tag{3-37}$$

where P_{out} is the *real* power out at a specified load, P_{Fe} is the constant iron loss and P_{Cu} is the copper loss which varies as the square of the current.

An example is in order: suppose that we have a transformer rated as

25 kVA, 7200-240 V, 60 Hz, single phase

---

*As given in Eq. (3-35), the efficiency is in per-unit. Very often the expression is multiplied by 100, and the efficiency is thus in percent.

and the open- and short-circuit tests have revealed:

$P_{Fe} = 160$ W   at rated 240 volts input to the low voltage and,

$P_{Cu} = 450$ W   at rated current of 3.47 A on the high-voltage winding

The efficiency at rated output and unity power factor will thus be

$$\eta = \frac{25000 \times 1.0}{25000 \times 1.0 + 160 + 450}$$

$$\eta = 0.976 \quad \text{or} \quad 97.6\%$$

On the other hand, if the transformer is loaded to rated 25000 voltamperes, but at 0.8 power factor, the efficiency will be

$$\eta = \frac{25000 \times 0.8}{25000 \times 0.8 + 160 + 450}$$

$$\eta = 0.970$$

As yet another case, suppose that the transformer is loaded to one-half voltampere load at 0.8 power factor. Then the efficiency will be

$$\eta = \frac{25000 \times 0.5 \times 0.8}{25000 \times 0.5 \times 0.8 + 160 + 0.5^2 \times 450}$$

$$\eta = 0.974$$

where it will be noted that the copper loss is multiplied by 0.5^2 because the current is only half the rated current at which the original data of 450 watts was based.

The example above serves to illustrate the statement that the efficiency of a transformer is a function of the load and the power factor of the load. This statement is, of course, based on the assumption of a load at rated voltage and frequency. These matters may be better studied by forming an equation for the efficiency in terms of the voltampere load $\overline{S}$, the power factor (p.f.), the core (or iron) losses, $\overline{P}_{Fe}$, and the full load I^2R (copper) losses, $\overline{P}_{Cu}$. For convenience, the values will be given in per-unit, and on that basis the copper losses at any given voltampere load will be given by the simple expression $\overline{S}^2\overline{P}_{Cu}$. The efficiency then is

$$\eta = \frac{\overline{S}(\text{p.f.})}{\overline{S}(\text{p.f.}) + \overline{P}_{Fe} + \overline{S}^2\overline{P}_{Cu}} \tag{3-38}$$

Typical plots of efficiency versus load at various power factors will be found to be like those of Figure 3-36. The changes in efficiency with load and power factor are exaggerated to show more clearly. It will be noted that the efficiency is always zero at zero load, since there is an input as long as the transformer is energized — the core losses at least. The efficiency increases with output and has a maximum value at some load or other. The load at which maximum efficiency is realized is easily found from the calculus by taking

$$\frac{d\eta}{d\overline{S}} = 0$$

from which we find that efficiency is at a maximum when

$$\overline{S} = \sqrt{\overline{P}_{Fe}/\overline{P}_{Cu}}$$

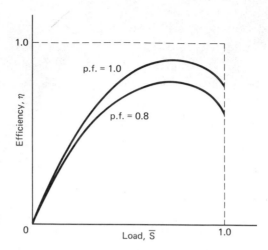

Figure 3-36   Variation of efficiency with voltampere load at two sample power factors.

or,

$$\overline{P}_{Fe} = \overline{S}^2 \overline{P}_{Cu}$$

or in other words at that load where the constant losses and the variable losses are equal.*

The ratio of the losses is under the control of the designer. For example, if the cross section of the core is increased at the same maximum flux density, the iron losses will be increased, since the loss per pound is a constant. On the other hand, the volts per turn will be increased because the maximum flux, ϕ, is increased and there will be fewer turns and less copper loss. Thus there is a trade-off between iron and copper losses and the point of maximum efficiency may be set as the designer wills.

It might at first be thought that the point of maximum efficiency would always be set at the rated full load but this is not necessarily so. For example, consider a pole-mounted distribution transformer feeding a residential load. During the nighttime hours there is little load on the transformer but the iron losses go on, and then at the time of peak load the copper losses are at their maximum value. Greatest economy will probably be realized by proportioning the iron and copper so that peak efficiency occurs at something less than full load. On the other hand, consider a large power transformer at a hydro station which is a "run of the river" plant; that is, there is little storage, and the installed equipment is run at full load almost all the time. Such a transformer would be best designed to have maximum efficiency at full load. Decisions on the advisability of various loss ratios, and points of maximum efficiency, are often settled on the basis of so-called *all day efficiency* which is given by

$$\text{All day efficiency} = \frac{\text{Total } \textit{energy} \text{ out}}{\text{Total } \textit{energy} \text{ in}}$$

*This is a particular example of a principle often encountered in engineering economics known as "Kelvin's Law".

where the total energy, in say kWh, is calculated on the basis of a certain load pattern. An example load pattern might be one-half load for eight hours, three-fourths load for ten hours, and full load for six hours. Further consideration of this matter will be reserved for the problem section.

ADDITIONAL READING MATERIAL

1. Brown, D. and E. P. Hamilton III, *Electromechanical Energy Conversion,* New York: MacMillan Publishing Company, 1984.
2. Chapman, S. J., *Electric Machinery Fundamentals,* New York: McGraw-Hill Book Company, 1985.
3. Chaston, A. N., *Electric Machinery,* Reston, Virginia: Reston Publications Company, Inc., 1986.
4. Del Toro, V., *Electric Machines and Power Systems,* Englewood Cliffs, New Jersey: Prentice-Hall, Inc., 1985.
5. Elgerd, O. I., *Basic Electric Power Engineering,* Reading, Massachusetts: Addison-Wesley Publishing Company, 1977.
6. Elgerd, O. I., *Electric Energy Systems Theory: An Introduction,* New York: McGraw-Hill Book Company, 1982.
7. Fitzgerald, A. E., C. Kingsley, Jr., and S. D. Umans, *Electric Machinery,* New York: McGraw-Hill Book Company, 1983.
8. Gross, C. A., *Power System Analysis,* New York: John Wiley & Sons, 1979.
9. Lindsay, J. F., and M. H. Rashid, *Electromechanics and Electric Machinery,* Englewood Cliffs, New Jersey: Prentice-Hall, Inc., 1986.
10. Nasar, S. A., *Electric Machines and Electromechanics,* Schaum's Outline Series in Engineering, New York: McGraw-Hill Book Company, 1981.
11. Nasar, S. A., *Electric Energy Conversion and Transmission,* New York: MacMillan Publishing Company, 1985.
12. Shultz, R. D. and R. A. Smith, *Introduction to Electric Power Engineering,* New York: Harper & Row, Publishers, 1985.
13. Slemon, G. R., and A. Straughen, *Electric Machines,* Reading, Massachusetts: Addison-Wesley Publishing Company, 1980.

STUDY EXERCISES

1. Consider a toroid like that of Figure 3-1, to which the following data apply:

$$r_1 = 8 \text{ cm}, \ r_2 = 10 \text{ cm}, \ h = 2 \text{ cm}$$
$$N = 200 \text{ turns}, \ \mu = \mu_r\mu_0 = 3000 \times 4\pi 10^{-7}$$

It is desired to establish a flux of 0.0006 Weber in the toroidal steel core.
a. What is the average flux density, B, in the core?
b. How much current, i, is required to establish the desired flux in the core if we use the approximate method of Eq. (3-1)?
c. What is the percent error for this example case if we use Eq. (3-1) instead of the "exact" method first described in this section?

d. What is the reluctance of this core?

e. What is the inductance of the coil?

2. a. Find the voltage per turn of a transformer having a cross-sectional core area of $A_c = 0.002$ m² The core is operated $B_{max} = 1.5\ T$. The frequency is 60 Hz. Assume the transformer to be idea

b. If the transformer of (a) above is rated at 50 kVA, 2400-240 volts, 60 Hz, how many turns mus be provided in the high-voltage and the low-voltage windings?

c. If a 40 kVA load at 0.8, current lagging, is connected to the secondary, what are the current drawn by the two sides of the transformer?

d. What is the actual load impedance in complex ohms? What is the value referred to the high voltage side?

3.

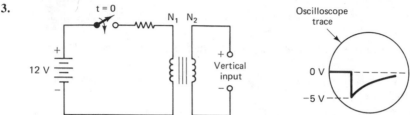

The circuit diagram above shows a test being performed on a transformer of unknown properties When a 12-volt battery is suddenly connected to one winding, a cathode ray oscilloscope shows trace as shown to the right of the circuit diagram.

a. Place polarity dots on a sketch of the transformer windings to be consistent with the experimenta results observed.

b. What is the turns ratio of the transformer?

4.

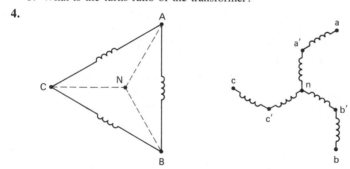

The diagram above illustrates a three-phase transformer connection known as *delta to zig-zag*. The usual conventions of orientation of the coil symbols, their polarity marks, and the relation to voltage phasors has been used in drawing the connection diagram. Each of the low-voltage (zig-zag) coil has the same number of turns for the first part of this problem. The transformers have a rating a a three-phase bank of 5000 kVA, 115-15 kV.

a. What must be the voltage and current rating of each coil? Mark these values on a sketch of the transformer connections.

b. What is the phase shift between E_{AN} and E_{an}?

c. What is the magnitude of the voltage from point a to point b'?

d. Suppose that the bank voltage magnitude ratio is to be as above, but there is to be a 10° phase shift between E_{AN} and E_{an}, what should the voltage ratings of the individual coils on the zig-zag side be?

5. The figure below illustrates an ideal transformer with three windings — with the number of turns for each winding shown. The 500 turn and the 200 turn windings are connected to a pi section, and winding polarities are given by the dots. The pi section impedances are given alongside the drawing. What will be the impedance in (complex) ohms looking into the 100 turn winding from a source connected to that winding?

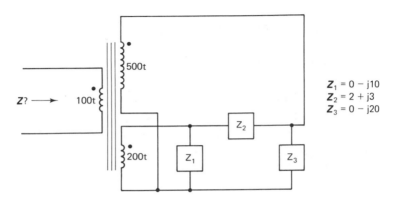

$$Z_1 = 0 - j10$$
$$Z_2 = 2 + j3$$
$$Z_3 = 0 - j20$$

6. The exciting current of a transformer is often given as a percentage of the rated current, and likewise, the core loss may be given as a percentage of rated voltamperes. Suppose that we have a transformer rated 500 kVA, 13.8-2.4 kV, and we know the exciting current to be 5 percent and the core loss as 1 percent.
   a. What are G_c and B_c in siemens of the equivalent circuit of Figure 3.21 (as viewed from the high-voltage side)?
   b. What are the corresponding values viewed from the low-voltage side?
   c. Suppose, for analysis purposes, someone wishes to use a series R-X branch instead of the parallel G-B of part (a) above. What are the values of R and X?

7. Repeat Problem 6, working in per-unit on the transformer's rating as base.

8. One authority gives expressions for core loss components as:
   $$\text{Hysteresis loss} = K_h f B_{max}^2$$
   $$\text{Eddy Curr. loss} = K_e f^2 B_{max}^2$$
   Suppose that the transformer of Problem 6 is operated at 10 percent overvoltage and 5 percent overfrequency, and we know that the hysteresis loss is twice that of the eddy currents at rated voltage and frequency.
   a. What is B_{max} in percent of normal under the above conditions?
   b. What is the total core loss in watts?
   c. What is the core loss if computed using the equivalent circuit (Figure 3.25)?

9. A 50-kVA, 7200-240–volt transformer is short-circuited on the low-voltage side and sufficient voltage applied to the high-voltage side to make one-half rated current flow. A wattmeter reads a power input of 250 watts under these conditions. A varmeter reads a reactive voltampere input of 1000 vars.
   a. What are the series R, X, and Z (a phasor), both in ohms referred to the high-voltage side and in per-unit on the transformer's own rating as base?
   b. If the transformer were short circuited on the high-voltage side, how many volts would have to be applied to the low-voltage winding to make rated current flow in that winding?

SOLUTIONS TO STUDY EXERCISES

1. a. $B = \phi/A = \dfrac{0.0006}{(10-8)2\times 10^{-4}} = \underline{1.5\text{ T}}$ ⟵ a

b. $\mathcal{F} = Ni = \dfrac{\phi}{\mu A}\, 2\pi r_{ave}$ from Eq. (3-1)

$= \dfrac{0.0006 \times 2\pi(10+8)/2}{3000 \times 4\pi 10^{-7}(10-8)2} \times 10^2 = 225$ A-turns

$i = \mathcal{F}/N = 225/200 = \underline{1.125\text{ A}}$ ⟵ b

c. $Ni = \dfrac{\phi 2\pi}{\mu h\, \ln(r_2/r_1)} = \dfrac{0.0006 \times 2\pi}{3000 \times 4\pi 10^{-7} 2 \times 10^{-2}\,\ln(10/8)}$

$= 224.07$ A-turns

$i = 224.07/200 = 1.1204$ A

$\text{Error} = \dfrac{1.125 - 1.1204}{1.1204} \times 100 = \underline{0.41\%}$ ⟵ c

d. $\mathcal{R} = \dfrac{l}{\mu A} = \dfrac{2\pi(10+8)10^{-2}/2}{3000 \times 4\pi 10^{-7}(10-8)2 \times 10^{-4}}$

$= \underline{3.75 \times 10^5}$ ⟵ d

e. $L = N^2/\mathcal{R} = 200^2/3.75 \times 10^5 = \underline{0.107\text{ H}}$ ⟵

2. $\phi_{max} = 1.5 \times 0.002 = 0.003$ W

$e/\text{turn} = d\phi/dt = d(0.003\cos \omega t)/dt = 1.131 \sin \omega t$

$E_{max}/\text{turn} = 1.131$

$E_{rms}/\text{turn} = 1.131/\sqrt{2} = \underline{0.8\text{ V/turn}}$ ⟵ a

$\left.\begin{array}{l} N_1 = 2400/0.8 = 3000 \text{ turns} \\ N_2 = 240/0.8 = 300 \text{ turns} \end{array}\right\}$ ⟵ b

$\left.\begin{array}{l} I_1 = 40 \times 10^3/2400 = 16.67 \text{ A} \\ I_2 = 40 \times 10^3/240 = 166.67 \text{ A} \end{array}\right\}$ ⟵ c

$\left.\begin{array}{l} Z_2 = 240\underline{/0°}/166.67\underline{/-\cos^{-1}0.8} = 1.44\underline{/36.8°} = 1.152 + j0.864 \\ Z_1 = (N_1/N_2)^2 Z_2 = 10^2 Z_2 = 144\underline{/36.8°} = 115.2 + j86.4 \end{array}\right\}$

3.

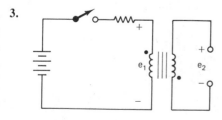

at $t = 0^+$ (just after switch closes),

$e_1 = +12$ with polarity shown

$e_2 = -5$ from 'scope trace

therefore bottom end of side 2 is positive and the dots must be placed as shown.

ratio $N_1/N_2 = 12/5$

4. a.

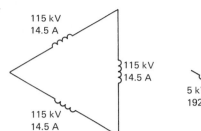

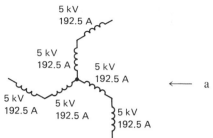

High-voltage side: $E_{coil} = E_{l-l} = 115$ kV
$$I_{line} = 5000/\sqrt{3} \times 115 = 25.1 \text{ A}$$
$$I_{coil} = I_{line}/\sqrt{3} = 14.5 \text{ A}$$
Low-voltage side: $E_{l-n} = 15/\sqrt{3} = 8.66$ kV
$$E_{coil} = E_{l-n}/(2 \cos 30) = 5 \text{ kV}$$
$$I_{coil} = I_{line} = 5000/\sqrt{3} \times 15 = 192.5 \text{ A}$$

b. Note that, either from the connection diagram which serves as a quasi phasor diagram, or by sketching a phasor diagram,

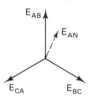

we see that $\underline{/E_{an}} = \underline{/E_{AN}}$; the voltages are *in phase* ← b
$|E_{ab'}| = 5 \sin 30 + 5 + 5 \sin 30$
$= \underline{10 \text{ kV}}$ ← c

c.

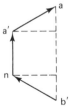

d. Two answers are possible:
  i. E_{an} is 10° behind E_{AN}, or
  ii. E_{an} is 10° ahead of E_{AN}
  Considering the first case:

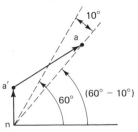

$E_{a'n}\underline{/90°} + E_{aa'}\underline{/30°} = 15/3\underline{/60° - 10°}$
Equating real parts
$E_{aa'} \cos 30 = (15/3) \cos(60° - 10°)$
$E_{aa'} = 6.43$
Equating imaginary parts
$E_{a'n} \sin 90 + 6.43 \sin 30 = (15/3) \sin 50$
$E_{a'n} = \underline{3.42 \text{ kV}}$
By similar means for case ii, we have
$E_{aa'} = 3.42$ and $E_{a'n} = \underline{6.43 \text{ kV}}$ ← d

5.

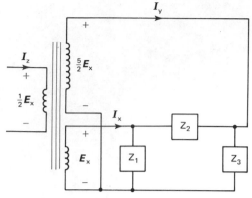

Define one of the voltages as E_x, for example. The other voltages follow from the turns ratios.

$$I_x = \frac{E_x}{-j10} + \frac{E_x - (5/2)E_x}{2 + j3}$$

$$= (-0.2308 + j0.4462)E_x$$

$$I_y = \frac{(5/2)E_x}{-j20} + \frac{(5/2)E_x - E_x}{2 + j3}$$

$$= (0.2308 - j0.2212)E_x$$

From ampere turn balance

$$100I_z = 500I_y + 200I_x$$

$$= (0.6923 - j0.2135)E_x$$

$$Z = \tfrac{1}{2}E_x/I_z = \underline{0.6595 + j0.2034} \quad \longleftarrow \quad 5$$

6. $\quad I_{1(\text{rated})} = 500/138 = 36.23$ A

$$I_{ex} = 0.05 \times 36.23 = 1.1815 \text{ A}$$

$$Y_{c1} = 1.1815/13800 = 0.0001313 \text{ S}$$

Core loss $= 0.01 \times 500 = 5$ kW

$$5000 = 13800^2 G_c$$

$$\left.\begin{array}{l} G_{c1} = \underline{2.63 \times 10^{-5} \text{ S}} \\ B_{c1} = \sqrt{Y_{c1}^2 - G_{c1}^2} = \underline{-1.29 \times 10^{-4} \text{ S}} \end{array}\right\} \quad \longleftarrow \quad a$$

$$Y_{c2} = (13.8/2.4)^2 Y_{c1}$$

$$\left.\begin{array}{l} G_{c2} = \underline{8.68 \times 10^{-4} \text{ S}} \\ B_{c2} = \underline{-4.25 \times 10^{-3} \text{ S}} \end{array}\right\} \quad \longleftarrow \quad b$$

$$R + jX = 1/(G_c + jB_c) = \underline{1523 + j7463} \quad \longleftarrow \quad c$$

7. $Y_{\text{base 1}} = I_{\text{base 1}}/E_{\text{base 1}} = \dfrac{500/13.8}{13.8 \times 10^3} = 0.00263$

$$\left.\begin{array}{l} \overline{G}_c = G_{c1}/Y_{\text{base 1}} = (2.63 \times 10^{-5})/2.63 \times 10^{-3} = 0.01 \\ \overline{B}_c = B_{c1}/Y_{\text{base 1}} = (-1.29 \times 10^{-4})/2.63 \times 10^{-3} = -0.049 \end{array}\right\} \quad \longleftarrow \quad a$$

Alternately we could work with side 2 quantities and arrive at *same* values in per-unit. $\quad \longleftarrow \quad$ b

As yet another (and better) alternative, we might recognize that:

$$\overline{G}_c = G_a/Y_{\text{base}} = G_c E_{\text{base}}/Y_{\text{base}} E_{\text{base}} = G_c E_{\text{base}}^2/S_{\text{base}}$$

$$= \text{Core loss/Rating} = 0.01 \text{ (as given)}$$

$$\overline{Y}_c = Y_c/Y_{\text{base}} = Y_c E_{\text{base}}/Y_{\text{base}} E_{\text{base}}$$

$$= I_{ex}/I_{\text{rated}} = 0.05 \text{ (as given)}$$

$\overline{B}_c = 0.05^2 - 0.01^2 = -0.049$

$\overline{R} + j\overline{X} = 1/(\overline{G} + j\overline{B}) = \underline{4.0 + j19.6}$ ⟵ c

8. a. $E_1 = K\phi_1 f_1$

    $1.1 E_1 = K\phi_2 1.05 f_1$

    $\phi_2/\phi_1 = 1.1/1.05 = 1.0476$

    But, $B = \phi/A$, so $B_{max} = \underline{104.76\% \text{ of normal}}$ ⟵ a

  b. Core loss (from Problem 6) $= 0.01 \times 500 = 5$ kW

    $5 = P_h + P_e = 2P_e + P_e = 3P_e$

    $P_e = 5/3 = 1.66$

    $P_h = 2 \times 1.66 = 3.33$

    New $P_h = 3.33 \times 1.05 \times 1.0476^2 = 3.84$

        $P_e = 1.66 \times 1.05^2 \times 1.0476^2 = 2.016$

        $P_{core} = P_h + P_e = \underline{5.86 \text{ kW}}$ ⟵ b

  c. $P_{core} = E^2 G_c = (1.1 \times 13.8)^2 \times 2.63 \times 10^{-5} \times 10^3$

           $= \underline{6.06 \text{ kW}}$ ⟵ c

9. a. $I = \frac{1}{2}(50000/7200) = 3.47$

      $I^2 R = 250, R = 250/3.47^2 = 20.73$

      $I^2 X = 1000, X = 1000/3.47^2 = 82.94$

    $R + jX = 20.73 + j82.94 = 85.5\underline{/75.96°}$

      $Z_{base} = 7200^2/50000 = 1036$

        $\overline{R} = R/Z_{base} = 20.73/1036 = \underline{0.02}$

        $\overline{X} = X/Z_{base} = 82.94/1036 = \underline{0.08}$ ⟵ a

        $\overline{Z} = \overline{R} + j\overline{X} = 0.02 + j0.08 = \underline{0.0825/75.96°}$

  b. $Z_{low} = 85.5(240/7200)^2 = 0.095$

    $I_{rated} = 50000/240 = 208$

    $E_{test} = 208 \times 0.095 = 19.8$ V

    Alternately, since $\overline{Z} = E_{test}/E_{base}$

    $E_{test} = \overline{Z}E_{base} = 0.0825 \times 240 = \underline{19.8 \text{ V}}$ ⟵ b

HOMEWORK PROBLEMS

1.

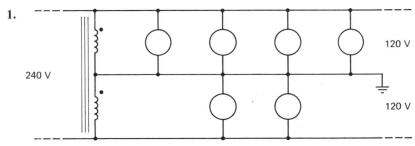

240 V

120 V

120 V

In a certain area of a factory we have a three-phase, 240 volts supply fed by a delta-connected transformer bank. We now wish to provide for some 120 volts single phase lighting load. To do this we take a transformer of 1:1 ratio and connect as shown above, thus providing a center tap with 120 volts on each side. We supply 1000 watts of lighting on one side and 500 watts on the other.

  a. What must be the current rating of each winding of the transformer?

  b. If any of the lights may be off at any given time how does this affect your answer to (a) above?

2.

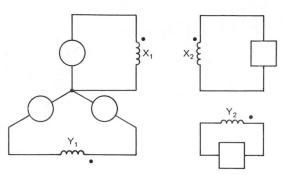

We have a balanced three-phase four-wire, 4160 volts source available and we wish to feed a balanced two-phase load of 50 kVA per phase at 240 volts by means of the transformer connection shown above.

The two transformers above are marked X and Y with their high and low voltage windings designated by subscripts 1 and 2, respectively.

If you were ordering these transformers from the manufacturer, what voltage and current ratings would you specify for each of the four windings?

3.

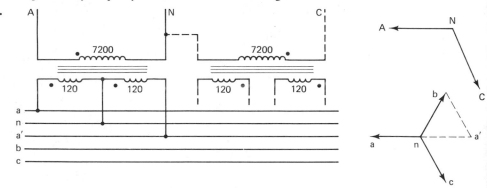

A certain utility supplies single-phase 120/240 volts power by running one-phase wire and neutral (E_{AN}) from the three-phase substation to a single-phase transformer supplying the load. This is the transformer at the left above. It is now desired to supply a new load of 50 kVA balanced three-phase at 120/208 Y volts. It is proposed for economic reasons to run only one additional wire (phase C) from the substation and supply one additional transformer (shown at the right) at the load site. The phasor diagram above illustrates how balanced three-phase voltage may be obtained in this way.

a. Show how the two 120-volt windings must be connected to the low voltage busses in order to obtain the desired phase relations.

b. What should be the current ratings of the 120-volt and 7200-volt windings of the new transformer?

4. We are to test a certain 50 kVA, 2400-480–volts transformer in the laboratory. If there is no phase-angle meter or wattmeter available we might use the connection shown in the following figure. We short-circuit the low-voltage winding and apply from a variable voltage source through a two-ohm resistor on the high-voltage winding with voltage readings taken at three locations as shown.

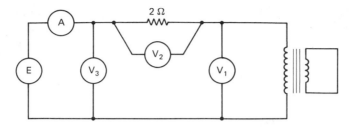

We pass rated current through the ammeter and find voltage readings at the three locations as follows:

$$V_1 = 86.59 \text{ volts}$$
$$V_2 = 41.66 \text{ volts}$$
$$V_3 = 115.03 \text{ volts}$$

a. What are R_{eq} and X_{eq} of this transformer as in Figure 3-27? It is assumed that the exciting current is negligible.

b. What would R_{eq} and X_{eq} be if measured from the low-voltage side?

c. What is the per-unit impedance of the transformer on its own rating as base?

A utility wishes to upgrade a 230 kV transmission line to 345 kV by supplying transformers between the 230 kV bus and the 345 kV line. Two methods of using Y-connected transformers are under consideration: (1) a conventional transformer sketched on a per-phase basis below, or, (2) an autotransformer shown on the right in the figure. In each case the transformer is to supply 100 MVA per phase to the line.

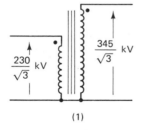

(1)

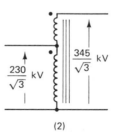

(2)

a. What must be the voltage and current rating of each coil in each case?

In comparing two alternatives as above it is significant to compute the total "MVA of parts" defined as:

$$\text{MVA of parts} = \sum_{i=1}^{i=n} E_i I_i$$

which is simply the sum of the voltampere ratings of the coils of a given transformer construction.

b. What are the "MVA of parts" for each of the methods, (1) and (2), shown above?

c. If we estimate that the weight and cost of transformers of similar voltage ratings are proportional to the total MVA of parts, what is the cost ratio of the two methods?

When we use an *auto*transformer connection as in Problem 5, part 2, we not only result in a physically smaller unit to handle the power but also the series impedance, $\overline{R}_{eq} + j\overline{X}_{eq}$, is less.

Suppose that computation shows that, when fed by an infinite (zero internal impedance) 230 kV bus with the 345 kV side short-circuited we draw a short-circuit current of 26000 amperes.

Standards call for the short-time current rating of a transformer to be a maximum of 25 times norm
rated current.

How much total additional series impedance from the system or extra current-limitir
reactors must be provided to limit the current to a safe level for a short-time fault duration?

Give values in ohms per phase and assume that all impedances can be approximated by pu
inductive reactance.

7. Recall that *voltage regulation* is the change in voltage magnitude expressed as a fraction or
percentage of rated full-load voltage under two specified loading conditions, normally no-load ar
full-load at a given power factor. Thus:

$$\% \text{ Voltage regulation} = \frac{E_{nl} - E_{fl}}{E_{fl}} \times 100$$

The circuit below illustrates the concept. Suppose that the source voltage is adjusted to a value suc
that the load voltage, E_2, is held at the rated value (13.8 kV in this case) when rated current at th
specified power factor is drawn by the load. When the switch, S, is opened the voltage E_2 change
to the no-load value (source voltage times the transformer ratio) and we can compute the voltag
regulation.

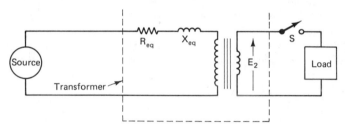

Transformer: 1000 kVA, single-phase,   Load: 1000 kVA

69-13.8 kV   13.8 kV

$R_{eq} = 48$ ohms   0.8 p.f.

$X_{eq} = 290$ ohms   current lagging

a. What magnitude of source voltage is required to supply rated voltage to the load? Note that thi
source voltage is maintained.
b. What is the voltage E_2 when the switch is opened?
c. What is the percent voltage regulation of this transformer at this load?
d. What power factor of the load would result in the largest possible magnitude of the voltag
regulation?
   Hint: Sketch a phasor diagram showing the voltage and current components referred to one sid
   of the transformer.
e. Is it possible for the voltage regulation of the transformer to be zero (or even negative) at som
power factor?

8. Suppose that we have a 25 kVA, 2400-240 volts, single-phase, 60 Hz transformer with $R_{eq} =$
0.03 ohm and $X_{eq} = 0.09$ ohm, referred to the 240-V side. The core loss of the transformer i
150 watts at the rated voltage and frequency.
a. If the transformer supplies a load of 25 kVA, 0.8 power factor, current lagging on the 240-vo
side, what is the percent voltage regulation and what is the efficiency?
b. As a second case, suppose that the transformer is connected as an autotransformer as in Fig
ure 3-8 and supplies rated current at an output voltage of 2400 + 240 = 2640 volts and wit

a power factor of 0.8, current lagging. What is the percent voltage regulation and the efficiency when used in this manner?

9.

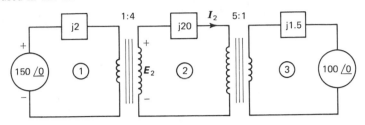

The diagram above shows two ideal transformers with ratios given, two voltage sources, and three impedances in ohms. It has been decided to deal with this circuit in per unit, using S_{1b} = 1000 VA and E_{1b} = 100 volts.

a. Complete the tabulation of base quantities in the respective regions:

Region	S_b	E_b	I_b	Z_b
①	1000	100		
②				
③				

b. Draw the circuit diagram with all quantities labelled in per unit.

c. What are I_2 and E_2, respectively, in amps and volts as well as in per-unit?

10. A certain large generator has a three-phase rating of 1000 MVA and 25 kV. The Thevenin equivalent or "internal" impedance of the machine is given as 0.01 + $j0.95$ per unit on the machine's own rating as base.

   a. What is the impedance in ohms per phase?

   b. What is the E_{l-n} in volts when the machine voltage is 1.1 per unit?

   c. If we have a short circuit on all three phases at the terminals of the machine when the internal voltage is 1.1 per unit, what current flows in amperes and in per-unit?

11. A certain three-phase transformer is rated 500 MVA, 25-500 kV (circuit voltages) and the manufacturer has given the impedance as 18 percent on the transformer's own rating as base.

   Suppose that, because of other units in the system, the base quantities have been chosen as 1000 MVA and 450 kV (on the 500-kV side of the transformer). What is the per-unit impedance of the transformer on this new base? What is base current on the low-voltage side on the new base?

12. Suppose that we have three different 100-kVA transformers that are being considered to supply a 480-volt load from a 7620 volt source. Loss data on the three transformers (marked A, B, and C) is available as follows:

Transformer	Iron Loss, P_{Fe}	Copper Loss, P_{Cu}
A	0.75 kW	0.75 kW
B	0.40 kW	1.10 kW
C	0.30 kW	1.20 kW

   a. What is the efficiency of each transformer at full load, 0.8 power factor, current lagging?

b. At what fraction of full load will each transformer have maximum efficiency and what is that maximum efficiency? Compute at both 0.8 p.f. and unity p.f.

c. If the daily load cycle over 24 hours is as follows:

Hours	Per-unit load
10	0.1
8	0.5
6	1.0

which of the three transformers will waste the least energy in a day's operation?

Chapter 4

Electromechanical

Energy

Conversion

- An introduction to the magnetic field concepts commonly used in analysis of electromechanical energy converters.
- The fundamentals of the conversion process from an energy function and other points of view.
- Analysis of some simple devices.

4.1 INTRODUCTORY EXAMPLE

Electrical engineers may be interested in electrical phenomena for the sake of processing and transmitting intelligence, as in the case of computers or communication links, or they may be interested in transmitting and utilizing energy for energy's sake. As an example, the hundreds of thousands of horsepower of the falling water at a large dam are transmitted across a state as a source of energy for lighting, heating, motors to drive mechanical apparatus, or for electrochemical processing. A key link in this process is the conversion of mechanical energy to electrical form (generators) or electrical energy to mechanical form (motors). This conversion between electric and mechanical form is the subject of this chapter.

As a first example, consider the case of two charged parallel metal plates as in Figure 4-1. This case is considered because the geometry of the field is simple and the fundamental process is not obscured by mathematical details. An edgewise view is shown in the figure illustrating the flow lines of the electric field that is set up when the charge of $+Q$ is applied to the top plate and a charge of $-Q$ to the lower plate (as by touching the plates momentarily by the terminals of the battery).* If the distance of separation of the plates, z, is small compared to the other dimensions of the plates, the D vector and the E vector set up by the charge can be considered to be uniform throughout the volume between the plates, and the flow lines are straight, parallel lines perpendicular to the plates. In other words, edge effects will be ignored for simplicity.

Under the above conditions the following simple equations relate the field quantities:

$$D = Q/A, \qquad E = D/\varepsilon_o = Q/A\varepsilon_o, \quad \text{and} \quad V = Ez = Qz/A\varepsilon_o$$

where A is the area of the plates, ε_o is the permittivity of free space $(1/36\pi\,10^9)$, and V is the potential difference between the plates.

Now suppose that we grasp the plates and pull them apart so the distance of separation is increased to $z + \Delta z$. We will do this with insulated gloves so that the

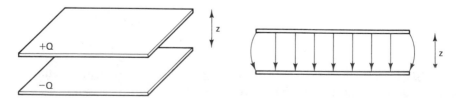

Figure 4-1 Parallel plate example.

*It is regrettable that, because of the finite number of letters in the alphabet, various letters are assigned different meanings in different contexts. The letter "Q" is commonly used for reactive volt-amperes and was so used earlier — but Q is also invariably used for electric charge and will be so used here. (In hydraulics we also use Q for flow in cubic feet per second.) Such dual usage will not usually cause trouble but the student is advised to note the change in meaning of letters like Q, E, and V in this chapter as contrasted with the usage of earlier chapters.

charge will not change. The voltage between the plates is now given by
$$V = Q(z + \Delta z)/A\varepsilon_o$$
and we see that the voltage has increased by $\Delta V = Q\Delta z/A\varepsilon_o$. The energy stored in the field is obtained from the relation $W = QV/2$ and hence the energy stored before and after is given by the equations:
$$W = Q^2 z/2A\varepsilon_o \text{ (before)} \quad \text{and} \quad W = Q^2(z + \Delta z)/2A\varepsilon_o \text{ (after)}$$
that is, the energy has increased by the increment $\Delta W = Q^2\Delta z/2A\varepsilon_o$ by virtue of the movement.

The source of the energy increment is easily seen; in pulling the plates apart we must do work against the electric force trying to pull the two unlike charges together. Mechanical work (by the mover) has been converted to electric energy stored in the field, and is available if, for instance, we now discharge the plates as by a resistor connected between them. The extra increment of energy added by mechanical means is used as well as the original energy supplied by electrical means. We have had an energy conversion from mechanical to electrical form!

The reverse action can occur. If, for example, we simply release a plate that has been held, the plates will be drawn together, assuming that gravitational forces have somehow been balanced out. As the plates are accelerated by force of the electric field, stored field energy is converted into mechanical energy in kinetic form. Alternately, we might allow the plate to be drawn against the force of a restraining spring, thus converting the energy to potential energy in mechanical form. The original example might be termed *generator action* and the second example be called *motor action*.

Suppose that we wish to write the differential equation of motion for the motor action case. We then wish to have an expression for the force of electric origin. An expression for the force may be found in various ways. We might conceivably use Coulomb's law to set up the force between differential elements of charge on the plates, and then sum the forces over the entire area of the plates. Such a summation would be a vector summation, which is usually more difficult than dealing with scalar sums. It is easier to use the scalar quantity of energy to derive the force equation. If we were to move the plates a small distance, dz, then energy balance gives us the necessary equation, $f\,dz = dW$ for the incremental work done by the externally applied force in the z direction. From this we obtain
$$f = dW/dz$$
This is the force we apply in the z direction to cause motion. For the case at hand
$$f = \frac{d}{dz}[Q^2 z/2A\varepsilon_o] = Q^2/2A\varepsilon_o$$
from the preceding energy relation. It should be noted that whether we actually move or not, the force exists. This is sometimes called the principle of virtual work in theoretical mechanics. The force of electric origin opposing the motion is the negative of the above with respect to the z direction.

The expression for the field energy stored could be obtained in various ways. We might in some cases wish to use the expression for the energy stored in a capacitor, $W = Q^2/2C$. Alternately, we might wish to use the expression for energy stored per unit volume, $w = DE/2$ and sum this throughout the volume. (In the case of this uniform field, simply multiply by the total volume.)

Clearly, the ability of a given apparatus to develop force is related to the ability of the field to store energy. It is interesting to compare the electric field case with the magnetic field case in this connection. The breakdown strength of air limits the E vector to about 3×10^6 volts per meter. On the other hand, for the magnetic field we can (by using steel in the flux paths outside the air region where motion takes place) attain a B vector of 1 weber per square meter fairly easily. The energy stored per-unit volume in each case would then be

Electric Field	Magnetic Field
$w = DE/2 = E^2 \varepsilon_o/2$	$w = BH/2 = B^2/2\mu_o$
$= (3 \times 10^6)^2/2(36\pi 10^9)$	$= 1^2/2(4\pi 10^{-7})$
$= (1/8\pi)10^3$ joules$/m^3$	$= (1/8\pi)10^7$ joules$/m^3$

By comparing the possible energy densities for the two fields it is small wonder that in devices intended to develop considerable force, the magnetic field is almost universally the coupling medium. Electric field transducers such as the condenser microphone or capacitor voltmeter are sometimes encountered where only a trivial amount of energy is involved. Electrostatic machines have been of historical interest and such apparatus as the Van de Graaff generator even now is used for special applications. For even fractional horsepower motors, however, magnetic field devices are invariably used. For this reason, the following pages will be devoted almost entirely to magnetic field devices.

4.2 MAGNETIC CIRCUITS WITH AIR GAPS

We have already discussed magnetic fields in relation to the transformer, where the high permeability of the steel core dominated the field picture. When we come to electromechanical devices, a further complication arises in that we must have air gaps in the magnetic pathways in order that motion may be possible. The H vector in the air gap is very large in comparison with that in the steel, and often has the dominant effect in analyzing the field.

We now return to the toroidal core and coil example of Chapter 3 (Fig. 3-1), except that we now consider the effect of an air gap, l_a meters long as shown in Figure 4-2. Ampère's law is still valid for this case, but H in the steel and H in the air gap are now different and we must integrate in two segments

$$\oint \vec{H} \cdot \vec{dl} = \int_{l_{st}} \vec{H}_{st} \cdot \vec{dl} + \int_{l_a} \vec{H}_a \cdot \vec{dl} = Ni$$

where l_{st} is the length of the path in steel and l_a is the length of the path in air. By assuming constant H in each region we arrive at

$$H_{st}l_{st} + H_a l_a = Ni$$

but we now have two unknowns and but one equation. The second equation is

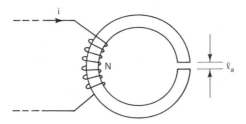

Figure 4-2   Elementary magnetic circuit.

obtained from the continuity of the B vector at the air-steel interface ($\vec{\nabla} \cdot \vec{B} = 0$).

$$B_{st} = B_a \quad \text{and therefore} \quad \mu_r \mu_o H_{st} = \mu_o H_a$$

Note that μ_r for air is unity. If we substitute this relation in the earlier equation we have

$$H_a\left[\frac{l_{st}}{\mu_r} + l_a\right] = Ni$$

If we use the information that $B_a = \mu_o H_a$ and $\phi = B_a A$ we obtain

$$\phi\left[\frac{l_{st}}{A\mu_r\mu_o} + \frac{l_a}{A\mu_o}\right] = Ni = \mathcal{F}$$

The quantity in brackets is the reluctance of the magnetic circuit (the reciprocal of permeance, $\mathcal{P}$).

$$\mathcal{R} = \left[\frac{l_{st}}{A\mu_r\mu_o} + \frac{l_a}{A\mu_o}\right]$$

and we make the first observation that reluctances of two series segments of a path combine by addition, since the two terms are separately the reluctance expressions for the steel and air segments of the total path. Also we observe that the reluctance of the steel is typically much smaller than that of the air. If the length of the steel were for example 50 times that of the air and the relative permeability of the steel were 5,000, the reluctance of the steel would be only 1 percent of that of the air! For this reason we often neglect the reluctance of the steel portion when there is a significant air gap. Under these conditions reluctance becomes a more useful computational tool, since it is a constant.

As in the case of the transformer, the voltage at the terminals of a coil wound around the magnetic circuit is given by Faraday's law

$$e = d\lambda/dt$$

where a positive sign infers a voltage drop in respect to that particular current direction, which would cause positive flux linkages, lambda. Also, as in the case of the transformer, it will be usual in the magnetic circuits of this chapter to make the simplifying assumption that the total flux or flux linkages can be given by the expression $\lambda = \phi N$, where ϕ is the flux of the B vector across the cross-sectional area of the steel core given by $\phi = BA$ and N is the number of turns of the coil.

If we now refer again to the toroid with an air gap in Figure 4-2, the flux is given by $\phi = (\mu_o A/l_a)Ni$ when the reluctance of the steel portion is negligible. It follows that $\lambda = N\phi = (\mu_o A / l_a)N^2 i = (\mathcal{P}N^2)i$. We see that for this case the flux linkages are proportional to the current, and the constant of proportionality is $\mathcal{P}N^2$ (or $N^2/\mathcal{R}$), which is recognized as the *inductance, L*, hence

$$L = N^2\mathcal{P} = N^2/\mathcal{R} \quad \text{and} \quad \lambda = Li$$

as was developed in Chapter 3.

4.3 COMPUTATIONAL DETAILS

The simple toroidal coil of Figure 4-2 is not very representative of most actual magnetic circuits of concern in practical cases. In many practical cases, the cross section of steel may vary at different places along the circuit as illustrated in the segment shown in Figure 4-3(a). To analyze such a case, imagine that a Gaussian surface includes the transition as shown by the dotted lines in the figure. The zero divergence of the B vector requires that $\oint \vec{B} \cdot \vec{dS} = 0$ over any closed surface. The B vector is very small everywhere except where the steel pierces the surface. If the B is assumed uniform across the area, $B_l A_l = B_r A_r$, where the subscripts refer to left and right, respectively; that is, the flux ϕ entering one side of the transition is equal to the flux leaving through the other side. The details of flux distribution right at the transition are complex and the assumption of uniform flux distribution across the area is of course in error, but this is usually ignored for simplicity.

In an air gap, the flux pattern actually bulges or "fringes" as shown in Figure 4-3(b). An analysis could be made by solving Laplace's equation numerically but if the length of the gap is small compared to other dimensions, one often ignores the fringing (assume $B_{\text{air}} = B_{\text{steel}}$) or makes an empirical correction by assuming the air gap area to be increased by adding the length of the air gap to the linear dimensions of the steel cross section. Thus if the steel cross section is a times b the air gap area is taken as $(a + l_a)(b + l_a)$. This is simply an old rule of thumb which often gives reasonable checks with test values.

In the analysis of the simple toroid, the permeability of the steel was given as $\mu_{st} = \mu_r\mu_o$ and the relative permeability μ_r was used in the mathematics as though it were known. Actually, the permeability of the steel varies with the magnitude of the B vector (and sometimes with its direction), and one must have data on the

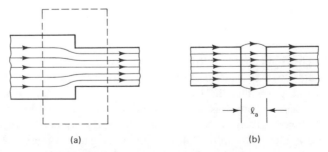

(a)   (b)

Figure 4-3   The behavior of flux at a transition.

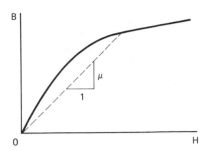

Figure 4-4   A typical *B-H* curve for steel.   0

particular alloy being used. Such information is often available in the form of *B-H* curves like those of Figure 4-4. This curve was first introduced as Figure 3-14. At any particular level of *B* or *H*, a line from the origin to the curve has the slope of μ ($= \mu_r \mu_o$) of the steel. It is to be noted that as the level of *B* increases, the permeability of the steel falls off.

It is not common to use the *permeability* for practical computations. Instead, computations follow the outline below, which will be illustrated for the hypothetical magnetic circuit of Figure 4-5. In this case there are several different sections in series. Each is identified by a number, and it is assumed that the length and cross-sectional area of each section is known as, for example, A_1 and l_1 for the long section around which the coil of N turns is wound. We begin by taking a value for the flux ϕ which is the same in all sections.

$$\phi_1 = \phi_2 = \phi_3 = \phi_4$$

The magnitude of the *B* vector for each section is found next

$$B_1 = \phi/A_1, \qquad B_2 = \phi/A_2, \qquad B_3 = \phi/A_3, \qquad B_4 = \phi/A_4$$

The *H* magnitude may now be found from the magnetization curves for each value of *B*, and the line integral $\oint \vec{H} \cdot \vec{dl} = Ni$ be evaluated in sections as before.

$$H_1 l_1 + H_2 l_2 + H_3 l_3 + H_4 l_4 = Ni$$

and from this sum the current required to produce the given value of flux may be found. If the problem is approached from the other direction (given a current, find the flux), then it is necessary to use cut-and-try to guess the flux. A few guesses will

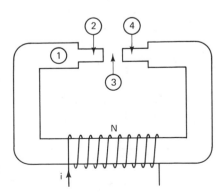

Figure 4-5 Example for magnetic circuit computations.

bracket the desired value and a rough curve will permit interpolation to find the answer.

One would not ordinarily refer to a curve to find H for section 3, which is an air gap in the example. Instead, one merely finds B from the air gap area (correcting for fringing as above if desired) and then uses the equation $B_{air} = \mu_o H_{air}$ where $\mu_o = 4\pi 10^{-7}$.

We might generalize the above example for any series circuit by the equations

$$Ni = \sum_k H_k l_k \quad \text{or} \quad i = (1/N) \sum_k H_k l_k$$

In many practical cases we will wish to compute a succession of points of flux and current and plot either the flux ϕ or flux linkages $\lambda = N\phi$ versus the current i, as in Figure 4-6. It is convenient for understanding to plot the current required for the air gap alone at any given value of ϕ or λ as $i = H_a l_a / N = \mathcal{F}_a / N$. This forms a straight line as shown in the figure. For low values of B, the ampere turns required by the steel portions of the path (sum of all other $H_k l_k$ terms) are small compared to those required by the air, and the net curve follows the "air gap line" closely. When B approaches the knee of the B-H curve, the quantity $\mathcal{F}_{steel}/N$ becomes noticeable, and the curve of ϕ vs. i or λ vs. i falls away from the air gap line as shown. Curves such as those in Figure 4-6 are known as *magnetization curves*.

Along the air gap line the flux linkages are proportional to the current, and the inductance parameter is seen to be equal to the slope of the line since $\lambda = Li$ is the defining equation for inductance. On the curved sections, the use of inductance is no longer valid as a description of this relation. Even in this case, however, we sometimes use the term *incremental inductance* to describe the slope of the curve at a certain point and relate small increments in flux linkages to an increment in current according to the equation $\Delta\lambda = L \Delta i$.

Sometimes a magnetic circuit may involve parallel branches as well as series branches, as is illustrated for two typical cases in Figure 4-7. In such cases the continuity of flux requires that the flux entering a branch point shall be equal to the flux leaving. The analogy may be drawn to the electric circuit and Kirchhoff's

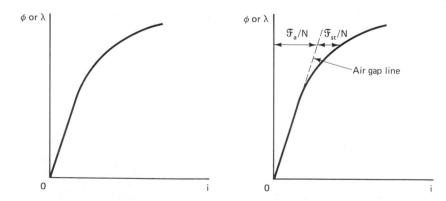

Figure 4-6 Typical magnetization curves.

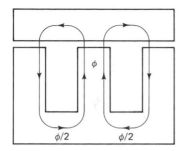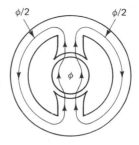

Figure 4-7   Magnetic circuits involving parallel branches.

current law at a junction. Because of the symmetry of the cases illustrated in Figure 4-7, half of the flux follows each of the parallel paths. Cases not involving symmetry may involve a more complex distribution, although for cases where an air gap dominates, it may be possible to use the reluctance concept and solve according to the laws of the equivalent electric circuit. Without symmetry or dominant air gaps, it may be necessary to find the flux distribution by cut-and-try.

4.4 ENERGY CONVERSION

Now we are ready to study cases involving energy conversion using the magnetic field as the coupling medium. Figure 4-8 illustrates a simple device wherein a magnetic field attracts a movable steel member called the *armature*. It is assumed that the movable armature is constrained by frictionless guides so that it moves parallel to itself at all times and a single mechanical variable x serves to describe its position. The force developed by the field will be given by f and taken as positive in the x direction. It will be noted that the references call for electric power *in* as positive and the mechanical work done by the device as positive. This then is the "motor action" case. Generator action could be covered by the same references if numerically negative power is indicated.

The electric power in is given by $p = ei$, where $e = d\lambda/dt$ from Faraday's law. Note that any resistance of the coil is assumed to be external to the diagram and power consumed in such resistance is to be accounted for elsewhere. As in the

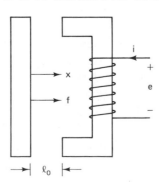

Figure 4-8   First example of energy conversion with magnetic field.

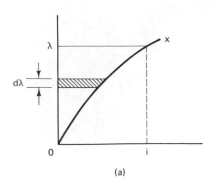

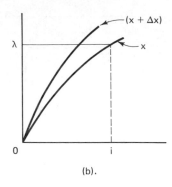

(a)   (b).

Figure 4-9 Magnetization curves for device of Figure 4-8.

discussion of the transformer, the power into the coil may be given by $p = i(d\lambda/dt)$ and the differential amount of *energy* supplied in a time dt by $dW = p\,dt = i\,d\lambda$. If the magnetization curve of the device is as shown in Figure 4-9(a) for a given x, then this increment in energy is shown by the shaded horizontal wedge at some value of λ. The total energy put into the device by holding x fixed and allowing the flux linkages to grow from zero to some value λ is the summation

$$W = \int_o^\lambda i\,d\lambda$$

and this energy is represented by the area to the left of the curve between zero and λ..* If the limits of the integral are reversed, W is the negative of the original value. This means that the energy originally put in is stored in the field and is completely returned to the circuit when the field collapses.

Energy storage in a magnetic field was discussed in Chapter 3 in connection with the transformer core along the lines given above, but the emphasis in connection with electromechanical energy conversion is different. First, let it be pointed out that the devices in this chapter involve a mechanical displacement variable: x in the example at hand, θ in later examples involving rotary motion. The presence of an additional variable in the form of mechanical displacement complicates the picture, but on the other hand, the inevitable air gap in the magnetic circuit makes certain approximations feasible. In considering the transformer core, it was pointed out that hysteresis and eddy current effects cause the magnetization curve of λ versus i to have a double-valued character on the λ-i plane. Such relations are inadmissible in the type of mathematical treatment used for the devices of this chapter. Fortunately, because most of the ampere turns are consumed in the air gap, the hysteresis loop is very slender, representing only a small increment around the air gap line, which was illustrated in Figure 4-6. In consideration of all these things, we will represent the magnetization curve as a single valued function of the variables. If indeed it is desired to include the energy loss of hysteresis and eddy

*The actual energy is, of course, the area times the scale factors of the plot. We will assume unity scale factors in the following discussion, and thus be able to be rather simplistic in referring energy to an area and vice versa.

currents, we might approximate the loss by means of a fictitious shunting resistor, as was done to represent transformer core losses.

Now consider the curve for a new value of x as $(x + \Delta x)$ in part (b) of Figure 4-9. The energy put into the field of this value of x, and available on collapse of the field, is evidently different for the same upper limit of λ. The field energy is thus a function of both λ and x, and this dependence on the two variables may be indicated by the functional notation

$$W_f = W_f(\lambda, x)$$

The "state" of the field energy is determined by the state variables, x and λ.

The nature of the energy function $W_f(\lambda, x)$ might be visualized by imagining a vertical axis for W_f erected over a plane with coordinate axes λ and x. Each point in the λ-x plane corresponds to a particular value of the energy W_f and the totality of points of W_f forms a surface over the λ-x plane. If we imagine that we travel from one point (λ_1, x_1) to a second point (λ_2, x_2) in the λ-x plane, there is a perfectly definite change in energy that occurs. Reversing the change in (λ, x) results in the negative of the original energy change. Further, it clearly cannot make any difference what path or trajectory is followed between the two points in the λ-x plane so far as net energy change is concerned. These are all properties of a *conservative* system, and the reader by now may be thinking of another similar conservative system, such as the electrostatic potential function in a two dimensional field, which is a good analogy to the energy function under discussion. More will be said of this later.

We can account for system energy changes by examining the magnetization curves for various hypothetical changes. Figure 4-10 illustrates some changes that could take place. Since the system is conservative, we can write equations based upon conservation of energy as

Elec. energy in = Increase in field energy + Mechanical energy out.

Part (a) of Figure 4-10 shows a motion from point a to point b along a line of constant flux linkages. If flux linkages are constant, $d\lambda/dt = 0$, $e = 0$, and no

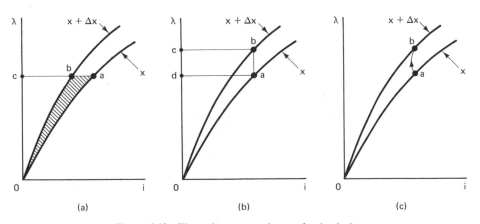

(a)   (b)   (c)

Figure 4-10 Illustrating energy changes for the device.

electrical energy is put in. The field energy at x is represented by the area oac and the field energy at $x + \Delta x$ is given by obc. We then write

$$0 = obc - oac + \text{Mech. energy out}$$

and,

$$\text{Mech. energy out} = oac - obc = oab$$

in terms of the areas, where oab is shaded to indicate the mechanical work done. The student may wish to note that motion from $x + \Delta x$ to x along the same line results in mechanical work $= -oab$; that is, generator action. Note that energy is inherently positive but the *increment* in energy may be negative.

If, as in part (b) of the figure we move from point a to point b along a line of constant current, electrical energy is added in the amount $\int i d\lambda$ represented by the rectangle $dabc$ and we can write

$$dabc = obc - oad + \text{Mech. energy out}$$

or

$$\text{Mech. energy out} = oad + dabc - obc = oab$$

a similar wedge to the first case but not shaded for fear of confusion of lines!

A more realistic motion might ensue from a situation where the coil is fed from a voltage source through a resistor, thus requiring a steady-state current of E/R for both the x curve and the $x + \Delta x$ curve. During a transition the flux linkages increase and cause a momentary dip in the current. The trajectory in the λ-i plane may then be shown in part (c). Time is a parameter along the trajectory, and specific points on that trajectory would have to be found from the dynamics of the system. The student may note that the mechanical work done is again given by the area oab.

To solve the system dynamics or for other purposes, we will wish to find the mechanical force. This we may do as follows. Conservation of energy requires that

$$dW_{\text{elect}} = i d\lambda = dW_f + f dx$$

or

$$dW_f = i d\lambda - f dx \tag{4-1}$$

but

$$dW_f = \frac{\partial W_f}{\partial \lambda} d\lambda + \frac{\partial W_f}{\partial x} dx \tag{4-2}$$

which is the expression for the total differential of the function $W_f(\lambda, x)$ by the rules of the calculus.* The two variables, λ and x are independent; that is, the equations must be true no matter how λ and x vary, specifically if $d\lambda$ and dx are separately considered to be zero, we conclude that it must be true that

$$i = \frac{\partial W_f}{\partial \lambda} \tag{4-3}$$

---

*This expression might be compared with the analogous expression for the increment in electrostatic potential as the dot product between the gradient and the displacement. The partial derivatives of W_f are the components of a gradient vector, and the increments in λ and x are the components of a displacement vector in the λ-x plane.

and

$$f = -\frac{\partial W_f}{\partial x} \tag{4-4}$$

Equation (4-4) in particular is an equation of great utility and we will illustrate by finding the force of electromagnetic origin for the device in Figure 4-8. If we assume that the air gap dominates the magnetic circuit, the field energy

$$W_f = \frac{\lambda^2}{2L} \quad \text{where } L = \frac{N^2}{\mathcal{R}} = \frac{N^2 A \mu_o}{2(l_o - x)}$$

In the expression for inductance the symbol l_o stands for the length of the air gap when the armature is at $x = 0$. From the energy expression, Eq. (4-4)

$$f = -\frac{\partial}{\partial x} \frac{\lambda^2(l_o - x)}{N^2 A \mu_o} = \frac{\lambda^2}{N^2 A \mu_o}$$

which was the objective.

For many purposes it is more convenient to have the force expressed in terms of current rather than flux linkages. We note that $\lambda = Li$ and make the substitution to find

$$f = \frac{N^2 A \mu_o i^2}{4(l_o - x)^2}$$

In applying this device to a practical situation, we may wish to find displacement or current as a function of time. For example, we may wish to know how long it would take this device to close a pair of contacts, or to open a valve. Such a solution requires that we write the differential equations and solve. Suppose, for example, that the device is connected mechanically and electrically to other devices, as in Figure 4-11. The mechanical elements are taken to be a spring without initial tension when $x = 0$, and a viscous friction element or dash pot. The electrical elements involve a voltage source, E, and resistance, R, which accounts for all resistance of coil and circuit. If we sum mechanical forces on the armature we have

$$\frac{N^2 A \mu_o i^2}{4(l_o - x)^2} = M \frac{d^2 x}{dt^2} + D \frac{dx}{dt} + kx \tag{4-5}$$

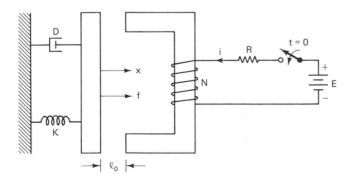

Figure 4-11   Application of the device of Figure 4-8.

Around the electrical loop we have

$$E = Ri + \frac{d\lambda}{dt}$$

where

$$\lambda = Li \quad \text{and} \quad \frac{d\lambda}{dt} = L\frac{di}{dt} + i\frac{dL}{dt}$$

The second term above is a voltage term attributable to the motion and may not be familiar to the student used to dealing in circuits, where L is constant. Now substituting and evaluating

$$E = Ri + \frac{N^2 A\mu_o}{2(l_o - x)} \frac{di}{dt} + \frac{iN^2 A\mu_o}{2(l_o - x)^2} \frac{dx}{dt} \qquad (4\text{-}6)$$

We now have the two simultaneous differential equations (4-5) and (4-6) necessary to solve for x and i subject to the initial conditions that at time $t = 0$, $x = 0$, $dx/dt = 0$, and $i = 0$. Unfortunately, the equations are nonlinear and cannot be solved by usual techniques such as Laplace transformation. We may still find a numerical solution by integrating on the digital computer, or we may use model or analog simulation techniques. The details of solution will not be discussed further in this treatment, however.

4.5 COENERGY

In the preceding work, the force was obtained in terms of λ, the flux linkages, and later the substitute variable i was inserted. This was done because we would not have known how to perform the partial differentiation of the field energy in such a way as to hold λ constant if the field energy had originally been expressed in terms of current. An alternate approach uses a different energy function defined as *coenergy*, W_f', according to the equation

$$W_f' = i\lambda - W_f \quad \text{or} \quad W_f + W_f' = i\lambda$$

from which

$$dW_f + dW_f' = i \, d\lambda + \lambda \, di$$

A simple graphical interpretation as in Figure 4-12 will aid in the understanding.

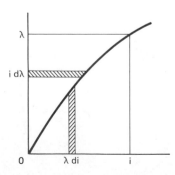

Figure 4-12 The relation between energy and coenergy.

Whereas energy is the summation of $i\,d\lambda$, coenergy is the summation of $\lambda\,di$,

$$W_f = \int_o^\lambda i\,d\lambda \quad \text{but} \quad W_f' = \int_o^i \lambda\,di$$

That is, *energy* is represented by the area at the upper left in the figure, and *coenergy* is represented by the area at the lower right. The sum of the two is simply the total area of the rectangle, or $i\lambda$. It should be noted that, if λ vs. i is linear, energy and coenergy are equal. Energy and coenergy both have the dimensions of joules, and though energy has a very real physical meaning, coenergy is just a function defined for mathematical convenience. Since coenergy is given by

$$W_f' = \int_o^i \lambda\,di$$

for a given x and would be different for different values of x, coenergy is also a function of both i and x, or in functional notation

$$W_f' = W_f'(i, x)$$

Proceeding with the mathematics since

$$dWf + dW_f' = i\,d\lambda + \lambda\,di$$

and

$$dW_f = i\,d\lambda - f\,dx$$

subtracting

$$dW_f' = \lambda\,di + f\,dx \tag{4-7}$$

but again the total differential of a continuous function is given by

$$dW_f' = \frac{\partial W_f'}{\partial i}\,di + \frac{\partial W_f'}{\partial x}\,dx \tag{4-8}$$

The variables i and x are independent and so the coefficients of di and dx in the two equations (4-7) and (4-8) above must be equal to give

$$\lambda = \frac{\partial W_f'}{\partial i} \tag{4-9}$$

$$f = \frac{\partial W_f'}{\partial x} \tag{4-10}$$

Of these two equations Eq. (4-10) is perhaps the most important for our present purposes.

The difference between the force equations using energy and coenergy functions deserves further emphasis.

$$f = -\frac{\partial W_f}{\partial x} \quad \text{or} \quad f = +\frac{\partial W_f'}{\partial x}$$

Note first the minus sign as a difference but also the fact that energy is a function of flux linkages λ and x, whereas coenergy is a function of current i and x. This means that using an energy function, λ is held constant during the differentiation and, using coenergy, current i is held fixed in order to find the force. This may perhaps be clarified by Figure 4-13. As x changes from an initial value of x to $x + \Delta x$ along a line of constant flux linkages, we move from a to b in the figure.

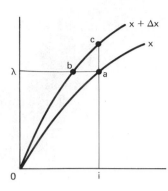

Figure 4-13 Relation between energy and coenergy increments.

As we move along a line of constant current, we move from a to c. The increments in the energy functions are

$$\Delta W_f = -oab \quad \text{and} \quad \Delta W_f' = oac$$

As Δx approaches zero, the two areas approach equality in magnitude, since the difference area (the small triangle bac) is a second-order infinitesimal. The two increments are negatives of each other, so the difference in the two force equations is easily seen.

If we generalize for a moment, we note that both coenergy and energy for a *linear* case are equal and given by

$$W_f = W_f' = Li^2/2 = \lambda^2/2L$$

and force in terms of the current is

$$f = (i^2/2)\, dL/dx \qquad (4\text{-}11)$$

that is, for a linear single coil case at least, the production of a mechanical force is dependent upon a variation of inductance with mechanical displacement. If we return to the example of Figure 4-8 as an application of the coenergy formulation for force, we see

$$W_f' = \frac{Li^2}{2} = \frac{N^2 \mu_o A}{4(l_o - x)} i^2$$

and

$$f = \frac{N^2 \mu_o A}{4(l_o - x)^2} i^2$$

which checks with the original expression, except that it was not necessary to first hold λ fixed while differentiating, and later substitute current as the variable.

4.6 ANOTHER EXAMPLE: ROTATIONAL MOTION

Now let us consider the device sketched in Figure 4-14 with a rotating armature or *rotor* under the influence of a stationary field or *stator*. The mechanical variables are now torque, T, and angle, θ, and the differential increment in mechanical work done by the device is $T\, d\theta$ when torque and angle are considered positive in the same direction (i.e. motor action again). It would now be possible to derive the expression

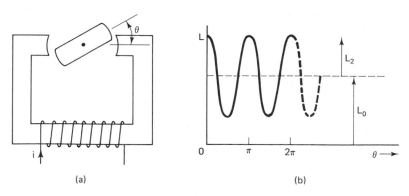

(a)   (b)

Figure 4-14   The reluctance motor.

for torque as was done for force but let it simply be said that everywhere in the development torque would replace force and angle would replace displacement and we would have

$$T = \frac{\partial W_f'}{\partial \theta}$$

Since torque is evidently dependent upon variation of inductance with angle, it is necessary to consider L vs. θ as in part (b) of Figure 4-14. The inductance is a periodic function of 2θ and, as such, can be represented by a Fourier series as

$$L = L_0 + L_2 \cos 2\theta + \text{(higher order terms)}$$

If we ignore the higher order terms then the coenergy is

$$W_f' = \frac{1}{2}i^2(L_0 + L_2 \cos 2\theta)$$

and the torque developed is

$$T = \frac{\partial W_f'}{\partial \theta} = -i^2 L_2 \sin 2\theta$$

It is instructive to study this torque relation for some specific currents.

First, suppose that the current is given by $i = I$, a constant (dc).

$$T = -I^2 L_2 \sin 2\theta$$

A plot of this torque against θ is shown in Figure 4-15. It will be noted that, without external forces there are four equilibrium points in each revolution that are at $\theta = \pi/2$, π, $3\pi/2$, and 2π or 0. If the rotor is brought to any one of these points the torque is zero. At $\theta = \pi/2$ and $3\pi/2$, however, the equilibrium is unstable; that is, a small perturbation sets up a torque, tending to move the rotor still farther from the point. At the stable points 0 and π it can be seen that the electromagnetic torque tends to line the rotor up with the stator.

If we would now attach a helical spring to the shaft at such a position as to hold the rotor at 45° ($\pi/4$) with zero current, the deflection at non-zero current would be a function of that current; the device could be used as an ammeter. To illustrate, a dotted line giving the opposing torque of the spring has been added to the figure.

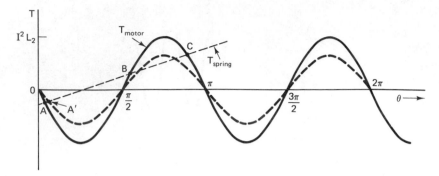

Figure 4-15   Torque variation with angle.

Where $T_{\text{motor}} - T_{\text{spr}} = 0$ or, in other words, $T_{\text{motor}} = T_{\text{spr}}$, we have equilibrium (again subject to stability considerations). If $T_{\text{motor}} - T_{\text{spr}} \neq 0$, the net torque becomes an accelerating torque and motion ensues.

It will be noted that, at point A in the figure, a small displacement toward the right results in a negative accelerating torque, $T_{\text{motor}} - T_{\text{spr}}$, and the rotor moves back toward A; that is, we are stable. A similar analysis shows point B to be unstable and point C to be stable.

If we are operating at point A and the magnitude of the current, I, is changed then the intersection point of the motor torque and the spring torque moves to a new point A'. By placing an indicator on the shaft we could make calibration marks on a scale and we would have an elementary ammeter. If, on the other hand, we would pass an ac current through the coil as

$$i = I_m \cos \omega_s t$$

where the subscript s refers to stator, then the torque becomes a function of both angle, θ, and time.

$$T = -I_m^2 \cos^2 \omega_s t L_2 \sin 2\theta$$
$$= -I^2(1 + \cos 2\omega_s t)L_2 \sin 2\theta$$

We see that the torque is made up of an average (constant) component

$$T_{\text{ave}} = -I^2 L_2 \sin 2\theta$$

and a time-varying component of double frequency

$$T(t) = -I^2 \cos 2\omega_s t L_2 \sin 2\theta$$

Because of the mass of the rotor, the device is unable to respond to the double-frequency sinusoidal component and takes a position corresponding to the average. Since this component works out to be the average of the square of the current, the scale can be calibrated in terms of the square root, and we have an rms reading ammeter for ac currents.

If we remove the spring and allow the rotor to rotate at a mechanical velocity ω_m, we have some interesting possibilities of using the device as a motor. For example, suppose

$$\theta = \omega_m t + \delta$$

The torque equation becomes more complex with these modes of the variables

$$T = -i^2 L_2 \sin 2\theta$$
$$= I_m^2 L_2 \cos^2 \omega_s t \, \sin(2\omega_m t + 2\delta)$$

The significance of this equation is difficult to grasp, but by application of familiar trig identities we may obtain

$$T = -\frac{I_m^2 L_2}{2} \sin(2\omega_m t + 2\delta)$$
$$-\frac{I_m^2 L_2}{4}\left\{ \sin[2(\omega_m + \omega_s)t + 2\delta] + \sin[2(\omega_m - \omega_s)t + 2\delta] \right\} \quad (4\text{-}12)$$

We see from the above equation that the torque function is composed of a sum of sinusoids of various frequencies. For most cases, the average torque over a period of time would be zero, and thus the device could not act as a motor to supply a given torque load on its shaft. If, however, it should happen that $\omega_m = \omega_s$, the last term in the equation becomes a constant and the average torque becomes

$$T_{ave} = -\frac{I_m^2 L_2}{4} \sin 2\delta$$

The total torque as a time function contains pulsating components even under this condition ($\omega_m = \omega_s$), but for typical magnitudes of the parameters the heavy steel rotor is completely unable to respond to the pulsating components to any measurable degree, and the average torque is the only significant quantity. In other words, the mass of the rotor forms a "low pass filter."

The average torque is seen to be a function of the angle δ which appeared in the expression for the mechanical angle as a seemingly arbitrary initial angle of the rotor. Note that the time $t = 0$ is taken at the instant that the stator current is at its maximum. If we were to time a strobe light to fire at each positive current crest, the light flash would "stop" the rotor motion and the eye would see the rotor lying at the angle δ.

The average torque is sketched as a plot against δ in Figure 4-16. A couple of possible load torque levels are also indicated on the sketch. If we consider a positive load torque, there are two possible equilibrium points at negative angles.* Consideration of small perturbations will reveal that point A is stable, whereas point B is unstable. The load torque is seen to determine the angle δ. A higher load torque demand will cause the rotor to fall back to a more negative value of δ. There is a power limit (note that power and torque are proportional to constant speed), and at loads beyond the crest of the curve the motor will stall.

For small positive angles, we see that the load torque is negative numerically; if we were to drive the shaft from a prime mover, the angle δ would advance and the device would absorb torque and power and deliver electrical power as a generator.

Two or three comments should be made about details before leaving this device for other examples. Since the motor runs with the mechanical speed in

---

*By "load" torque we refer to a torque in opposition to the rotor motion. Total torque is thus $T_{(magnetic)}$-$T_{(load)}$ in the forward direction.

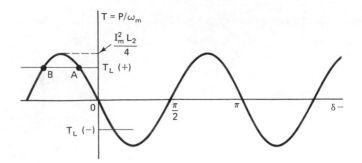

Figure 4-16   Average torque of a reluctance motor.

radians per second equal to the stator source angular frequency, the motor is said to be a *synchronous* motor. Note that a mechanical speed $\omega_m = -\omega_s$ would also result in a non-zero average torque. A motor such as this will not start by itself, but will run in whatever direction it is started.

4.7 MULTIPORT DEVICES

In the examples of the previous sections there was one electrical input or *port* and one mechanical output or *port*. There might be more of either type of ports although the usual multiport devices consist of only one mechanical output and two or more coils forming the electrical inputs. We speak of "inputs" and "outputs" in terms of the motor action references in this case.

A device having two coils and a single mechanical variable, taking θ as the variable for the rotary motion case, will be used in the following discussion. The extension to a larger number of coils would follow along similar lines. With two coils in a given position we may write the equations

$$\lambda_1 = L_{11}i_1 + L_{12}i_2$$
$$\lambda_2 = L_{21}i_1 + L_{22}i_2$$

The terms L_{12} and L_{21} are mutual inductance terms and are written in the form L_{ij} rather than using the symbol M, thus making it more convenient to handle a case of still more coils than the present example. For any given mechanical configuration, the values of the mutual terms L_{ij} may be positive or negative numerically (or zero). In general, all the inductance coefficients will be functions of the mechanical variable, θ, for devices such as those under discussion.

If we differentiate the above equations for a given position, θ, we have

$$e_1 = \frac{d\lambda_1}{dt} = L_{11}\frac{di_1}{dt} + L_{12}\frac{di_2}{dt}$$

$$e_2 = \frac{d\lambda_2}{dt} = L_{21}\frac{di_1}{dt} + L_{22}\frac{di_2}{dt}$$

and if we proceed as in the single coil case, we have for incremental energy changes over a time interval of dt

$$dw_1 = i_1 \, d\lambda_1 = L_{11}i_1 \, di_1 + L_{12}i_1 \, di_2$$
$$dw_2 = i_2 \, d\lambda_2 = L_{21}i_2 \, di_1 + L_{22}i_2 \, di_2$$

By summation (integration) with a fixed mechanical position we find the total energy, equal to the coenergy for the linear case, as

$$W_f = W'_f = \frac{L_{11}i_1^2}{2} + \frac{L_{12}i_1i_2}{2} + \frac{L_{21}i_2i_1}{2} + \frac{L_{22}i_2^2}{2}$$

The summation may be made over any convenient trajectory from an initial value of $i_1 = i_2 = 0$ to a final state of i_1 and i_2.*

Since from the previous pages we know that the torque is given by

$$T = \frac{\partial W'_f}{\partial \theta}$$

we have

$$T = \frac{i_1^2 \, dL_{11}}{2 \, d\theta} + \frac{i_1i_2 \, dL_{12}}{2 \, d\theta} + \frac{i_2i_1 \, dL_{21}}{2 \, d\theta} + \frac{i_2^2 \, dL_{22}}{2 \, d\theta} \qquad (4\text{-}13)$$

and this equation is the multiport analogy to the single-coil case.

4.8 AN ILLUSTRATION OF A MULTIPORT DEVICE

As an illustration of the application of Eq. (4-13), consider the device shown in Figure 4-17. There is one coil with current i_1 on a stationary member of *stator* and another coil with current i_2 on a rotating member or *rotor*. It might be assumed that the self inductances, L_{11} and L_{22}, are relatively constant with angle θ, but the mutual inductances, $L_{12} = L_{21}$, vary with θ. Since the variation is periodic in θ, a first approximation to the variation would be to use the first term of a Fourier series, thus let us say

$$L_{12} = L_{21} = M \, \cos \theta$$

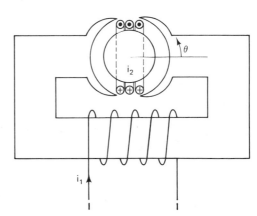

Figure 4-17 A device with two electrical ports and one mechanical port (shaft angle θ).

*Students will recall from earlier circuits courses that the freedom of choice of trajectory requires that $L_{12} = L_{21}$ if the total energy stored at a given i_1, i_2 point is to be independent of the particular integration path!

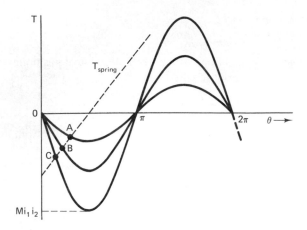

Figure 4-18　Torque variation with θ for various values of the product $i_1 i_2$.

where M is a constant, indeed this is the maximum value of the mutual inductance, and would appear to occur in the position shown in the figure where $\theta = 0$.

With the assumption of constant L_{11} and L_{22}, Eq. (4-13) gives

$$T = -i_1 i_2 M \sin \theta \tag{4-14}$$

A plot of Eq. (4-14) for three sample values of the product $i_1 i_2$ is shown in Figure 4-18. It may be noted that this figure is remarkably similar to Figure 4-15 except that the torque is periodic in θ rather than 2θ. Again, suppose that a restraining spring is attached to the shaft such that it exerts zero torque at $\theta = \pi/2$ or 90°. The spring torque is shown as a straight line (Hooke's law) on the figure.

As an application of the device, suppose that the current i_1 is proportional to a current in a circuit being measured, as $i_1 = K_i i$ and that i_2 is proportional to a voltage as $i_2 = K_e e$. The product $i_1 i_2 = K_i K_e e i$ would thus be proportional to the power of the circuit being measured, and the rotor would take positions in angle such as A, B, and C where the spring and device torque add to zero. Such a device might be calibrated and used as a wattmeter (compare with Figure 1-28).

A larger energy application of the device using a rotor coil and one or more stator coils would be a synchronous motor. Such devices are treated in Chapter 5, but further discussion will be omitted at this point.

4.9 FORCE-CURRENT AND VELOCITY-VOLTAGE RELATIONSHIPS

The examples in the preceding sections were analyzed using energy and coenergy functions W_f and W_f', and this method was found to be effective for those particular problems. There are other means of finding forces, torques, and so forth, and for some classes of problems these other methods may be more effective. It is surely part of the *art* of engineering to find the most suitable approach for a given problem.

In the discussion of the force on the plates of a capacitor in Section 4.1, it was indicated that one might perhaps proceed in the direction of using Coulomb's law to find the force between infinitesimal elements of charge on the plates, in which case the total force would be the sum of the infinitesimal components. The sum is a vector sum, however, and such a process may be very tedious unless there are certain symmetries present.

In the case of magnetic field problems, a very basic relation from field theory gives the force on an infinitesimal length of conductor, $\vec{dl}$, carrying a current i, and located in a field, $\vec{B}$, as

$$\vec{df} = i\,\vec{dl} \times \vec{B} \tag{4-15}$$

The direction of the force is obtained from the cross product of the vectors $\vec{dl}$ and $\vec{B}$.* The right-hand rule for the cross product might be stated as: Let the fingers extend along the first vector, $\vec{dl}$, and curl into the direction of the second vector, $\vec{B}$, in which case the thumb points in the positive direction of the force.

Application of Eq. (4-15) to forces on conductors of arbitrary shape would be very complex because of the vector summation necessary to get the total force. One special case works out simply, however and, since this case is commonly encountered, it is worthy of consideration. That is the case of a long, straight, conductor in a uniform field. For this case we have as the total force on a conductor of length, $\vec{l}$, in a uniform field, B,

$$\vec{f} = i\vec{l} \times \vec{B}$$

It will be recalled that the cross product can be expressed in terms of the sine of the angle between the vectors. For the special, but most common, case where $\vec{l}$ and $\vec{B}$ are perpendicular, the sine of 90° is unity and we have a common expression given in terms of the *magnitudes* of the vectors as

$$f = Bli \tag{4-16}$$

A pictorial representation of this relation is given in Figure 4-19(a), where arrows represent the direction of the components (as usual, a cross indicates the tail of a receding arrow and a dot the head of an approaching arrow). The three elements, f, l, and B form a mutually orthogonal set with directions following from the original cross product formulation.[†]

The developed force is shown in part (a) of the figure. If, owing to the force, motion ensues, then mechanical work may be done. In part (b) of the figure a plan view of the conductor is shown with the direction lines of the B field shown as receding arrows. If we assume that the conductor moves with velocity, v, then mechanical work is being done at the rate $p_m = fv$. This is an example of energy conversion, the energy comes from the electric circuit supporting the current. A voltage, e, will be generated along the conductor as it passes through the B field and,

---

*It will be remembered that current is a scalar. It is assumed, however, that the reference direction for the current is taken in the same sense as the direction of the vector $\vec{dl}$. A short discussion of the vector notation is included as Appendix A for those who have not previously been introduced to this symbolism.

[†]The student may have encountered other memory aids (Fleming's rule, etc.) and it is immaterial which is found the easiest to remember and apply, but it is necessary to have some way of finding the directional relations.

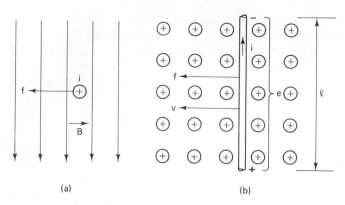

(a)   (b)

Figure 4-19   A straight conductor in a uniform magnetic field.

from the conservation of energy we see that the electrical input must be equal to the mechanical output. In equations

$$ei = fv$$
$$= Bliv$$

from which*

$$e = Blv \qquad\qquad (4\text{-}17)$$

The student should note the references in terms of which these quantities are expressed. Specifically we have mechanical power delivered and electrical power entering the conductor; that is, we have the *motor action* case with these references. It will be recalled that the expression for force in terms of energy functions was also given in the reference frame of motor action. Reversing any one reference will add a minus sign to the equations.

Equations (4-16) and (4-17) are convenient for analysis of many devices *where the equations are applicable.* It must be remembered that the equations were based on a straight conductor in a uniform field with orthogonal relations between the vector quantities. The next section illustrates the application of these equations to a simple example.

4.10 AN ILLUSTRATIVE EXAMPLE OF THE METHODS OF SECTION 4.9

Perhaps one of the most common examples of the application of Eqs. (4-16) and (4-17) is the dc motor or generator; however, in the interests of simplicity we will choose a related but simpler structure for our illustration. The structure chosen is that of the D'Arsonval galvanometer movement, a structure much used in dc indicating instruments (and, with a rectifier, for ac instruments as well). Figure 4-20(a) shows a "phantom" view of the magnetic structure. A permanent magnet with poles shown

---

*The student may have seen this same relation developed from field theory starting with the force developed on a moving charge, leading to an electric field intensity vector which, when integrated along the length of the conductor leads to the voltage, e.

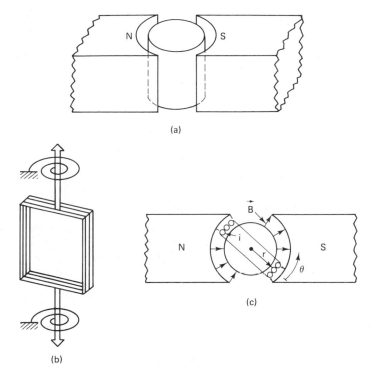

(a)

(b)

(c)

Figure 4-20   A D'Arsonval galvanometer movement.

as N and S forms a magnetic field in an air gap between the pole pieces and a stationary slug of steel. The air gap is uniform, and it will be assumed that the B field is uniform and directed across the gap normal to the steel structure as indicated in part (c) of the figure. A coil of N turns is built as shown in part (b) of the figure. The coil is mounted on a shaft and supported in the air gap on jeweled bearings and fed with a current, i, through two hair springs which serve both as conducting leads and as a mechanical restraint on motion of the coil. A cross section of the coil in the gap is given in part (c), where the usual conventions are used to show directions.

If the active length of conductor in the magnetic field is h, then from Eq. (4-16) each conductor will experience a force

$$f = Bhi$$

and the total torque produced by the N turns (each of which has two sides) will be given by

$$T = 2BhrNi$$

We might choose to simplify this expression by recognizing that the quantity $2hr$ is the active area of the coil, A, hence

$$T = BANi$$

If the stiffness constant of the springs is taken as K, where the spring torque is given as

$$T_{sp} = K\theta$$

and if the current is a constant, I, then in the steady state the coil will take an angular position where the electromagnetic torque is balanced by the spring torque, that is

$$K\theta = BANI$$

or

$$\theta = BANI/K \qquad (4\text{-}18)$$

In words, the angular deflection is directly proportional to the current in the coil and this is the basis for using the device to measure current, or, when placed in series with a resistor, for using the device as a voltmeter.

When the angle θ changes with time, we must consider the dynamics of the system as described by differential equations if we are interested in other than the steady state with a constant current in the coil. As an example, consider the circuit of Figure 4-21 where a voltage source is seen feeding the coil through a resistor, R. It is assumed that the coil is at rest at the angle zero when the switch, S, is closed. Around the loop Kirchhoff's voltage law gives us

$$E = Ri + e_m$$

where e_m stands for the "motional" voltage generated in the coil as in Eq. (4-17).* Since the velocity of the conductors through the B field is given by $v = r(d\theta/dt)$ we have for e_m

$$e_m = 2BhrN(d\theta/dt)$$
$$= BAN(d\theta/dt)$$

or

$$E = Ri + BAN(d\theta/dt) \qquad (4\text{-}19)$$

The mechanical torque developed by the current in the coil is absorbed by the spring torque mentioned previously, and also by the inertial torque owing to the mass of the rotating coil and other parts. If we express the polar moment of inertia of the moving system as J we have the equation

$$BANi = K\theta + J(d^2\theta/dt^2) \qquad (4\text{-}20)$$

Now solving (4-19) for i and inserting into (4-20) we have

$$\frac{EBAN}{R} = J\frac{d^2\theta}{dt^2} + \frac{B^2A^2N^2}{R}\frac{d\theta}{dt} + K\theta \qquad (4\text{-}21)$$

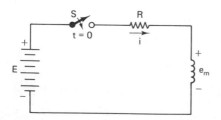

Figure 4-21   The galvanometer fed by a voltage source.

*There may also be a small $L(di/dt)$ voltage in the loop, but this is ordinarily neglected compared to the motional voltage.

This equation is seen to be a linear, second-order differential equation with constant coefficients and should be well known to students of circuit theory, where it is often met for the first time in connection with the analysis of a series RLC circuit with constant voltage excitation applied. We can solve by any of the well-known methods, subject to finding the initial conditions from the physics of the situation. As stated, the system was assumed to be at rest when the switch was closed at $t = 0$ and therefore at that time $\theta = 0$ and $d\theta/dt = 0$.

By whatever means we use for solution we find that the steady state response is

$$\theta_{ss} = \frac{EBAN}{KR} \tag{4-22}$$

which is seen to be consistent with Eq. (4-15) since the steady-state current is given by E/R.

The nature of the transient behavior is determined from the zeros of the characteristic equation, which is

$$Js^2 + \frac{B^2A^2N^2}{R}s + K = 0$$

in the s-domain. The roots are

$$s = -\frac{B^2A^2N^2}{2JR} \pm \sqrt{\frac{B^4A^4N^4}{4R^2J^2} - \frac{K}{J}} \tag{4-23}$$

and, as is well known, the shape of the response curve is dramatically affected by the radical term known as the *discriminant*. For values of terms under the radical which are positive, zero, or negative we have behavior that is described as overdamped, critically damped, and oscillatory, respectively. We see these three cases sketched in that order as (a), (b), and (c) in Figure 4-22 below. If all of the system parameters are constant but we can vary the series resistance, that resistance determines which of the three cases we have. A high resistance leads to oscillatory behavior and a low resistance leads to overdamped behavior. If the term under the radical in Eq. (4-20) is zero then we have a value for R of

$$R = \sqrt{\frac{B^4A^4N^4}{4KJ}} \tag{4-24}$$

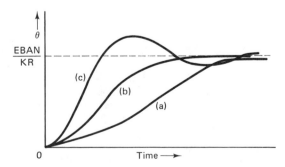

Figure 4-22 The step function response of the galvanometer.

which is called the critical damping resistance. The choice of a value for R or of the other parameters of the device by the designer must be a compromise by the designer between sluggish response and oscillatory behavior leading to overshoot, as in the curve (c) of Figure 4-22.

In practice, most devices of this nature have their coils wound on an aluminum bobbin, which forms a closed one-turn winding and gives additional damping owing to the generated voltage and consequent currents. This extra damping is particularly necessary in the case of a device used as a milliammeter, for example, which might be regarded as being fed from a current source. The current source could be approximated from the foregoing analysis by considering the case when both E and R are allowed to be very large. Further analysis of this device will be reserved for the problem exercises.

ADDITIONAL READING MATERIAL

1. Brown, D., and E. P. Hamilton III, *Electromechanical Energy Conversion,* New York: MacMillan Publishing Company, 1984.

2. Chapman, S. J., *Electric Machinery Fundamentals,* New York: McGraw-Hill Book Company, 1985.

3. Chaston, A. N., *Electric Machinery,* Reston, Virginia: Reston Publications Company, Inc., 1986.

4. Del Toro, V., *Electric Machines and Power Systems,* Englewood Cliffs, New Jersey: Prentice-Hall, Inc., 1985.

5. Fitzgerald, A. E., C. Kingsley, Jr., and S. D. Umans, *Electric Machinery,* New York: McGraw-Hill Book Company, 1983.

6. Lindsay, J. F., and M. H. Rashid, *Electromechanics and Electric Machinery,* Englewood Cliffs, New Jersey: Prentice-Hall, Inc., 1986.

7. Nasar, S. A., *Electric Machines and Electromechanics,* Schaum's Outline Series in Engineering, New York: McGraw Hill Book Company, 1981.

8. Nasar, S. A., *Electric Energy Conversion and Transmission,* New York: MacMillan Publishing Company, 1985.

9. Shultz, R. D., and R. A. Smith, *Introduction to Electric Power Engineering,* New York: Harper & Row, Publishers, 1985.

10. Slemon, G. R., and A. Straughen, *Electric Machines,* Reading, Massachusetts: Addison-Wesley Publishing Company, 1980.

STUDY EXERCISES

1. A certain choke coil for an electronic power supply is found to have a very constant inductance of 5 henries over the desired range of current. The device is constructed with a steel core and a series air gap. We wish to consider various alterations of the design—each alteration by itself, leaving all else unchanged at the original value. What will be the inductance if:
   a. The length of the air gap is doubled?
   b. The number of turns of the coil is doubled?
   c. The length of the steel portion of the magnetic circuit is doubled?
   d. The cross-sectional area of the magnetic circuit is doubled?

2. The sketch at the right shows a magnetic circuit with a tapered slug of steel supported between two pole pieces. There are two air gaps with dimensions as shown. It may be assumed that the steel consumes negligible magnetomotive force and that fringing in the gaps may be ignored. The coil has 500 turns and $\mu_o = 4\pi 10^{-7}$.

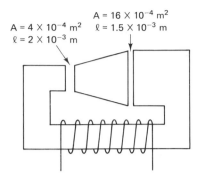

$A = 16 \times 10^{-4}\ m^2$
$\ell = 1.5 \times 10^{-3}\ m$
$A = 4 \times 10^{-4}\ m^2$
$\ell = 2 \times 10^{-3}\ m$

   a. If B in the left-hand air gap must be 1 weber/m^2, how much current is required in the coil?
   b. What is the inductance of this device?

3. Consider the device illustrated in Problem 2. If there is a dc current of I amperes in the coil, what is the net force on the steel slug between the poles? Be sure to specify the direction you call positive.

4. The figure below shows a magnetic circuit with an air gap. A small block of steel is located asymmetrically in the air gap — supported by nonmagnetic materials. The air gap is thus divided into three regions, 1, 2, and 3, and the physical dimensions of each region are given beside the figure. It is assumed that the permeability of the steel portions is practically infinite compared to that of the air, and that the flow lines of the B vector pass straight across the gap with negligible fringing. The areas given are those normal to the B vectors and the lengths given are those colinear with the B vectors.
   a. What is the reluctance of each region?
   b. What is the total net reluctance between the pole faces?
   c. If a total magnetic flux of 0.0004 weber is required, then what magnetomotive force in ampere-turns is required in the coil?

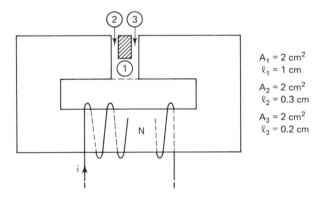

$A_1 = 2\ cm^2$
$\ell_1 = 1\ cm$

$A_2 = 2\ cm^2$
$\ell_2 = 0.3\ cm$

$A_3 = 2\ cm^2$
$\ell_3 = 0.2\ cm$

5. A rotational device is arranged with a single coil passing 2 amperes. It is found that, over the region of interest, the inductance of the coil varies with the angle of rotation, θ, according to the function

$$L(\theta) = 2\theta - \frac{1}{16\pi}\theta^2 - \cos\theta$$

   a. What is the torque developed in the θ direction?
   b. Suppose that we have a spring that develops a constant torque in opposition to the θ direction and in the amount of 2 newton-meters:

   Sketch the curves of electromagnetic torque developed and the spring torque and indicate the point(s) of *stable* equilibrium.

6.

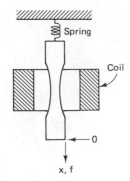

At the left is shown a coil (which has a dc current I) and a movable iron core of dumbbell shape. The core is suspended by a spring such that $x = 0$ when the current is off; that is, the spring force $= mg$. The initial spring force and gravity can therefore be ignored.

a. Sketch L (inductance of the coil) vs. x.
b. Sketch f (electromagnetic force) vs x.
c. Superpose a curve of spring force on (b) and show points of stable equilibrium.

7. In the analysis of the reluctance motor it was assumed that the stator current was $i = I_m \cos \omega_s t$ and the mechanical angle was $\theta = \omega_s t + \delta$.

a. Under these conditions what is the stator voltage, $v(t)$?
b. Check your answer to part (a) by showing that the average electrical power into the device is equal to the average mechanical power out.

8. The elementary linear motor shown below is often used to illustrate the application of Eqs. (4-16) and (4-17). The sketch shows two parallel horizontal rails 0.5 meter apart with a crossbar at right angles supported by the rails but free to move in the x direction with negligible friction. A current source of 100 amperes passes current through the crossbar because of contact with the rails. A vertical magnetic field of uniform $B = 0.8$ tesla, then reacts with the current to cause a force on the crossbar.

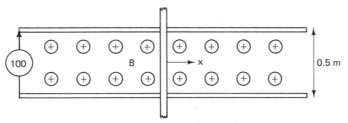

a. What is the force on the bar, and in what direction does it act?
b. If the bar has a mass of 200 grams, what is the acceleration of the bar? Assume Newtonian mechanics, $f = ma$, are adequate.
c. If the bar starts from rest, at what velocity will it be moving 2 seconds after the current is applied?
d. What is the kinetic energy stored in the bar at the end of 2 seconds?
e. Show that the electrical input to the system (all resistance ignored) over the acceleration period is equal to the kinetic energy stored in the bar.

9.

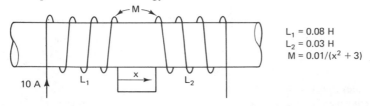

$L_1 = 0.08$ H
$L_2 = 0.03$ H
$M = 0.01/(x^2 + 3)$

Ten amperes are passed through the two coils shown above. Since the coils are in series, the total inductance is $L_0 = L_1 + L_2 \pm 2M$

a. What is the force, f, developed between the two coils?
b. Does the force tend to move the coils apart — or together?

SOLUTIONS TO STUDY EXERCISES

1. If we assume that the reluctance of the steel is negligible (as it must be if inductance is constant with current)

$$L = N^2 \mathcal{P} = \frac{N^2 \mu_o A_g}{l_g}$$

a. With l_g doubled, L is *halved*.
b. With N doubled, L is *quadrupled*.
c. With l_{st} doubled, L is *unaffected*.
d. With A doubled, L is *doubled*.

2. a. If $B_{\text{left}} = 1$, $B_{\text{right}} = 1 \times 4 \times 10^{-4}/16 \times 10^{-4} = 0.25$ (Since $\vec{\nabla} \cdot \vec{B} = 0$)

$$H_{\text{left}} = \frac{B_{\text{left}}}{\mu_o} \qquad H_{\text{right}} = \frac{B_{\text{right}}}{\mu_o}$$

$$\mathcal{F}_{\text{total}} = NI = H_{\text{left}} l_{\text{left}} + H_{\text{right}} l_{\text{right}}$$

$$500I = \frac{1}{4\pi 10^{-7}} \times 2 \times 10^{-3} + \frac{0.25}{4\pi 10^{-7}} \times 1.5 \times 10^{-3}$$

$$I = \underline{3.78A} \quad \longleftarrow \quad \text{a}$$

Alternately (or as a check)

$$\mathcal{R} = \mathcal{R}_{\text{left}} + \mathcal{R}_{\text{right}}$$

$$= \frac{l_{\text{left}}}{\mu_o A_{\text{left}}} + \frac{l_{\text{right}}}{\mu_o A_{\text{right}}}$$

$$\phi = \mathcal{F}/\mathcal{R}$$

$$BA = \frac{NI}{\mathcal{R}} \quad \text{or} \quad I = \frac{BA\mathcal{R}}{N}$$

$$= 3.78A$$

b. $L = N^2 \mathcal{P} = \dfrac{N^2}{\mathcal{R}}$

$$= \frac{500^2}{\dfrac{2 \times 10^{-3}}{\mu_o 4 \times 10^{-4}} + \dfrac{1.5 \times 10^{-3}}{\mu_o 16 \times 10^{-4}}}$$

$$= \underline{0.0529 \text{ H}} \quad \longleftarrow \quad \text{b}$$

3. Choose x to the *left* (and hence force to the left) as positive, an arbitrary choice, but the following must be written consistent with the choice.

$$f = \frac{\partial W_f'}{\partial x}; \quad W_f' = \frac{LI^2}{2} = \frac{N^2 I^2}{2\mathcal{R}}$$

$$= \frac{500^2 I^2}{2} \frac{d}{dx} \left[\frac{1}{\dfrac{(l_1 - x)}{A_1} + \dfrac{(l_2 - x)}{A_2}} \right]$$

at $x = 0$,

$$f = \underline{8.35 I^2} \quad \longleftarrow \quad 3$$

4. a. $\mathcal{R}_1 = \dfrac{l_1}{\mu_o A_1} = \dfrac{1 \times 10^{-2}}{4\pi 10^{-7} 2 \times 10^{-4}} = \underline{3.98 \times 10^7}$

$\mathcal{R}_2 = \dfrac{l_2}{\mu_o A_2} = \dfrac{0.3 \times 10^{-2}}{4\pi 10^{-7} 2 \times 10^{-4}} = \underline{1.19 \times 10^7}$ $\longleftarrow$ a

$\mathcal{R}_3 = \dfrac{l_3}{\mu_o A_3} = \dfrac{0.2 \times 10^{-2}}{4\pi 10^{-7} 2 \times 10^{-4}} = \underline{7.95 \times 10^7}$

b. Reluctances combine in series and parallel like resistances in an electric circuit. An analog:

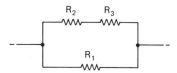

Hence, $\mathcal{R}_{tot} = \dfrac{\mathcal{R}_1(\mathcal{R}_2 + \mathcal{R}_3)}{\mathcal{R}_1 + \mathcal{R}_2 + \mathcal{R}_3} = \underline{1.33 \times 10^7}$ $\longleftarrow$ b

c. $\mathcal{F} = \phi\mathcal{R} = 0.0004 \times 1.33 \times 10^7 = \underline{5305\ \text{AT}}$ $\longleftarrow$ c

5. a. $W_f' = Li^2/2 = (2^2/2)[2\theta - (1/16\pi)\theta^2 - \cos\theta]$

$= 4\theta - (1/8\pi)\theta^2 - 2\cos\theta$

$T_m = +\partial W_f'/\partial\theta = \underline{4 - (1/4\pi)\theta + 2\sin\theta}$ $\longleftarrow$ a

b. Note, in the figure below, that the acceleration torque,

$$T_a = J\,d^2\theta/dt^2 = T_m - T_{spring}$$

We have equilibrium points at A, B, C, and D where T_a is zero.

At points A and C a small perturbation sets up an accelerating torque that restores the system to the equilibrium point and we are *stable*.

At points B and D the accelerating torque is disruptive and we are *unstable*.

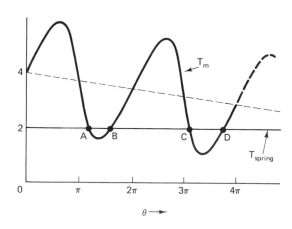

6. a. Note from the figure that the permeance, $\mathcal{P}$, is a minimum when $x = 0$ and a maximum when the thick portion of the dumbbell is in the coil. Ultimately, however, the permeance will fall off when the core is out of the coil.

Since $L = N^2 \mathcal{P}$, we have

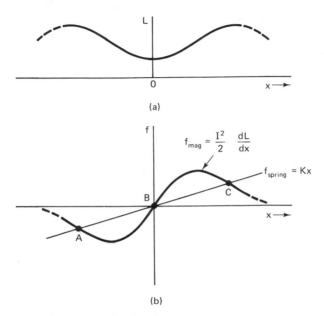

(a)

(b)

From the figure, note that the accelerating force

$$f_a = f_{mag} - f_{spring}$$

is zero at points A, B, and C. If we consider small perturbations from these equilibrium points, we see that A and C are stable points and B is unstable by reasoning similar to that of the preceding problem.

7. **a.** $v = d\lambda/dt$, where $\lambda = Li$

$$= [L_0 + L_2 \cos(2\omega_s t + 2\delta)]I \cos \omega_s t$$

so:

$$v = d\lambda/dt = -\omega_s L_0 I \sin \omega_s t - \frac{3}{2}\omega_s L_2 I \sin(3\omega t + 2\delta)$$

$$- \frac{\omega_s L_2}{2} \cos 2\delta \sin \omega_s t + \frac{\omega_s L_2 I}{2} \sin 2\delta \cos(\omega t + \pi)$$

b. The electrical input is given by $p = vi$, hence the equation above must be multiplied by $i = I_m \cos \omega_s t$ and then we take the average.

The product, vi, is thus a long complex expression, but we may note that, in computing the average, it will be only the product of the current times the last term of the voltage expression that contributes a non-zero value to the average. All other product terms are between orthogonal functions and integrate to zero over a complete period in taking the average, hence

$$P_{elect} = \frac{V_m I_m}{2} \cos(\theta_v - \theta_i) = -\frac{\omega_s L_2 I^2}{4} \sin 2\delta$$

On the other hand, the mechanical power is found from

$$P_{mech} = \omega_s T_{ave} = -\frac{\omega_s L_2 I^2}{4} \sin 2\delta$$

8. a. $f = Bli = 0.8 \times 0.5 \times 100 = \underline{40 \text{ Newtons}}$ ⟵ a
   since $\vec{f} = i\vec{l} \times \vec{B}$ (or by any other favorite memory aid)
   we see that the force is positive in the x direction, toward the right.

   b. $40 = 200 \times 10^{-3} a$
      $a = \underline{200 \text{ m/sec}^2}$ ⟵ b

   c. $v = at = 200 \times 2 = \underline{400 \text{ m/sec}}$ ⟵ c

   d. $W = \dfrac{1}{2} mv^2 = \dfrac{200 \times 10^{-3} \times 400^2}{2} = \underline{16\,000 \text{ joules}}$ ⟵ d

   e. $e = Blv = 0.8 \times 0.5 \times 200t$
      $= 80t \text{ volts}$
      $p = ei = 80t\,100 = 8000t$
      $W = \displaystyle\int_0^2 p\,dt = \int_0^2 8000t\,dt = \dfrac{8000t^2}{2}\bigg|_0^2 = \underline{16\,000 \text{ joules}}$ ⟵ e

9. a. $L = 0.08 + 0.03 - 2 \times 0.01/(x^2 + 3)$
      $W_f' = Li^2/2 = L\,100^2/2 = 50L$
      $\quad = 4 + 1.5 - 1/(x^2 + 3)$
      $f = \partial W_f'/\partial x = 0 + 0 - (-1)(x^2 + 3)^{-2} 2x$
      $\quad = +2x/(x^2 + 3)^2$ ⟵ a

   b. Since the force is positive in the x direction, the coils are forced *apart*. ⟵ b

HOMEWORK PROBLEMS

1. A certain magnetic actuator similar to that of Figure 4-8 is tested at the two extremes of travel to find the λ vs. i functions. The curves are shown below and functions given that fit the curves reasonably well.

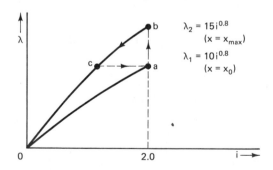

Three points, a, b, and c are identified on the curves, together with paths between the points (trajectories on the λ-i plane).

For each of the trajectories listed below, find the various energy increments for the table which correspond to each movement. For sake of definiteness in reference, take electrical *input* and mechanical *output* as references — the motor action case. Note that, on this basis, some increments may be numerically negative.

Energy Increments in Joules

Path	$\Delta W_{field}(\pm)$	$\Delta W_{elec}(\pm)$	$\Delta W_{mech}(\pm)$
a → b			
b → c			
c → a			
b → c → a			

2. Suppose that the reluctance motor has an inductance given by

$$L = L_0 + L_2 \cos 2\theta + L_4 \cos 4\theta$$

   a. If the stator current is $i = I_m \cos \omega_s t$, at what speeds could this machine run as a motor?
   b. What is the ratio of the pull-out (max) torque at the different speeds?

3. In the simple reluctance motor (Fig. 4-14) it would perhaps be more realistic to consider a voltage source connected across the coil rather than assuming a known current. Let us assume that the voltage $v = -V_m \sin \omega_s t$ is applied to the coil. If the coil resistance is negligible, the flux linkages λ are given by $\lambda = \int v\, dt$. Derive an expression for the average torque if the rotor runs at a mechanical velocity ω_m such that $\theta = \omega_m t + \delta$ and $\omega_m = \omega_s$. Similar simplifying assumptions may be made to the case in the notes; in particular it may be assumed that the inverse inductance, $1/L = \Gamma$, can be represented by two terms of a Fourier series, $\Gamma = \Gamma_0 - \Gamma_2 \cos 2\theta$.

4.

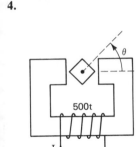
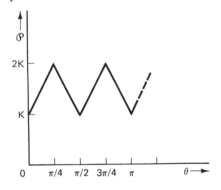

In the figure above, an almost square cross-section rotor is placed between two pole pieces. As the angle, θ, varies, the permeance (reciprocal of reluctance) varies approximately as shown in the figure at the right. The figure shows the variation over π radians; symmetry suggests that the curve merely repeats, over and over. K is just a constant that accounts for specific dimensions and units.

Sketch the plot of developed torque as a function of angle, θ, assuming a constant current of 5 amperes in the coil. Label the magnitudes of angle and torque at each point of discontinuity.

5. The figure below shows a "black box" arranged with a coil passing a constant current of 5 amperes. The device acts on an iron rod to produce a force in the x direction.

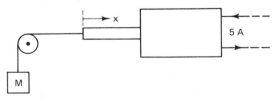

It is found that the inductance is the following function of x:

$$L(x) = 10 + 4x + 2(x - 1)^3$$

a. What is the electromagnetic force developed on the iron bar as a function of x?

b. There is also a restraining force, Mg, of 150 newtons from the mass, M, arranged as shown. Sketch a plot of the electromagnetic force and the force Mg opposing the motion and show where the *stable* point(s) of equilibrium are.

6. One possible usage of a device such as that of Figure 4-17 is that of a synchronous motor as described below.

A dc current, I, is passed through the rotor coil (marked i_2 in the figure). An ac current, $i_1 = I_m \cos \omega_s t$ is passed through the stator coil. The rotor rotates at a constant velocity such that $\theta = \omega_m t + \delta$.

a. Derive an expression for the torque developed from the interaction of the currents.

b. For what speed(s), ω_m, will there result a non-zero average torque?

c. For the speed(s) of (b) above, what is the average torque?

d. Sketch the plot of the average torque as a function of δ.

e. Compare your sketch of (d) with Figure 4-16 for the reluctance motor and comment on similarities and differences.

7. Consider the magnetic field of the earth, which might be approximated as one gauss (0.0001 tesla) and oriented toward the north magnetic pole in the northern hemisphere.

A dc overhead line runs from east to west at right angles to the earth's field, and it carries a short circuit (fault) current of 10,000 amperes under some conditions. We will consider the conductor, which carries positive current from east to west.

What is the force (and in what direction) on a mile of conductor?

8.

B = 0.9 tesla
r = 0.2 meter
I = 100 amperes
ℓ = 0.25 meter

One class of rotating machine involves a relation between current-carrying conductors and magnetic field devices as symbolized above where one sample conductor is shown on the surface of a cylindrical steel rotor. The rotor is under the influence of the pole pieces shown in part, and the result is that the conductor lies in a region of $B = 0.9T$. There is a length of 0.25 meter of conductor exposed to the uniform field. The conductor axis is parallel to the axis of the cylindrical rotor.

a. What torque in newton-meters is contributed by the sample conductor toward rotation of the machine?

b. If the machine rotates at 1200 revolutions per minute, what voltage is induced along the length of the sample conductor?

c. How many watts of electric power are converted by this conductor?

Chapter 5

Synchronous

Machines —

Classical Theory

- A description of the principles of the synchronous machine.
- The presentation of an equivalent circuit for the machine.
- The behavior of the machine with various loads and excitations.
- The use of the machine as a motor.

5.1 INTRODUCTION

An elementary example of a synchronous machine was presented in Chapter 4 in the form of a small reluctance motor. The torque (or power) was shown plotted as a function of the angle δ in Figure 4-16. It was noted that the angle δ corresponds to the position of the rotor at time $t = 0$, when the current (and hence the magnetic field) is at its maximum strength. With a small negative value for the power angle δ the magnetic forces tend to pull the rotor forward, and thus support a positive load torque in opposition to the magnetic torque. The case of negative δ thus corresponds to motor action.

If a numerically negative load torque is supplied — a *driving* torque — the angle δ advances and the rotor pulls against the restraint of the magnetic field. We thus *supply* mechanical energy and *generate* electric power. It is a characteristic of most machines that the action is reversible, so we group generators and motors under one heading: that of electrical machines. Analysis may proceed by assuming reference directions for the variables to describe either motor or generator action as suits the convenience of the moment, and the opposite effect will be described in terms of negative numerical values for the variables.

The reluctance motor (or generator) has the advantage of simplicity, but suffers some serious disadvantages in terms of a practical machine for handling any reasonable amount of power. First we might note that a magnetic field is required for operation and hence reactive voltamperes, Q, must be supplied to support the magnetic field. This Q must be drawn from the line in the case of the motor and *also* in the case of generator action. Other types of apparatus must then be associated with the generator in order to make it function.

A second difficulty with the simple motor of Chapter 4 lies in the complex variation of torque with time as shown in equation (4-12). The plot of Figure 4-16 shows average torque and we rely on the inertia of the rotating mass system to smooth out the instantaneous time variations.

The first point, the need for magnetic excitation, can be met by winding a coil on the rotor and supplying the magnetic field from a dc source connected to the rotor or *field* winding. The second point, the variation of torque with time can be met by going to a polyphase structure with two or more coils on the stator or *armature*. Typically, then, at least four coils are involved (for a three-phase machine) and the energy or coenergy functions will involve ten terms considering self- and mutual induction between the four coils. Development of the machine properties then becomes quite tedious. Fortunately, if we exploit the symmetry of a balanced three-phase system we can analyze the machine by simpler methods — and these are methods that lend themselves to physical understanding of the machine operation. We will proceed in the following sections to develop the properties of actual polyphase synchronous machines by the classical method inferred.

5.2 THE GENERATOR

The general arrangement of a typical generator was introduced in Chapter 2 in order to explain how polyphase voltages are generated in a machine. A similar structure

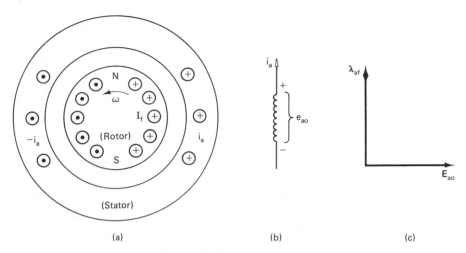

Figure 5-1 Elementary generator.

is shown in Figure 5-1 to begin a more detailed analysis of the properties of such machines.

Only one group of stator coils is shown in Figure 5-1(a), that of phase a. The position of the other coils forming phases b and c will be indicated later. The rotor has a winding with a dc current, I_f, flowing and as a result magnetic flux linkages λ_{af} are established with the stator or *armature* coils.

The flux linkages λ_{af} are roughly proportional to the field current I_f. Because of the *B-H* characteristics of the steel, the relation is not exactly linear, but as a first approximation we might assume domination of the flux path by the air gap as in Figure 4-6 (page 146).

The stator coils are so arranged in the slots that the flux linkages of the stator coils established by the field current vary sinusoidally as the rotor turns through the angle ωt. The methods of accomplishing this result are quite involved and are treated under the subject of armature winding in more detailed texts; for our present purposes we will merely assume that it is possible to arrange the coils properly to accomplish sinusoidal variation.

If we relate the reference direction for flux linkages to the current by the right-hand rule, we see that, at the instant shown in the figure the flux linkages of phase a are at a positive maximum and, if expressed as a sine function then*

$$\lambda_{af}(t) = \lambda_{af}^m \sin(\omega t + 90°)$$

From Faraday's law we have the voltage generated in phase a as

$$e_{ao} = -d\lambda_{af}/dt = -\omega\lambda_{af}^m \cos(\omega t + 90°) = \omega\lambda_{af}^m \sin \omega t$$

The minus sign is included because we wish to use reference polarities of generator action as in part (b) of the figure; that is, a voltage *rise* in the current direction.

---

*The subscripts on λ as λ_{af} are chosen to show that this symbol stands for the flux linkages of phase a owing to the field current. Other components of flux linkages will be discussed in the following pages. The subscript o on the voltage symbol stands for *open circuit* voltage.

As usual, it is more convenient to deal with the sinusoids by means of phasor representation, and a phasor diagram of a flux linkage phasor and a voltage phasor is shown in part (c) of Figure 5-1. Note that the flux linkage phasor, λ_{af} and the voltage phasor E_{ao}, are shown without the superscript m; in other words, they are rms phasors. Further note that the operation of differentiation with respect to time becomes multiplication by $j\omega$ when the time functions are described by phasors, that is

$$E_{ao} = -j\omega\lambda_{af}$$

For the particular time reference shown in the figure the two phasors are at angles of 90° and zero degrees as in

$$E_{ao} = E_{ao}\underline{/0°} \quad \text{and} \quad \lambda_{af} = \lambda_{af}\underline{/90°}$$

If we describe the three-phase generator by showing the other two phases as b and c in Figure 5-2 we note that the flux linkages of phases b and c lag those of phase a by 120° and 240°, respectively. Likewise the voltage phasors lag by similar angles and form a balanced three-phase set as in part (b) of the figure. The phasor diagram becomes much more cluttered by the inclusion of all three phases and therefore we will concentrate on phase a as the reference phase for discussion and consider that the other phasors form elements of a balanced set.

It is interesting to note that, if we study phase a as in Figure 5-1(a) and (c) the orientation of the field structure and the phasor λ_{af} are the same. As time proceeds, the field structure rotates counterclockwise in (a) whereas the phasor λ_{af} when multiplied by $\varepsilon^{j\omega t}$ also rotates counterclockwise to form the time function

$$\lambda_{af}(t) = \text{Im } \sqrt{2}\,\lambda_{af}\varepsilon^{j\omega t}$$

The particular choice of orientation for these quantities is convenient for graphical visualization although of course not necessary since the quasi-physical sketch of (a) and the phasor diagram of (c) are two different things!

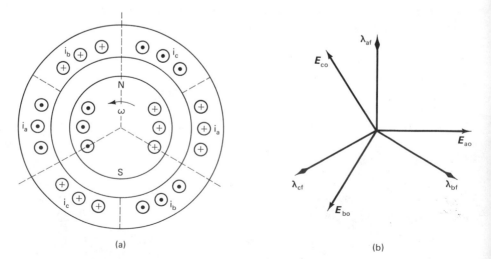

(a)   (b)

Figure 5-2   Three-phase generator.

5.3 RELATION OF POLES, SPEED, AND FREQUENCY

The diagram of Figure 5-1 illustrates a so-called two-pole structure. The effect of the field current is to produce a north magnetic pole on top of the rotor and a south magnetic pole on the bottom of the rotor as viewed in the position shown. As the field rotates, the flux linkages of the stator coils go through one complete cycle for each revolution, so the generated voltage does likewise; the number of cycles per second or frequency is equal to the number of revolutions per second of the rotor. If the speed of the rotor is given in revolutions per minute (r/min) as is common then the relation becomes

$$f = (r/min)/60$$

Many machines are constructed with more than two magnetic poles on the rotor and with the stator coils grouped accordingly. Figure 5-3 illustrates a couple of possibilities — a four-pole and a six-pole structure. If we have more than two poles, we must note that each coil would have a complete cycle of voltage generated every time a *pair* of poles passed by. The relation between rotor speed and frequency must then be modified. If p is the number of poles, then $p/2$ is the number of pairs of poles, and we have

$$f = \frac{p \ (r/min)}{2 \times 60} = \frac{p \ (r/min)}{120} \tag{5-1}$$

Thus for a given frequency the speed of the rotor and the number of poles are inversely proportional.*

At usual power frequencies and ratings, centrifugal force becomes a problem with generators of two- and four-pole construction because of their relatively high

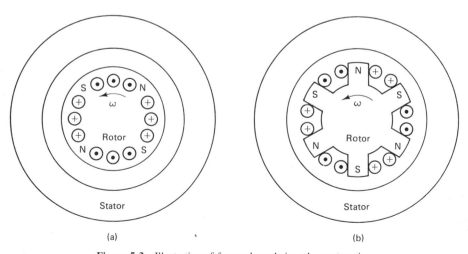

Figure 5-3   Illustration of four-pole and six-pole construction.

*Specifically, for a 60-Hz machine, a two-pole machine rotates at 3600 r/min, a four-pole machine at 1800 r/min, a 60-pole machine at 120 r/min, etc. Steam turbines run best at high speed and such generators are usually either two- or four-pole structures. Hydro turbines, on the other hand, run best at slower speeds and the generators will have many poles.

speed. Such machines are usually built as shown in Figure 5-3(a) with a so-called *round* or smooth rotor construction since this lends itself to greater mechanical strength in withstanding centrifugal forces. The smooth rotor construction becomes difficult for larger numbers of poles, however, and such generators are usually built with construction like that of (b) of the figure, which is called *salient* pole construction.

When we deal with construction of more than two poles, the factor $p/2$ creeps into the equation as in equation (5-1). Since this is just a scale factor and forms a bit of a nuisance, we will use the two-pole machine as our illustration to develop the characteristics of the generator in the work to follow.

5.4 EFFECT OF ARMATURE CURRENT: THE REVOLVING FIELD

If the coils of the stator are connected to loads such as *impedances,* current flows in the stator windings, and these currents affect the net flux linkages of the stator coils—thus changing the generated voltage. The voltages of the preceding figure must, then, be regarded as open-circuit voltages, and we must look further to analyze the effect of the armature currents as well as to find the net generated voltage.

The magnetic field set up by the armature currents depends strongly on the distribution of the armature conductors. Consider first the situation of Figure 5-4, where the armature coils are shown concentrated in a small region on the horizontal axis. If we assume that the permeability of the steel is very great compared to that of the air gap, the flow lines of the B and H vectors cross the gap at right angles to the steel surfaces. Likewise, because of the high permeability of the steel, almost all of the ampere-turns of the armature magnetomotive force are consumed in crossing the air gap, so the H vector and the B vector are everywhere the same

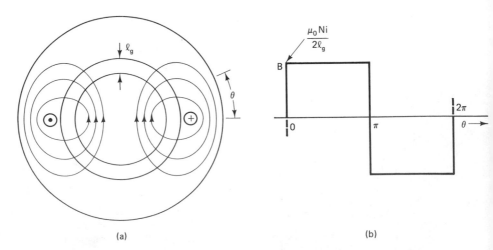

(a)   (b)

Figure 5-4   Effect of current in the armature coils—concentrated coils, phase *a* shown.

around the periphery of the air gap, except right *at* the armature coil. Part (b) of the figure shows the distribution of the B vector as a function of the angle θ around the periphery. If we have N turns in phase a and a current i_a, the magnitude of B is given by

$$B_a = \mu_o H = \mu_o \frac{Ni_a}{2l_g}$$

where l_g stands for the length of the air gap.

It was mentioned before that the armature coils are actually distributed to generate a sine-shaped wave. As an example, suppose the coils are divided into three parts as in Figure 5-5. The effect of this distribution on the air gap flux density set up by the armature current is shown in part (b) of the figure. We see that this shape begins to approximate a sine wave. By further distribution we may approach very closely to a sinusoidal distribution, and we will assume that such has been done and we can describe the B magnitude owing to i_a by the equation

$$B_a = \frac{\mu_o N}{2l_g} i_a \sin \theta$$
$$= K_1 i_a \sin \theta \tag{5-2}$$

Constants will be carried forward in the simplified form like K_1 above in the work to follow in order to simplify the equations.

It is to be noted that the B_a of Eq. (5-2) is a function of time and θ both since i_a varies sinusoidally with time. The plot of Figure 5-5(b) is shown for some positive value of i_a. If we take i_a as given by $i_a = I^m \sin \omega t$ and plot the B magnitude versus time for several sample instants of time, we obtain a family of sine functions as illustrated in Figure 5-6. People who have studied wave phenomena call this kind of variation a *standing wave*.

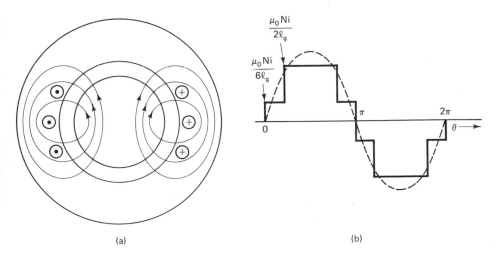

(a)   (b)

Figure 5-5   Effect of distributing the coils.

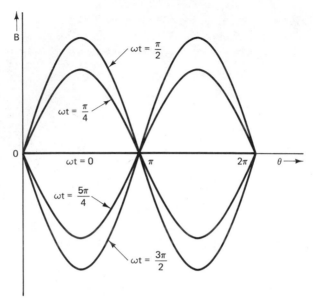

Figure 5-6   The standing wave of B owing to the current in one phase of the armature.

If we consider the case of balanced three-phase currents in the stator (arma-ture) windings, the other two phases will set up similar B waves except displaced by 120° and 240° in time and space.* Specifically, we have

$$B_b = K_1 i_b \sin(\theta - 120°) \tag{5-3}$$

$$B_c = K_1 i_c \sin(\theta - 240°) \tag{5-4}$$

If we now take i_a as some rms magnitude I, and phase angle α we have as the time function descriptions of the three balanced currents:

$$i_a = \sqrt{2}I \, \sin(\omega t + \alpha) \tag{5-5}$$

$$i_b = \sqrt{2}I \, \sin(\omega t + \alpha - 120°) \tag{5-6}$$

$$i_c = \sqrt{2}I \, \sin(\omega t + \alpha - 240°) \tag{5-7}$$

We now substitute equations (5-5) through (5-7) into equations (5-2) through (5-4) and sum to find the total flux density function B_o in the air gap as a function of the time, t, and position θ.

$$\begin{aligned} B_o = K_1\sqrt{2}I[&\sin(\omega t + \alpha) \sin \theta \\ &+ \sin(\omega t + \alpha - 120°) \sin(\theta - 120°) \\ &+ \sin(\omega t + \alpha - 240°) \sin(\theta - 240°)] \end{aligned} \tag{5-8}$$

Equation (5-8) may be greatly simplified by trigonometric identities to give

$$B_o = (3/2)K_1\sqrt{2}I \, \cos(\theta - \omega t - \alpha) \tag{5-9}$$

*We will restrict ourselves to the balanced case entirely in the work to follow. The unbalanced case is far more complex and requires a change of variable operation such as that known as *symmetrical components*. Any discussion of this is deferred to more specialized courses.

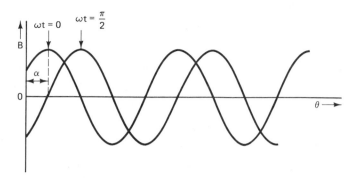

Figure 5-7   The B wave in the air gap as a traveling wave.

Equation (5-9) represents yet another kind of wave equation known as a *traveling wave*. To illustrate the meaning of this term, consider the waves sketched in Figure 5-7 where B_o is shown as a function of θ for varying time instants. Note that, as time increases by increments Δt, for each successive instant of time the curve of B vs. θ is the same except advanced by the angle $\omega(\Delta t)$. The wave is thus seen to preserve constant amplitude and advance around the periphery through the angle θ at an angular velocity ω.

The variation of B with θ and time is said to constitute a *revolving field* and this concept is one of the most important concepts in understanding the action of rotating ac machinery. The student will do well to study how it is that the three-phase belts of conductors, displaced in angle, and with currents displaced in time cause a succession of B values to travel around the periphery of the machine with a constant amplitude and a fixed velocity ω.

5.5 EFFECT OF ARMATURE CURRENT: SYNCHRONOUS REACTANCE

As the traveling wave of B passes around the stator (the revolving field) flux linkages are set up with the stator coils, which vary with time and therefore which generate voltages in these stator coils.* If we consider one turn of the stator winding lying on the horizontal axis of the machine in the phase a belt, then the flux linkages owing to the revolving field may be found by an integration process as

$$\lambda_{a'} = \int_o^\pi B_o lr\, d\theta = \int_o^\pi \frac{3}{2} K_1 \sqrt{2} I \cos(\omega t + \alpha - \theta) lr\, d\theta$$

In the integral the term l is taken as the active length of the conductor and r is the radius, thus the product $lr\, d\theta$ is a differential amount of area.

---

*The B wave also sets up flux linkages with the rotor coil, but these mutual flux linkages are constant, since the B wave is of constant magnitude and revolves at the same speed as the rotor. No voltage is generated by constant flux linkages, so we ignore this effect in the work to follow. If we did not have balanced steady-state ac in the stator coils, there would be an interaction with the rotor coil and the analysis would be much more complicated.

$$\lambda_{a'} = -\frac{3}{2}K_1 lr \sqrt{2} I \, \sin(\omega t + \alpha - \theta)]_o^\pi$$

$$= 2 \times \frac{3}{2}K_1 lr \sqrt{2} I \, \sin(\omega t + \alpha)$$

$$= K_2 I \, \sin(\omega t + \alpha) \tag{5-10}$$

The other turns of phase a winding also have flux linkages set up by the revolving field. The flux linkages of the other turns will have a displaced phase angle if the turns are located above or below the horizontal axis of the machine. If the turns are symmetrically distributed around the axis, the net flux linkages will have the same phase angle as the central axis turn does. The total flux linkages of the phase a belt will then be given by an expression such as

$$\lambda_{ar} = K_3 I \, \sin(\omega t + \alpha) \tag{5-11}$$

The ratio of the constants K_3/K_2 depends upon the number of turns and their distribution. This would be of interest to us if we were designing the machine, but for our present purposes it is enough to know that the ratio is a constant. The notation λ_{ar} is chosen in accordance with the usual custom of calling the effect of the armature currents the *armature reaction*. The windings of phases b and c will have similar flux linkages set up by the revolving field of armature reaction but those flux linkages will be delayed by 120° and 240°, respectively.

Since the armature reaction flux linkages of phase a vary sinusoidally with time, they can be represented by a phasor (rms magnitude)

$$\boldsymbol{\lambda}_{ar} = (K_3/\sqrt{2})I \underline{/\alpha} = K_4 I \underline{/\alpha} = K_4 \boldsymbol{I}_a \tag{5-12}$$

It is to be noted that the flux linkages of phase a are in phase with the current in phase a even though the flux linkages are due to all the currents in the balanced three-phase system and not to the current in phase a alone. This is evident from a comparison of Eq. (5-5) and (5-11). The phasor $\boldsymbol{\lambda}_{ar}$ of Eq. (5-12) is then parallel to a phasor representing the current $\boldsymbol{I}_a$; and these relations may be shown graphically by extending the phasor diagram of Figure 5-1(c) along the lines of Figure 5-8. An example current $\boldsymbol{I}_a$ is shown together with the resultant phasor for the flux linkages of phase a owing to $\boldsymbol{I}_a$ (acting in concert with its companions $\boldsymbol{I}_b$ and $\boldsymbol{I}_c$). The phasor $\boldsymbol{\lambda}_{ar}$ is

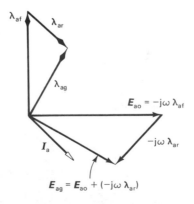

Figure 5-8 Phasor representation of armature reaction.

shown added to the phasor $\boldsymbol{\lambda}_{af}$ to give a resultant phasor representing the net flux linkages of phase a. The net generated voltage of phase a is shown as $\boldsymbol{E}_{ag}$ and is as before simply $-j\omega$ times the flux linkage phasor. The effect of the armature reaction is then such as to change the net generated voltage from the value $\boldsymbol{E}_{ao}$, which is generated under open-circuit conditions.

In mathematical form we note

$$\boldsymbol{\lambda}_{ag} = \boldsymbol{\lambda}_{af} + \boldsymbol{\lambda}_{ar}$$
$$-j\omega\boldsymbol{\lambda}_{ag} = -j\omega\boldsymbol{\lambda}_{af} + (-j\omega\boldsymbol{\lambda}_{ar})$$
$$\boldsymbol{E}_{ag} = \boldsymbol{E}_{ao} + (-j\omega\boldsymbol{\lambda}_{ar})$$

These voltages are the components sketched in the phasor diagram.

In connection with the phasor diagram of Figure 5-8, it is interesting to observe (as in the case of Figure 5-1) that the particular orientation of each of the phasors could also be given an interpretation as the orientation of the axis of symmetry of a flux package, and the resultant phasor $\boldsymbol{\lambda}_{ag}$ represents the orientation of the net flux across the air gap of the machine. We see that, depending on the phase angle of $\boldsymbol{I}_a$ the armature reaction may be directed with, against, or across the action of the main field. (More on this later.)

Although $\boldsymbol{E}_{ag}$ represents the net generated voltage component owing to the air gap flux (sometimes called the air gap voltage) not all of this voltage is seen at the terminals of the machine. Some of the voltage is lost in the resistance of the windings as $R_a\boldsymbol{I}_a$ and some more is lost in a reactance drop $jX_l\boldsymbol{I}_a$ where X_l is known as the *leakage reactance* of the armature winding and corresponds to stray flux paths linking the coils and not passing across the air gap. Linkages around the end turns of the coils are a principal source of this component. The net voltage at the terminals is then the voltage $\boldsymbol{E}_{ag}$ minus the drops $R_a\boldsymbol{I}_a$ and $jX_l\boldsymbol{I}_a$. This total picture is illustrated in the phasor diagram of Figure 5-9.*

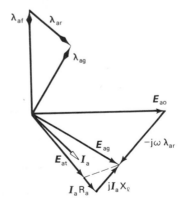

Figure 5-9   The effect of R_a and X_l added to the phasors of Figure 5-8.

*It should be stated that the relative length of the voltage drop $R_a\boldsymbol{I}_a$ is exaggerated in this phasor diagram and in many of those which follow in order to aid clarity of the drawing. Actually, R_a is proportionately very small in most generators of any significant size, perhaps as little as one one-hundredth of the reactance defined in the following.

Mathematically the phasor sum illustrated graphically in the figure can be given as

$$E_{at} = E_{ao} + (-j\omega\lambda_{ar}) - jX_l I_a - R_a I_a$$

Since λ_{ar} is given as $K_4 I_a$ by Eq. (5-12), the above equation can be rewritten as

$$E_{at} = E_{ao} - j\omega K_4 I_a - jX_l I_a - R_a I_a$$

and this equation in turn can be regrouped as

$$E_{at} = E_{ao} - (R_a + jX_l + j\omega K_4)I_a$$

The coefficient of I_a is simply a complex impedance and is known as the *synchronous impedance* of this machine. We will symbolize this impedance by Z_s where Z_s can be further expressed as

$$Z_s = R_a + j(X_l + \omega K_4)$$

We now define

$$X_s = (X_l + \omega K_4) \tag{5-13}$$

where X_s in turn is designated as the *synchronous reactance,* a quantity which is in part due to the leakage reactance X_l and in part due to the effects of armature reactance as made visible by the term ωK_4. A careful retrace of the physical quantities absorbed in the constant K_4 will show that ωK_4 does indeed have the dimensions of *ohms*. In terms of the quantities defined the relation between the open-circuit voltage E_{ao}, the current I_a and the terminal voltage E_{at} becomes

$$E_{at} = E_{ao} - (R_a + jX_s)I_a \tag{5-14}$$

and this relation is illustrated in the phasor diagram of Figure 5-10.

For purposes of analysis and discussion of the properties of this type of machine, it is convenient to note that the relations of Eq. (5-14) could also be obtained from the circuit of Figure 5-11, which is thus an equivalent circuit for phase *a* of the machine. The phasor diagram such as Figure 5-9 is helpful in visualizing the internal relations of the machine but the circuit equivalent is more convenient in analysis of machine behavior in combination with other equipment. Both modes of description will be used in the discussion of machine properties in the following pages.

The model developed for the generator is quite simple and will be used in the discussions to follow, but it should be emphasized once more that simplifying assumptions were made. In particular, the armature (stator) currents were assumed to be balanced, steady-state, sinusoidal currents. If the currents are not balanced or are not in the steady state, other factors enter the analysis, and conclusions based on

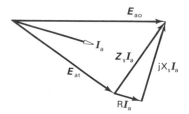

Figure 5-10 Phasor relations between open-circuit voltage, load current, and terminal voltage.

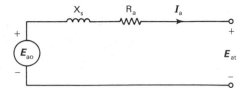

Figure 5-11 Equivalent circuit of the synchronous generator.

the simple model of Figure 5-11 may not be valid. Also, in adding or superposing the flux linkage components, as in Figure 5-8 we tacitly assume linearity. The steel in the machine may approach saturation, particularly in the narrow cross section of the stator teeth between the coil slots, so the superposition is not strictly correct. The assumption of linearity is usually made, however, for ease of analysis, and then corrections are made later to account for the effects of saturation.

5.6 THE SYNCHRONOUS GENERATOR WITH AN IMPEDANCE LOAD

Almost all of the ac power generated in this country is produced by generators of the type under discussion, and ordinarily the output of many generators is pooled into a vast power network or grid. Seldom does one machine act by itself, but, for the sake of simplicity, let us consider the case of a single generator feeding a balanced load with an impedance of Z_L per phase. The circuit diagram of Figure 5-12 illustrates this simple connection. For purposes of discussion, we will consider various impedance angles; that is, various power factors of the load. It is instructive to first consider two hypothetical (but unlikely) loads: zero power factor, current lagging, and zero power factor, current leading. These cases would correspond to pure inductive and capacitive reactances for Z_L respectively. The phasor diagrams of Figure 5-13(a) and (b) illustrate the two cases.

If we assume the same magnitude of terminal voltage, E_{at}, and load current I_a in the two cases we note that an open-circuit voltage, E_{ao} required to supply the terminal voltage under load is much higher for the current lagging case than for the current leading case. The open-circuit voltage is proportional to the field flux linkages, λ_{af}, and the flux linkages in turn are proportional to the field current in the rotor that causes them. It will be remembered from a diagram like Figure 5-8 that the flux linkages set up by the armature reaction (λ_{ar}) point in the same direction as the current phasor I_a. As a result, the lagging current in (a) of the figure tends to demagnetize the field, where the leading current in (b) tends to magnetize the field,

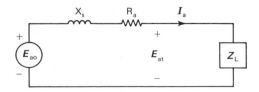

Figure 5-12 Generator with an impedance load.

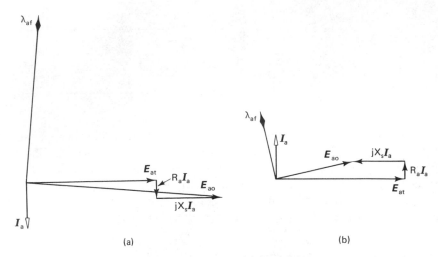

Figure 5-13 Effects of zero power factor load current.

thus accounting for the difference in the field current required to support a given terminal voltage in each case.*

The diagrams of Figure 5-13 show a voltage drop component $X_s I_a$ that is large with respect to the terminal voltage, E_{at}. This relative magnitude is typical of actual machines. A magnitude of X_s of 1.0 per-unit based on the machine's own rating is quite possible; with rated or base current the voltage drop in X_s may be as much as the full rated or base voltage! The magnitude of the synchronous reactance is strongly affected by the length of the air gap, as may be seen by following the developments of Eqs. (5-2) through (5-13). If the air gap is made large, the synchronous reactance may be made small but the penalty is economic—the larger air gap will require more field ampere turns to produce the flux linkage of the stator and the machine will be bigger, heavier, and more costly. In practice, a designer usually accepts a large magnitude of X_s and the field current is adjusted by an automatic device to hold the terminal voltage constant under loads of varying magnitude and varying power factor. Such a device is called an *automatic voltage regulator*. The theory of such devices is covered in courses in automatic feedback control systems.

More realistic loads at the position Z_L might be pure resistive (unity power factor) or partly inductive (current lagging) loads. Simple phasor diagrams for these cases are shown in Figure 5-14(a) and (b). For the same magnitude of terminal voltage and load current in these two cases, the change in flux linkages or open-circuit voltage required is not as great as shown in the previous cases. The axis of the armature reaction flux linkages is mainly across the main field axis for the unity

---

*An easy way of remembering this relation is to recall that positive reactive voltamperes, Q, must be *supplied to* an inductive load. Positive Q is magnetizing Q, so we must supply more field current to accomplish our result. On the other hand, capacitors *supply* positive Q or magnetizing energy, so we do not need to have as much field current!

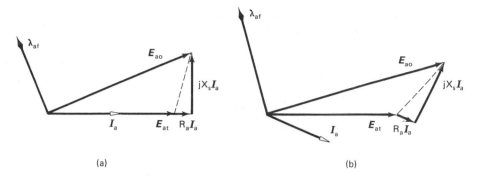

Figure 5-14   Generator with (a) unity power factor and (b) current lagging power factor loads.

power factor case, whereas that of an intermediate lagging power factor case is partly demagnetizing and partly across the main field axis. For discussion purposes of the effect of the armature reaction, we sometimes speak of the *direct* and the *quadrature* axes of the main field. In this terminology the zero power factor current has its reaction along the direct axis of the field and the unity power factor has a strong component on the quadrature axis of the field. More advanced treatments recognize a difference in magnetic reluctance on these two axes and divide the armature reaction into two components along their respective axes. We will not pursue this refinement here, but it would be well to note that sometimes the X_s of the preceding is given as X_d or X_q, (depending on whether the reaction is along the direct or quadrature axis).

A numerical example may be helpful in illustrating the foregoing relations. Suppose we have a small hydro generator feeding an isolated community in the mountains. The generator is a three-phase 1000 kVA unit rated at 13.8 kV, and the manufacturer's test report shows that $Z_s = 2.5 + j180$ ohms per phase. The load on the plant is 500 kW at 0.8 power factor, current lagging, and it is desired to feed the load at 13.8 kV at the terminals of the generator. We know that the generator produces its rated voltage of 13.8 kV with a field current of 40 amperes. We will assume the relation between field flux linkages, λ_{af}, and field current I_f is linear (this is not too good an assumption, but if someone would give us a curve of λ_{af} vs. I_f we would not have to make the linear assumption). The problem is to find the open circuit or "internal" voltage and the field current required to supply the load as specified.

The per-phase circuit diagram will look like Figure 5-12, with the following data on a per-phase basis

$$E_{at} = \frac{13.8}{\sqrt{3}} = 7.97 \text{ kV}$$

$$I_a = \frac{500}{\sqrt{3}\, 13.8 \times 0.8} = 26.15 \text{ amperes}$$

If we use E_{at} as the reference (angle zero)

$$E_{at} = 7.97\underline{/0}\text{ kV}$$

$$I_a = 26.15\underline{/-\cos^{-1}0.8} = 26.15\underline{/-36.87°}$$

From the circuit of Figure 5-12

$$\begin{aligned}
E_{ao} &= E_{at} + Z_s I_a \\
&= 7.97 + j0 + (2.5 + j180)26.15 \times 10^{-3}\underline{/-36.87°} \\
&= 10.84 + j3.73 = 11.47\underline{/18.96°}\text{ kV}.
\end{aligned}$$

From the linear assumption

$$I_f = 40\left(\frac{\sqrt{3}\,11.47}{13.8}\right) = 57.56\text{ amperes}$$

It might be of interest to treat this problem on a per-unit basis. For this purpose, we need to choose the base quantities and we will choose $S_b = 1000$ kVA and $E_b = 13.8$ kV; that is, the rated three-phase quantities. From Chapter 3 we recall that, in terms of three-phase bases, we have $Z_b = E_b^2/S_b = (kV_b)^2/MVA_b$ or

$$Z_b = 190.44\text{ ohms}$$

therefore

$$\begin{aligned}
\overline{Z}_s &= (2.5 + j180)/190.44 \\
&= 0.013 + j0.945 = 0.945\underline{/89.20}\text{ per-unit}
\end{aligned}$$

If the terminal voltage is 13.8 kV, this is one per-unit and the current, in per-unit is

$$\overline{I} = \frac{(500/1000)}{1.0 \times 0.8}\underline{/-\cos^{-1}0.8} = 0.625\underline{/-36.87°}$$

The open circuit, or "internal" voltage is then

$$\overline{E}_{ao} = 1 + j0 + (0.945\underline{/89.20})(0.625\underline{/-36.87}) = 1.43\underline{/18.9°}$$

The field current base would be 40 amperes and, with a linear assumption we would require a field current of $\overline{I}_f = 1.43$ per-unit, and last, the field current in amperes would be

$$I_f = \overline{I}_f I_{f(\text{base})} = 57.56\text{ amperes}$$

5.7 THE SYNCHRONOUS MACHINE CONNECTED TO A POTENTIAL SOURCE

It is most common to find synchronous machines connected to other machines or networks of other machines rather than an individual impedance load. It is from this usage that the name *synchronous* machines is derived. A machine such as is described in this chapter must run in step at the same frequency as the other machines with which it is connected; it must run *synchronously*.* As an instructive case, we will consider a situation like that shown in Figure 5-15, where a machine is con-

---

*The actual mechanical speed at which the machine must run to generate a given frequency is of course dependent on the number of poles in the structure, as given by the relation of Eq. (5-1).

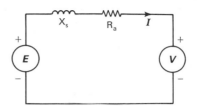

Figure 5-15 A model for a synchronous machine connected to an infinite bus.

nected to a perfect potential source at its terminals. In the parlance of a power engineer, such a source is called an *infinite bus*.

The voltage of the infinite bus is labeled V where $V = V\underline{/0°}$, that is, the bus voltage is used as the reference phasor. The internal voltage of the machine is given as E where $E = E\underline{/\delta}$. The angle δ of the voltage phasor is called the power angle for reasons that will be discussed in the following. From Kirchhoff's law we have

$$I = \frac{E\underline{/\delta} - V\underline{/0}}{R_a + jX_s} \tag{5-15}$$

The magnitude and phase angle of I vary widely depending upon the magnitude of E and the angle δ. Several cases will be discussed below.

First, let it be supposed that $E = V$, that is the angle δ is zero and the magnitude of E is adjusted by means of variation of the field current such that E equals the bus voltage V. From Eq. (5-15) we see that the current is zero, so no power is transferred. This is the situation when a machine is first connected to another voltage source because the operator carefully adjusts the machine so that excessive currents will not flow when the connection is made.

If the voltage magnitude remains fixed and we increase the driving torque of the prime mover driving the generator, the angle δ will increase. This may be understood by considering a diagram such as Figure 5-1(c), where the flux linkage phasor λ_{af} could be regarded as lying along the direct axis of the rotor and the voltage E_{ao} is seen to lie 90° behind the flux linkage phasor. If the rotor is driven harder mechanically the rotor will accelerate and advance in angle, and the "internal" voltage advances as well. Suppose that somehow the angle has been advanced a small amount; there is now a difference voltage applied to the impedance and a current flows. This current is illustrated in the phasor diagram of Figure 5-16. As the angle δ advances the difference voltage $(E - V)$ grows and the current increases. The current lags the difference voltage across the impedance $(R_a + jX_s)$ by almost 90° since X_s is far greater than R_a in usual machines. For this case, the current is then almost in phase with the bus voltage V. Because of the reference directions chosen, we see that the machine is acting as a generator and causes power to flow

Figure 5-16 The machine of Figure 5-15 with the rotor angle advanced. (Phasor Diagram)

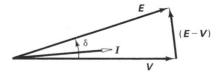

into the bus. If the angle δ is increased further by increasing the driving torque from the prime mover, the difference voltage, the current, and hence the power all increase until the additional power supplied by the driving source is transferred out of the generator and into the bus. There is a limit to this action, as will be discussed below.

If we consider another case, where the angle δ is left fixed at zero and the magnitude of the voltage E is increased by increasing the field excitation of the machine, a phasor diagram such as Figure 5-17 results. For this case, the difference voltage lies at an angle of zero and the resultant current I lags both V and E by almost 90°. From an energy viewpoint, we note that there is very little real power transferred but instead there is a transfer of positive reactive voltamperes, Q, *into* the bus. This action is another illustration of the principle that positive Q represents magnetizing effect, and when the machine is supplied with extra magnetizing effect by increasing the field current, the excess is delivered to the load bus.

In a similar fashion to part (a) of the figure we note in part (b) the consequences of reducing the rotor excitation and hence the magnitude of the voltage E. The current still *lags* the *difference* voltage $(E - V)$ but the position of this difference causes the current to *lead* the voltages V and E by almost 90°. The energy picture is then just the opposite of that of part (a)—the machine *supplies* current *leading* or negative reactive voltamperes, Q. We might then note that to supply current leading vars is the same as absorbing current lagging vars, and we can explain the action by noting that, if the machine does not get its necessary magnetization from the rotor field current then it will *draw* magnetizing Q from the bus voltage source V.

The difference between this case and that of Section 5.6 should be carefully noted. In the case of an impedance load on the terminals the consequence of a change of field excitation is a change in terminal voltage. In the case of a connection to a potential source of fixed voltage (an infinite bus) the terminal voltage cannot change and the consequence of a change in excitation is a flow of reactive voltamperes Q in the appropriate direction and sign to satisfy the excitation requirements of the machine.

We might continue this discussion by considering all manner of values for the magnitude and angle of E, but it will be more productive to derive, as in the next section, an expression for complex power flow in terms of the circuit parameters and study the situation in the light of that expression.

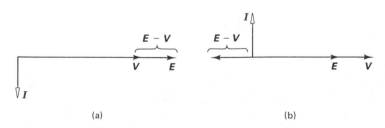

Figure 5-17   The machine of Figure 5-15 with the excitation varied. (Phasor Diagram)

As a numerical example let us once again consider the generator of the example in the preceding section rated 1000 kVA at 13.8 kV and with $\overline{Z}_s = 0.945 \underline{/89.2°}$ per unit.

Suppose we wish to supply 0.5 per unit power to the bus of Figure 5-15 with voltage $\overline{V} \underline{/0°} = 1 \underline{/0°}$ per unit. Further suppose that the field current is set at 60 amperes. The problem is to find the power angle δ and the reactive voltamperes supplied to the bus.

If

$$I_f = 60 \text{ amperes}, \quad E_{ao} = (60/40)13.8 \text{ kV} \quad \text{or} \quad 1.5 \text{ per unit}$$

$$\overline{I} = \frac{1.5 \underline{/\delta} - 1.0 \underline{/0°}}{0.945 \underline{/89.2°}} = 1.59 \underline{/\delta - 89.2°} - 1.06 \underline{/-89.2°}$$

$$\overline{S}_{bus} = \overline{V}\overline{I}^* = 1.59 \underline{/89.2° - \delta} - 1.06 \underline{/89.2°}$$

$$\overline{P}_{bus} = 0.5 = 1.59 \cos(89.2° - \delta) - 1.06 \cos 89.2°$$

Solving for δ,

$$\delta = 18.12°$$

$$\overline{Q}_{bus} = \text{Im } \overline{S}_{bus} = 1.59 \sin(89.2° - 18.12°) - 1.06 \sin 89.2°$$
$$= 0.443 \text{ per unit} = 443 \text{ kvars}$$

As a further extension, suppose that the field current were only 50 amperes — what would be the answer to the above problem?

If

$$I_f = 50 \text{ amperes}, \quad E_{ao} = (50/40)13.8 \text{ kV} \quad \text{or} \quad 1.2 \text{ per unit}$$

$$\overline{I} = \frac{1.2 \underline{/\delta} - 1.0 \underline{/0°}}{0.945 \underline{/89.2°}} = 1.27 \underline{/\delta - 89.2°} - 1.06 \underline{/89.2°}$$

$$\overline{S}_{bus} = \overline{V}\overline{I}^* = 1.27 \underline{/89.2° - \delta} - 1.06 \underline{/89.2°}$$

$$\overline{P}_{bus} = 0.5 = 1.27 \cos(89.2° - \delta) - 1.06 \cos 89.2°$$

Solving for δ,

$$\delta = 23.12°$$

$$\overline{Q}_{bus} = \text{Im } \overline{S}_{bus} = 1.27 \sin(89.2° - 23.12°) - 1.06 \sin 89.2°$$
$$= 0.103 \text{ per unit} = 103 \text{ kvars}$$

5.8 POWER TRANSFERRED THROUGH AN INDUCTIVE IMPEDANCE: A GENERALIZATION

The situation discussed in the previous section involving a synchronous machine connected to an infinite bus is really just a particular instance of the general case, where two ac voltages are connected through an inductive link and complex power flows as determined by the relative magnitudes and angles of the voltages. It so happens that most impedances in power systems are strongly inductive, so it is worthwhile to derive an expression for the power transfer for this case. We use the notation of Figure 5-15 and derive an expression for the complex power delivered to the bus of voltage $V = V\underline{/0°}$ connected to a voltage $E = E\underline{/\delta}$ through an impedance $Z = R_a + jX_s$. For convenience let us express Z in polar form as $Z = Z \underline{/\theta}$ and proceed as follows.

$$I = \frac{E\underline{/\delta} - V\underline{/0}}{Z\underline{/\theta}}$$

$$S_{bus} = VI^* = V\underline{/0}\left[\frac{E\underline{/-\delta} - V\underline{/0}}{Z\underline{/-\theta}}\right]$$

$$S_{bus} = \frac{EV}{Z}\underline{/\theta - \delta} - \frac{V^2}{Z}\underline{/\theta}$$

From this equation, we note that the real power and reactive voltamperes are given by

$$P_{bus} = \text{Re } S_{bus} = \frac{EV}{Z} \cos(\theta - \delta) - \frac{V^2}{Z} \cos\theta \qquad (5\text{-}16)$$

$$Q_{bus} = \text{Im } S_{bus} = \frac{EV}{Z} \sin(\theta - \delta) - \frac{V^2}{Z} \sin\theta \qquad (5\text{-}17)$$

If we are given values of the parameters we can of course substitute into equations (5-16) and (5-17) and obtain values for the real power P and the reactive voltamperes Q. For the sake of understanding, it is desirable to simplify these expressions by assuming that the impedance is purely inductive reactive, that is the angle θ is 90°. This is not far from the truth, for many power system elements where the impedance angles lie between 80° and 90° typically. If θ is 90° then Eqs. (5-16) and (5-17) become

$$P = \frac{EV}{Z} \sin\delta \qquad (5\text{-}18)$$

$$Q = \frac{EV}{Z} \cos\delta - \frac{V^2}{Z} \qquad (5\text{-}19)$$

Equation (5-18) illustrates the conclusions drawn from the phasor diagrams of Figure 5-16. If the voltage magnitudes E and V are held fixed, and we vary the power angle δ by varying the mechanical driving torque, we see that for $\delta = 0$ there is no power transferred. If the angle δ is increased slightly, power flows, and as δ increases the power at first increases almost linearly with δ because $\sin\delta \cong \delta$ for small angles. As the angle δ increases more, the power transfer follows the familiar sine loop shown in Figure 5-18. The mathematical analysis and the resultant plot of Figure 5-18 makes evident a property that might not at first be apparent from the simple phasor diagram analysis of the preceding section—the power transfer function of δ has a maximum of $P_{max} = EV/Z$. Physically speaking, we can describe the behavior of a synchronous generator as follows.

If we start with $\delta = 0$ and increase the driving torque the machine accelerates and δ increases, thus causing the machine to deliver electric power. At some value of δ the machine reaches equilibrium, where the electric power output balances the increased mechanical power owing to the increased driving torque. If we continue to increase the driving torque, however, a point is reached at the top of the sine loop

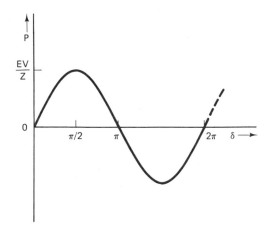

Figure 5-18   Power transfer through an inductive element.

where an increase in δ no longer increases the electric power output and the excess driving torque simply continues to accelerate the machine. We then say that the machine has lost synchronism, since it is running faster than it should to generate voltage at the frequency of the bus. Excessive currents will then flow and automatic equipment will break the connection between the machine and the bus.

If we revert now to consideration of the reactive voltamperes, Eq. (5-19) explains mathematically the relations observed in the phasor diagrams previously discussed. If δ is zero, or at least quite small, cos δ is nearly unity and Eq. (5-19) can be rewritten as

$$Q \cong \frac{V}{Z}(E - V) \qquad (5\text{-}20)$$

From Eq. (5-20) we see the previous conclusions verified. If $E = V$ there is no transfer of reactive voltamperes. If $E > V$, the reactive vars delivered to the bus are positive, and if $E < V$ the reactive vars delivered to the bus are negative; that is, the bus is supplying positive vars to the machine.

In summary of the ideas of this section we state:

1. The flow of real power requires a phase angle difference between the voltages on each end of an inductive reactive link and there is a limit to the amount of power that can be transferred at a given level of voltage and impedance.

2. The flow of reactive voltamperes is dominated by the difference in voltage magnitude at either end of an inductive link, with positive Q flowing from the higher voltage end to the lower voltage end of the link.

Of course the conclusions of (1) and (2) above are not independent of each other, as we may see by examining the Eqs. (5-16) and (5-17) carefully, but in gaining an understanding of machine behavior, it is helpful to consider the effects as outlined as the major effects at work, with only small cross-coupling between real power and reactive voltampere supply.

The relations studied above are most useful when considering a situation where the two voltage magnitudes are held at some particular value. In the case of the individual impedance load on a single machine, as covered earlier, the equations are still valid between the "internal" voltage, E_{ao} and the terminal voltage E_{at} as the voltages on either end of the impedance link but the terminal voltage E_{at} changes greatly as the load varies, so the equations are not quite so simple in interpretation for this case.

5.9 THE SYNCHRONOUS MOTOR

The reader may have already pondered the consequences of a negative value for the power angle δ in the preceding work dealing with a machine tied to a voltage source as in Figure 5-15. Suppose, for example, that the machine is connected to the voltage source with the voltage $E = V$. This of course implies the same frequency for E and V or else the angle between the two phasors would change constantly; the machine is brought to synchronous speed before the connection. If now the driving torque on the machine is reduced, the rotor will tend to slow down and E will fall back, resulting in a negative value for δ. From Eq. (5-18) we see that the power *delivered by* the machine will be numerically negative or in other words the machine *draws* positive power from the voltage source V and acts as a motor. As the driving torque becomes negative, it becomes more convenient to regard it as a *load torque* for this case of motor action. The machine is then described as a *synchronous motor*.

If motor action were originally considered as the basic action, the reference direction for current (and hence for power) in Figure 5-15 would have been reversed. Although we could continue to describe the motor action in terms of the original references by properly interpreting negative numerical values, it is more convenient conceptually to alter the references for power flow and restate the relations in terms of the new direction.

Figure 5-15 is redrawn with the reference direction for motor action chosen as Figure 5-19. The expression for the current now can be expressed by

$$I = \frac{V\underline{/0} - E\underline{/-\delta}}{R_a + jX_s} = \frac{V\underline{/0} - E\underline{/-\delta}}{Z\underline{/\theta}} \tag{5-21}$$

where the voltage phasor E is described by $E = E\underline{/-\delta}$ in which positive δ is now interpreted as the angle by which E lags *behind* the source voltage V, which is again taken as the reference phasor at an angle of zero. We now proceed to derive new equations like those of the generator but for the power *delivered by* the bus with

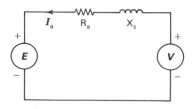

Figure 5-19 A synchronous machine with reference direction chosen for motor action.

voltage V.

$$S_{bus} = VI^* = V\underline{/0}\left[\frac{V\underline{/0} - E\underline{/\delta}}{Z\underline{/-\theta}}\right]$$

$$S_{bus} = \frac{V^2}{Z}\underline{/\theta} - \frac{EV}{Z}\underline{/\delta + \theta}$$

$$P_{bus} = \text{Re } S_{bus} = \frac{V^2}{Z}\cos\theta - \frac{EV}{Z}\cos(\delta + \theta) \tag{5-22}$$

$$Q_{bus} = \text{Im } S_{bus} = \frac{V^2}{Z}\sin\theta - \frac{EV}{Z}\sin(\delta + \theta) \tag{5-23}$$

Again the meaning of the equations is clarified if we consider that θ is closely approximated by 90° and we have

$$P_{bus} = \frac{EV}{Z}\sin\delta \tag{5-24}$$

$$Q_{bus} = \frac{V^2}{Z} - \frac{EV}{Z}\cos\delta \tag{5-25}$$

These equations are essentially the same as Eqs. (5-18) and (5-19); in fact, Eq. (5-24) is identical to Eq. (5-18), but it must be remembered that the significance of the angle δ is now opposite and represents the angle by which the rotor falls behind, rather than the angle of advance.

The curve of power versus delta is identical to Figure 5-18 for the generator except that P stands for the electric power *absorbed* by the motor from the source and the angle δ is the angle of lag of E and therefore of the rotor of the two-pole machine. The physical behavior is described in a similar fashion to that of the generator. As a mechanical load (which is now retarding torque) is placed on the motor shaft, the rotor slows down and δ becomes positive. Electric power then flows from the source V until the mechanical load is satisfied and the machine runs at the angle delta. It must again be pointed out that there is a maximum value for the load (when δ is 90°), and loading beyond this point causes the motor to drop out of synchronism resulting in excessive currents and ultimate disconnection of the motor from the source.

In the case of the motor we should note that, since the motor always runs at a fixed speed over its useful range of operation, the power and torque curves differ only by the scale constant ω. Figure 5-18 could therefore be relabeled in terms of torque versus power angle δ if so desired.

It should be mentioned that to bring the rotor up to speed (to make the original connection between E and V with each at the same frequency) implies a driving source. In the case of a generator, the source is simply the driving turbine or whatever prime mover is used. In the case of the motor, we occasionally have a load that can reverse and act as a driving source to bring the motor up to synchronous speed, but more often the motor is started by other means, such as auxiliary windings like those of an induction motor. Discussion of induction motor torque is deferred to the next chapter.

The flow of reactive voltamperes, Q, is given by Eq. (5-25) and again it is convenient to consider small values of δ and $\cos \delta \cong 1$ in order to emphasize the major effects. With this assumption

$$Q_{bus} = \frac{V}{Z}(V - E) \tag{5-26}$$

From Eq. (5-26) if $E = V$ there is no reactive drawn from the source V. If $V > E$, positive reactive vars are drawn from the source, which action can be explained by saying that insufficient field current has been supplied and additional magnetizing energy has to be drawn from the source in the form of positive (current lagging) Q.

If $E > V$, the motor draws negative vars from the source and acts like a capacitor. But to *draw* negative (current leading) vars from the source is the equivalent of *supplying* positive vars, and this action can be explained by the viewpoint that a large E means a large field current and excess magnetizing energy spills over to the source in the form of positive Q. We sometimes describe the field as over- or underexcited, depending on whether it draws current leading or lagging vars from the source.

An overexcited synchronous motor is a useful device since it may be used in the same role as static capacitors in order to improve the power factor of a load that might otherwise be predominantly inductive in a plant with lots of induction motors, which load thereby suffers a power factor penalty in the rates charged for electric energy. Sometimes utilities will install a synchronous motor with no shaft load at all and simply let the machine run in an over- or underexcited condition to act as a reactive load. Most often the machine runs overexcited, and therefore is known as a *synchronous condenser* (the word "condenser" was formerly used interchangeably with "capacitor").

As a further numerical example, suppose that the example machine of the preceding sections (rated at 1000 kVA, 13.8 kV, and with $\overline{Z}_s = 0.945\underline{/89.2°}$ per-unit) is to be used as a synchronous motor fed at 13.2 kV. It will be loaded mechanically to draw 800 kW. Other loads at the same plant total 1100 kW at 0.5 power factor, current lagging. It is desired to over-excite the machine to draw leading vars, and correct the overall plant power factor to 0.8, current lagging. What should be the field current setting, and is the machine capable of the task assigned without overloading on the basis of current rating?

A circuit diagram showing this situation is given in Figure 5-20. Rated values are used as bases.

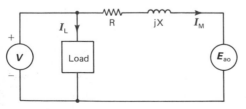

Figure 5-20   An example problem.

$$\overline{V} = \frac{13.2}{13.8} = 0.957 \text{ per unit}$$

$$\overline{S}_{\text{LOAD}} = \frac{1100}{1000} + j\frac{1100}{1000} \tan(\cos^{-1}0.5)$$

$$= 1.1 + j1.91 = 2.20\underline{/60°}$$

$$\overline{P}_{\text{TOTAL}} = P_{\text{LOAD}} + P_{\text{MOTOR}} = 1.1 + 0.8 = 1.9$$

$$\overline{Q}_{\text{TOTAL}} = 1.9 \tan[\cos^{-1}0.8] = 1.43$$

$$\overline{Q}_{\text{MOTOR}} = \overline{Q}_{\text{TOTAL}} - \overline{Q}_{\text{LOAD}} = -0.48$$

$$\overline{S}_{\text{MOTOR}} = \overline{P}_{\text{MOTOR}} + j\overline{Q}_{\text{MOTOR}} = 0.8 - j0.48$$

$$\overline{I}_{\text{MOTOR}} = \frac{\overline{S}_{\text{MOTOR}}^{*}}{\overline{V}*} = 0.98\underline{/30.98°}$$

$$\overline{E}_{ao} = \overline{V} - (0.945\underline{/89.2})\overline{I}_{\text{MOTOR}}$$

$$\overline{E}_{ao} = 1.63\underline{/-29.3°}$$

$$\overline{I}_f = \overline{E}_{ao} = 1.63$$

$$I_f = \overline{I}_f I_{f_{\text{base}}} = 1.63(40) = 65.13 \text{ amperes}$$

Since $\overline{I}_m = 0.98$ per unit (less than 1.0 per unit or rated) the motor is *not* overloaded.

5.10 A PHYSICAL PICTURE

The properties of the synchronous generator or motor have been developed by means of mathematical descriptions and phasor diagrams in the preceding pages. Engineers like to have a physical "feeling" for what is happening in devices and it is possible to give a picture of the operation of the machine in alternate terms that may be helpful to some in visualizing the operation.

Suppose we have a device that can detect the magnitude and direction of magnetic fields (Hall effect probes might be an example). If we probe the air gap region of the synchronous machine, we detect that a magnetic field is set up by action of the dc field winding on the rotor and by the revolving field action of the balanced ac currents in the stator windings. Without knowing the details of the windings which set up the fields, we could just as well conceive that the fields were due to a pair of concentric cylinders of magnetic material which are permanently magnetized in the proper polarity and magnitude. The two cylinders revolve together about their common axis. The fact that the stator iron is actually stationary, and that it is time variation of the currents in the coils that causes movement of the field, would not necessarily be detectable by means of the instrument probe in the air gap. Figure 5-21(a) and (b) attempts a picture of such a situation. The two permanently magnetized cylinders are rotating in the same (counterclockwise) direction but in case (a) the rotor is oriented ahead of the outer element and in (b) the opposite is true. The inner and outer cylinders each have two poles, a north pole and a south pole, and our memory tells us that opposite poles attract each other. A crude attempt has

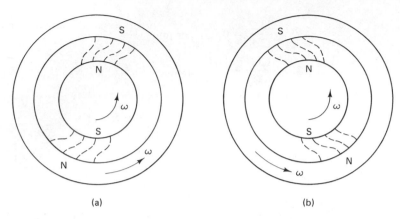

(a)   (b)

Figure 5-21   A magnetic field picture of the synchronous machine action.

been made to show the flow lines of the magnetic field in the air gap and to symbolize the fact that if the poles are not directly opposite each other there will be a tangential component of force tending to bring them into line.

In part (a) of the figure the inner cylinder is leading the outer and pulling it behind with a torque acting in opposition to some hypothetical resisting torque holding the outer cylinder back. Energy flow is from the inner to the outer cylinder. This is the situation in the generator where the rotor field advances by an angle δ ahead of the stator field. Power transfer in the amount $P = T\omega$ is accomplished by the interaction of the fields in the air gap where T is torque and ω is angular velocity. It is worthy of note that if the power and speed are constant the torque must be constant, and this helps us understand that the revolving field set up by balanced stator currents in the machine must be constant in magnitude, since the electric power output of a balanced polyphase generator is constant with time.

The opposite effect is symbolized in part (b) of the figure, where the outer cylinder is being driven ahead and is pulling the inner cylinder behind it against some resisting torque. This is motor action, where the energy flow is from the outer to the inner cylinder.

5.11 MACHINE RATINGS AND LOSSES

Each phase of the polyphase synchronous machine is capable of generating or absorbing just so much voltage because the speed is fixed and the flux density has a practical upper limit, owing to saturation of the steel. The windings have a current limitation because of I^2R losses and consequent heating effects. The stator is rated in kVA as a result of these limitations and as usual the nameplate kVA of a machine is the total kVA for all phases, and the voltage is the normal line-to-line voltage.

The rotor or field winding of the machine is also current limited by thermal effects, and this limitation may be severe, particularly in steam or gas turbine units where the speed is high and the rotor must be built compactly with small radius, to limit centrifugal force effects. Such rotors are sometimes liquid-cooled. It may be that rotor heating will limit the output of a given load when a generator is supplying a low power factor load (and the field excitation must be high to maintain the rated terminal voltage).

The output of a machine is determined by the load applied to the machine, but the input is always greater than the output because of energy losses in the machine. There are various mechanisms of energy loss in the machine. The armature resistance has already been mentioned in the discussion of the equivalent circuit of the machine. The power going into the rotor or field winding is also lost as I^2R of that winding.* The stator steel has a flux variation as the rotor revolves and there is a core loss component just as in the tranformer, comprised of both eddy-current and hysteresis losses. In a rotating machine there are also mechanical losses of a frictional nature, including air friction, which is known as *windage*. In large machines the total losses are so great that natural radiation and convection from the machine would be insufficient to keep the temperature within reasonable limits, so forced air is circulated through the windings and even cooling liquid circulation is sometimes used.

From the standpoint of the user, the most evident effect of the losses is that of the efficiency of the machine. As with almost all power apparatus the efficiency is defined as the ratio of the output to the input, multiplied by 100 if it is desired to express the efficiency in percent. Alternate mathematical expressions for efficiency are

$$\eta = \frac{\text{Output}}{\text{Input}} \times 100$$

$$\eta = \frac{\text{Output}}{\text{Output} + \text{Losses}} \times 100$$

$$\eta = \frac{\text{Input} - \text{Losses}}{\text{Input}} \times 100$$

ADDITIONAL READING MATERIAL

1. Brown, D., and E. P. Hamilton III, *Electromechanical Energy Conversion,* New York: MacMillan Publishing Company, 1984.

2. Chapman, S. J., *Electric Machinery Fundamentals,* New York: McGraw-Hill Book Company, 1985.

3. Chaston, A. N., *Electric Machinery,* Reston, Virginia: Reston Publications Company, Inc., 1986.

4. Del Toro, V., *Electric Machines and Power Systems,* Englewood Cliffs, New Jersey: Prentice-Hall, Inc., 1985.

5. Elgerd, O. I., *Basic Electric Power Engineering,* Reading, Massachusetts: Addison-Wesley Publishing Company, 1977.

*Some interesting development work is being done in using cryogenic techniques to cool the conductors into their superconductive region and eliminate this loss.

6. Elgerd, O. I., *Electric Energy Systems Theory: An Introduction*, New York: McGraw-Hill Book Company, 1982.

7. Fitzgerald, A. E., C. Kingsley, Jr., and S. D. Umans, *Electric Machinery*, New York: McGraw-Hill Book Company, 1983.

8. Lindsay, J. F., and M. H. Rashid, *Electromechanics and Electric Machinery*, Englewood Cliffs, New Jersey: Prentice-Hall, Inc., 1986.

9. Nasar, S. A., *Electric Machines and Electromechanics*, Schaum's Outline Series in Engineering, New York: McGraw-Hill Book Company, 1981.

10. Nasar, S. A., *Electric Energy Conversion and Transmission*, New York: MacMillan Publishing Company, 1985.

11. Shultz, R. D., and R. A. Smith, *Introduction to Electric Power Engineering*, New York: Harper & Row, Publishers, 1985.

12. Slemon, G. R., and A. Straughen, *Electric Machines*, Reading, Massachusetts: Addison-Wesley Publishing Company, 1980.

STUDY EXERCISES

1.

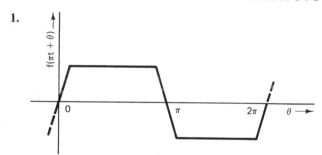

The curve shown above is a plot of a certain periodic function, $f(\pi t + \theta)$, for a specific value of t, $(t = 0)$. Sketch a plot of $f(\pi t + \theta)$ versus θ for values of:

a. $t = 1/4$

b. $t = 1/2$

c. In what direction is the wave traveling in relation to θ?

2. In a two-phase machine there are two sets of stator coils spaced 90 mechanical degrees apart (for a two-pole machine), and the currents under balanced conditions are equal and 90° apart in phase angle. As a result, the two-phase belts of conductors set up B waves as follows:

$$B_a = KI \sin(\omega t + \alpha) \sin \theta$$
$$B_b = KI \sin(\omega t + \alpha - 90°) \sin(\theta - 90°)$$

By manipulation of the trig identities, derive an expression for $B_0 = B_a + B_b$ analogous to Eq. (5-9).

3. At one time in Southern California, there were places served by 50-Hz power while other places were served by 60-Hz by the same utility company. The company tied the two portions of their system together by means of a pair of synchronous machines on the same shaft, one machine operating at 50 Hz and the other at 60 Hz.

What is the minimum number of poles that the individual machines could have for this type of operation and what would be the speed of the shaft in rpm?

4.

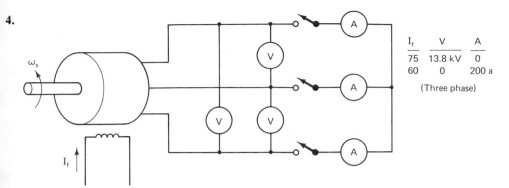

I_f	V	A
75	13.8 kV	0
60	0	200 a

(Three phase)

A certain synchronous machine is tested in the laboratory by driving it at constant rated speed and applying a small field current and slowly increasing it until it produces rated voltage on open circuit and then repeating until rated current flows on short circuit. Assuming linearity of the magnetic circuit, what is Z_s in ohms per phase and in per-unit on the machine's own rating as base? Test results are tabulated above.

5. A small synchronous ac generator has the following characteristics:

$$10 \text{ kVA, 208 volts, three-phase}$$
$$Z_s = 0.2 + j4.0 \qquad \text{ohms per phase}$$

For each of the three loads described below, the field current is set so that rated voltage is produced at the machine terminals with full rated current flowing. The field current is then unchanged when the load is reduced to zero.

What is the percent voltage regulation for loads of:
a. Unity power factor
b. 0.8 power factor, current lagging
c. 0.8 power factor, current leading

6. If the above machine is used as a motor and set to draw rated kVA from a 208-volt source at 0.7 power factor, current leading, what is the "internal" voltage E_{ao}?

7. A three-phase lossless synchronous generator is tied to a 15-kV three-phase bus of a utility that has such a small "internal" impedance that it may be regarded as an infinite bus. The machine itself has a synchronous reactance of 11 ohms per phase. The machine's field current is set so that E_{ao} is 120 percent of the bus voltage V_t; the machine is *over*excited. The prime mover torque is set at a value such that the machine *delivers* 12 MW (total) to the bus. Find the complex power, S, and the current magnitude, I.

8. If the machine described above operates as a motor drawing the same magnitude of current, $|I|$, as before, determine $S = P + jQ$. Assume that the machine's field current remains the same as before; that is, $|E_{ao}|$ is unchanged.

SOLUTIONS TO STUDY EXERCISES

1. a. $f(\pi/4 + \theta)$ advances by $\pi/4$ over $f(0 + \theta)$

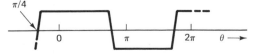

b. $f(\pi/2 + \theta)$ advances by $\pi/2$ over $f(0 + \theta)$

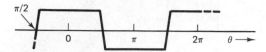

c. The wave is seen to be traveling in opposition to θ.

2. $B_0 = B_a + B_b = KI\{\sin(\omega t + \alpha) \sin \theta + \sin(\omega t + \alpha - 90) \sin(\theta - 90)\}$
$\qquad = (KI/2)\{\cos(\omega t + \alpha - \theta) - \cos(\omega t + \alpha + \theta)$
$\qquad\qquad + \cos(\omega t + \alpha - \theta) - \cos(\omega t + \alpha + \theta - 180)\}$
$\qquad = KI \cos(\omega t + \alpha - \theta)$ or alternately
$\qquad = KI \cos(\theta - \omega t - \alpha)$

which is seen to be a traveling wave proceeding in the θ direction.

3. $f = p(r/min)/120$ Eq. (5-1)

At 50 Hz	p	r/min		At 60 Hz	p	r/min
	2	3000			2	3600
	4	1500			4	1800
	6	1000			6	1200
	8	750			8	900
	10	600			10	720
	12	500			12	600

We note that if the 50-Hz machine has 10 poles and the 60-Hz machine has 12 poles, then they both run at 600-rpm and can be tied together mechanically.

4. At $I_f = 75$, $E_{ao} = 13.8/\sqrt{3} = 7.97$ kV/phase
At $I_f = 60$, $E_{ao} = (60/75)7.97 = 6.37$ kV/phase
$\quad Z_s = 6.37 \times 1000/200 = \underline{31.87 \text{ ohms per phase}} \longleftarrow$
$Z_{base} = E_b/I_b = 7.97 \times 1000/200 = 39.84$ ohms
$\quad Z_s = 31.87/39.84 = \underline{0.80 \text{ per unit}} \longleftarrow$

We might note as an aside that engineers dealing with synchronous machines sometimes use a quantity known as the *short-circuit ratio,* which is defined as

$$SCR = I_{f(open\ circuit)}/I_{f(short\ circuit)} \qquad \text{(At rated } V \text{ and } I)$$

In this case, we have SCR $= 75/60 = 1.25$, and we may further note that this is the reciprocal of the per-unit synchronous impedance, Z_s.

5.

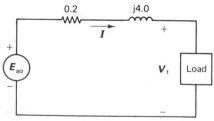

$I_{rated} = 10 \times 10^3/\sqrt{3} \times 208$
$\qquad = 27.76$ A

a. Unity power factor

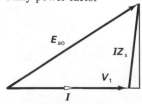

$V_t = 208/\sqrt{3}/0° = 120.1/0°$
$I = 27.76/0°$
$E_{ao} = V_t + IZ_s$
$E_{ao} = 120.1/0° + 27.76/0°(0.2 + j4.0)$
$\qquad = 167.7/41.47°$
%Reg $= [(167.7 - 120.1)/120.1] \times 100$
$\qquad = \underline{39.6\%} \longleftarrow$ a

b. 0.8 p.f., current lagging

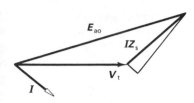

c. 0.8 p.f., current leading

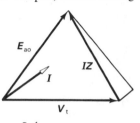

$V_t = 120.1\underline{/0°}$

$I = 27.7\underline{/-\cos^{-1}0.8}$

$= 27.7\underline{/-36.87°}$

$E_{ao} = 120.1\underline{/0°} + 27.76\underline{/-36.87°}(0.2 + j4.0)$

$= 209.3\underline{/24.10°}$

%Reg $= [(209.3 - 120.1)/120.1] \times 100$

$= 74.4\%$ ⟵ b

$V_t = 120.1\underline{/0°}$

$I = 27.76\underline{/+36.87°}$

$E_{ao} = 120.1\underline{/0°} + 27.76\underline{/36.87°}(0.2 + j4.0)$

$= 108.84\underline{/57.85°}$

%Reg $= [(108.34 - 120.1)/120.1] \times 100$

$= -9.4\%$ ⟵ c

6.

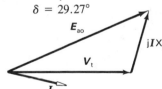

$E_{ao} = V_t - IZ_s$

$= 120.1\underline{/0°} - (27.7\underline{/\cos^{-1}0.7})$

$\times (0.2 + j4.0)$

$= 211.8\underline{/-22.6°}$

$E_{ao} = \underline{211.8 \text{ V}}$ ⟵

7. $V_t = 15/\sqrt{3}\underline{/0°} = 8{:}66\underline{/0°}$ kV $(l\text{-}n)$

$E_{ao} = 1.2 \times 8.66 = 10.39$ kV $(l\text{-}n)$

P/phase $= 12 \times 10^6/3 = 4 \times 10^6 = (E_{ao}V_t/X_s)\sin\delta$

$4 \times 10^6 = (10.39 \times 10^3 \times 8.66 \times 10^3/11)\sin\delta$

$\delta = 29.27°$

$I = (E_{ao} - V_t)/jX_s$

$= \dfrac{(10.39\underline{/29.67°} - 8.66\underline{/0°}}{j11}$

$= 0.4633\underline{/-4.56°}$

$S = V_tI^* = 4.00 + j0.319$ MVA/phase or $12.0 + j0.958$ MVA (total)

$S_{tot} = 12.04\underline{/4.56°}$ $S = \underline{12.04 \text{ MVA}}$, $Q = \underline{0.958 \text{ MVAr}}$ ⟵

8.

We *might* use the cosine law to find δ, knowing the three sides of the triangle:

$(IX_s)^2 = V_t^2 + E_{ao}^2 - 2V_tE_{ao}\cos\delta$

$\delta = -29.27°$

or, we might merely note that the phasor diagram is simply that of Problem 7 "flopped" over!

Proceeding:

$I = (V_t\underline{/0°} - E_{ao}\underline{/\delta})/jX_s = 0.463\underline{/+4.56°}$

$S = V_tI^* = 8.66\underline{/0°} \times 0.463\underline{/-4.56°}$

$= 4.00 - j0.3192$ MVA/phase

or $= \underline{12.00 - j0.9576 \text{ MVA (total)}}$ ⟵

HOMEWORK PROBLEMS

1. The figures below show a large magnet constructed for a research laboratory. The magnet has an iron frame with pole pieces of circular cross section with an air gap of 1.5 cm. There are windings of circular form embedded in the pole face. Each of the three coils has 12 turns and the conductors pass 400 amperes in each turn. We may assume that the iron core has practically infinite permeability compared to that of the air gap.

   a. Calculate the magnitude of the B vector as a function of the distance, x, radially outward from the center of the pole, and then plot these values versus x. Identify all points of discontinuity.

   b. If you were the design engineer and asked to make the B vector distribution approximate more closely a sinusoid, explain *qualitatively* what changes in winding distribution you would make to accomplish the objective?

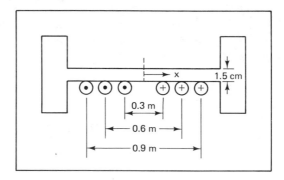

Cross-section A-A'

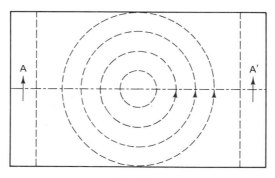

Top view

2. One of the original generators at Grand Coulee dam was rated

   108 MVA, 13.8 kV at 1.0 power factor, $\overline{X}_s = 64.5\%$

   a. At rated load, terminal voltage and power factor, what would be the "internal" voltage, E_{ao}, in kV per phase?

   b. If the machine is now asked to deliver rated VA at 0.8 power factor, current lagging, by what percent would the field current have to be increased (or decreased) if we assume a linear relation between flux linkages and field current?

   Note: Armature resistance, R_a, is to be neglected for this problem.

3. Suppose that two of the original eighteen generators at Grand Coulee are set aside to supply a local pumping load of 150 MW at 13.8 kV and 0.9 power factor, current lagging. Both machines have the characteristics given in Problem 2.

Now further suppose that the field current of machine 1 is fixed at a value that gives rated terminal voltage on open circuit, and the field current of machine 2 must be set to supply the load demands. The hydro turbines of each machine are set so that they divide the real power load equally.
a. What is the "internal" voltage of each machine, E_{ao1} and E_{ao2}?
b. What will be the current and power factor of each machine?

4. A small power system is composed of a synchronous generator, a transformer bank, and two transmission lines in parallel feeding a large metropolitan area which may be approximated by an infinite bus. The per-phase equivalent circuit is shown below with all reactances given in per unit on the machine's own rating as base. Resistances are to be assumed negligible.

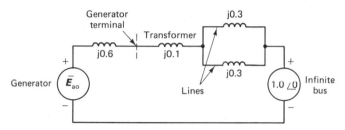

The infinite bus voltage is given as 1.0 per unit. The mechanical input to the machine and the machine excitation (field current) are set so that the machine delivers rated (1.0 per unit) volt-amperes at 0.8 power factor, current lagging, at the machine terminals and these adjustments of generator real power output and field current are maintained for all parts of the problem.
a. How much complex power, $\overline{S}$, is delivered to the infinite bus?
b. If one of the lines is disconnected owing to a lightning strike, how much complex power, $\overline{S}$, is delivered to the infinite bus?
c. In part (b), above, is the machine within its power limit for stable steady-state operation?

5. The text of this chapter discusses two cases of connection of a synchronous machine: (1) a purely passive impedance load and (2) connection to a potential source, fixed in magnitude and phase angle (an *infinite bus* in the language of the power engineer). An intermediate case would involve the connection of the machine to a system with a small "internal" impedance in the Thevenin's theorem type of representation.

The figure below shows a per-phase equivalent circuit of a generator connected to such a system. Voltages are in kilovolts per phase and impedances in ohms per phase. Resistances and other losses are to be assumed negligible. System frequency is 60 Hz.

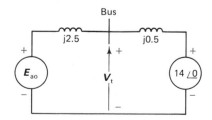

When the machine is first connected to the system, the operator brings the machine up to speed so that it is generating at 60 Hz and in step with the system. The field current is adjusted to bring the terminal voltage to 14 kV, and hence this is the value of E_{ao}.

a. If the prime mover input to the generator is now slowly increased, the electrical output increases in the same amount. How much prime mover driving power may be applied as a maximum without the machine's exceeding its stability limit?

b. If we now wish to deliver 200 MW to the bus at 0.8 power factor, current lagging, we must readjust the field current, thus changing the value of E_{ao}. What must the new value of E_{ao} be for the specified load? The value of 200 MW is the total load *or* the prime mover input to the machine.

c. With the value of field current set in part (b), what is the maximum prime mover drive to the machine that may be handled without loss of stability?

6. A certain 500-horsepower, 2400-volt, three-phase, 60-Hz, 6-pole synchronous motor has a synchronous reactance of 10.5 ohms per phase, and is assumed lossless for the purposes of this problem.

a. At what speed, in r/min, does the motor run?

b. If the motor is capable of running at 0.8 power factor, current *leading,* what is the rated current of the motor?

c. With rated voltage applied and the field current set for operation at unity power factor at full rated current load, what is the maximum torque of the motor in newton-meters?

d. Repeat part (c) for the case where the field current is set to cause the motor to operate at 0.8 power factor, current leading, at full rated current load.

7. Suppose that the motor of Problem 6 has an armature resistance, R_a of 0.25 ohm per phase. It also has mechanical losses of 20 kW (assumed constant) and a field loss of 10 kW when operated at full load, unity power factor, at rated voltage.

a. What is the full load efficiency at unity power factor?

b. If we assume that the field loss varies as $(E_{ao})^{1.4}$ what is the full load efficiency when operating at 0.8 power factor, current leading?

8. A certain industrial plant is supplied with electric power at 2400 volts (*l-l*, three-phase). The plant has 200 kW of unity power factor load such as the pure resistances of lighting and water heating. The plant also has 800 kW of motor load of a type which draws power at 0.707 p.f., current lagging. (The induction motors of the next chapter are a good example of this type of load.) All loads are balanced between the three phases — the values given above are *total* loads.

A new load on the plant requires a motor delivering 200 horsepower. If synchronous motors are available with a conversion efficiency of 0.85 (shaft output/stator input):

a. What must be the kVA rating of the new motor if we are to overexcite the field and cause the motor to draw power at a leading power factor, thus bringing the total plant load to an improved power factor of 0.90, current lagging?

b. If the new motor has $Z_s = 0 + j20$ ohms per phase, what must the internal voltage E_{ao} be set at to accomplish the desired objective?

9. A 12-pole, 60-Hz, three-phase synchronous generator has a synchronous impedance, $Z_s = 5 + j50$ ohms per phase. The generator is connected to an infinite bus (perfect potential source) of 15 000 volts, line-to-line. The field current (i.e., E_{ao}) is not changed from the value required to match the bus voltage when the machine is first connected to the bus. The prime mover throttle is then opened, thus causing the generator to advance. If the generator voltage phasor advances by 30°:

a. What is the complex power delivered to the bus?

b. By what mechanical angle does the generator shaft advance with respect to a synchronously rotating reference frame, as, for example viewed by a stroboscope?

10.

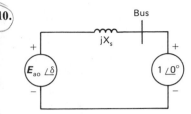

The diagram above shows a per-phase equivalent circuit of a synchronous machine of negligible losses and X_s of 0.80 per unit connected to an infinite bus of voltage $V_t = 1.0\underline{/0°}$. The controls are set such that the machine delivers 0.9 per unit power at 0.8 power factor, current lagging to the infinite bus.

a. What is the power angle, δ, of the machine's internal voltage, E_{ao}?

b. If the shaft connecting the machine to the prime mover driving source should break, what complex power, S, will then be delivered to the bus?

Chapter 6

The

Induction

Motor

- The general principles of the polyphase induction motor, leading to an equivalent circuit for the motor.
- Motor performance, in terms of the equivalent circuit, with a particular view of matching the motor to the load.
- Single-phase induction motors, briefly discussed.

6.1 INTRODUCTION

The synchronous motor described in Chapter 5 has some excellent characteristics such as constant speed (with constant frequency) and controllable reactive voltampere demand, which may be used for power factor correction. The motor does require a dc field supply for the rotor, however, and requires special attention to starting, all of which adds to operational and maintenance complications, as well as the expense of such a motor. Another type of ac motor is the *induction motor,* where the field excitation of the rotor circuits is accomplished by induced voltage from the stator windings, which are connected to the ac energy source.

The induction motor does not run at constant speed, although in usual applications the speed variation is very small. Since the magnetic field must be supplied from the ac source, the motor will draw current-lagging reactive voltamperes from the source and will operate at a power factor less than unity. Nevertheless, the simplicity of construction and operation of such motors leads to performance and cost advantages such that this class of motor is by far the most common ac motor in actual operation. Typically, in large plants the majority of motors will be of the induction type, with perhaps very few synchronous motors used in critical areas of application — and perhaps also used to improve the overall plant power factor by operating in the overexcited mode.

It has been said that the concept of a polyphase induction motor led the great electrical pioneer, Nikola Tesla, to develop polyphase systems in the first place. All the larger induction motors are of the polyphase type, but smaller units are supplied with single phase power at some sacrifice in performance characteristics.* In the following pages we will present the story of the induction motor in terms of a polyphase device first, and then later consider the differences in characteristics when single-phase devices are necessary.

6.2 THE ELEMENTARY PRINCIPLES OF THE INDUCTION MOTOR

The polyphase induction motor is provided with stator windings and a structure essentially identical to that of the synchronous machine. The rotor, on the other hand, may be supplied with another polyphase ac winding very similar to the stator winding, with the leads brought out of the motor through *slip rings* to an external resistance load. The rotor core is of the smooth cylindrical rotor type of construction. Alternate forms of windings for the rotor windings are very common and will be discussed later. A circuit diagram symbolizing the arrangements outlined above is given in Figure 6-1 where the three stator phases are shown connected in Y, and likewise, the rotor winding is represented in a Y connection. Variations in the form of delta connections and other numbers of phases are possible but the form shown will be used for discussion purposes.

*The single-phase induction motor is used in large numbers for domestic application. Many homes have a great number of these devices driving refrigerators, washers, dryers, and all manner of appliances. In numbers, the single-phase induction motor is surely the dominant type of all motors!

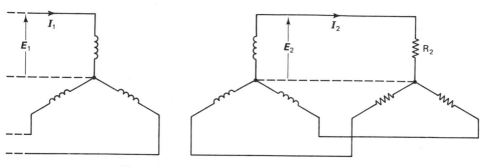

Figure 6-1   A circuit diagram of the induction motor.

By applying a balanced three-phase voltage to the stator windings we cause currents to flow and a revolving field is set up just as in the case of the armature reaction of the synchronous machine. The revolving field also links the rotor windings, and since these linkages change, a voltage is induced in the rotor coils. Currents then flow in the rotor and the interaction of these currents with the stator flux results in a mechanical torque, tending to cause the rotor to turn. By Lenz's law the direction of the effects will be such as to tend to cause the rotor to follow the revolving field of the stator; the currents are set up to *oppose* the change in flux linkages. A sketch symbolizing this action is given in Figure 6-2, where the stator flux is shown linking the rotor windings. The stator field is shown revolving at a speed ω_s and the rotor revolving at a speed ω_m. The subscript s may be thought of as standing for *stator* or *synchronous,* as desired, since the stator field will be excited from the basic energy supply and will revolve at a speed determined by the applied frequency and the number of poles.* See Eq. (5-1). The subscript m stands for *mechanical* and ω_m is ordinarily different from ω_s. Just as in the case of the synchronous machine, the number of poles enters the picture as a factor determining speed but we will use the two-pole case in sketches in order to simplify the presentation.

As mentioned above, the mechanical speed ω_m is normally different from the speed of the revolving field, ω_s, since otherwise the flux linkages with the rotor

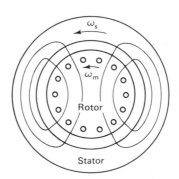

Figure 6-2   The revolving field linking the rotor conductors.

*See Section 5.3 of Chapter 5 for relationship of speed, frequency, and number of poles for the stator.

would not change and we would not have the induced voltages and resultant currents in the rotor windings upon which the motor depends for torque production. In practice, the rotor must lag behind the stator by some difference velocity $\omega_s - \omega_m$. This difference velocity is called the *slip,* and is usually symbolized by the letter S as in

$$S = \omega_s - \omega_m \quad \text{(radians per second)}$$

It is most common to normalize the slip using ω_s as the base, in which case we would have

$$s = \frac{\omega_s - \omega_m}{\omega_s} \quad \text{per-unit} \tag{6-1}$$

Since this normalized form is the most common, we will use the symbol s without any particular modification in the work to follow with the understanding that s means normalized or per-unit slip. Since the basic motor action occurs only when we have a non-zero value for s, it is not surprising that this quantity plays a leading role in the analysis of the motor.

One effect of the slip, s, in the motor performance can be illustrated by considering the frequency of the rotor currents. At standstill, the revolving field set up by the stator generates voltages in the rotor conductors at a frequency the same as that of the stator currents. The motor *could* be regarded as simply a three-phase transformer of unusual mechanical configuration. As the motor moves ahead, however, the relative velocity between the revolving field of the stator and the rotor conductors becomes less. The frequency of the rotor voltages and currents then becomes less in proportion to the reduction in relative velocity. If the rotor should travel at synchronous speed, ω_s, there would be *no* relative motion, the rotor flux linkages would not change, and there would be no generated voltage or consequent current. As the rotor speed approaches this condition, the rotor frequency approaches zero. The relation between rotor frequency f_r and stator frequency f_s can be stated succinctly as

$$f_r = sf_s \tag{6-2}$$

So long as there is some relative velocity between the stator field and the rotor (some positive slip), the rotor flux linkages change. The flux linkages in the three rotor coils vary sinusoidally, and between the same limits in all coils, but there will be a phase difference between them because of the space displacement of the rotor coils around the periphery of the rotor steel. If represented by phasors, the flux linkages would form a balanced three-phase set, and in turn the voltages generated would likewise form a balanced three-phase set. Last, the currents flowing as a result of the voltages form a balanced three-phase set too! By the same means as those employed to study the armature reaction of the synchronous machine in Chapter 5, it may be seen that the rotor currents cause a revolving field, which travels at a speed $\omega_r = 2\pi f_r$ *with respect to the rotor itself* (for the two-pole case).

We have now mentioned three different angular velocities: ω_s, the speed of the revolving field of the stator, ω_m, the speed of the rotor steel and copper (the shaft speed of the motor), and ω_r, the speed of the revolving field set up by the rotor currents *relative* to the rotor itself. These various speeds are related to each other and if the slip, s, is known, any one can be found from any of the others. For example since

$$\frac{\omega_s - \omega_m}{\omega_s} = s$$

then

$$\omega_m = (1 - s)\omega_s \qquad (6\text{-}3)$$

also since

$$f_r = sf_s$$
$$\omega_r = s\omega_s \qquad (6\text{-}4)$$

and

$$\omega_r + \omega_m = s\omega_s + (1 - s)\omega_s = \omega_s \qquad (6\text{-}5)$$

Equation (6-5) is particularly interesting, since it says that the combination of the mechanical speed ω_m, with the relative rotor field speed ω_r added to it, results in a revolving rotor field with a net angular velocity of ω_s, the same as the stator field velocity. Seen from the stator, the rotor *field* revolves along synchronously and the magnetic situation is not appreciably different from that of the synchronous machine with a dc excited rotor winding and a rotor traveling at ω_s. The slip frequency (rotor) currents make up the difference between the mechanical velocity of the rotor and synchronous speed. If we view the field from magnetic probes in the air gap itself, we might as well have the picture of Figure 5-21(b) where the revolving stator field is pulling the rotor field behind it!

The torque developed depends on the relative magnitudes and angle between the stator field and the rotor field. The angle, in turn, depends on the phase angle of the rotor currents in response to the induced voltages in the rotor. The rotor resistance and reactance determine the phase angle of the rotor currents, and we might pursue an analysis of the machine's characteristics by studying the magnitude and phase angle relations of the stator and rotor currents in detail. From the viewpoint of a simpler understanding of the motor, we prefer to view the rotor from the vantage point of the stator (where lab measurements would be made anyway in most machines), and predict the machine performance from a knowledge of the energy sent across the air gap by the stator. This is the approach of the following section.

6.3 MACHINE PERFORMANCE: AN EQUIVALENT CIRCUIT

The physical picture has been laid out, a revolving field is set up by the stator, the field induces voltages in the rotor, and the consequent currents interact with the revolving field by setting up a rotor field revolving synchronously with that of the stator. A convenient way to describe and analyze the motor is to find an equivalent circuit, which accounts for the various voltage and current components, and which shows the transfer of energy from stator to rotor to be converted into mechanical form.

When a voltage is applied to the stator a current flows and part of the voltage is absorbed in the stator winding resistance; part is absorbed in the stator leakage reactance, a reactance corresponding to flux linkages that do not cross the air gap to couple rotor and stator windings together. The remainder of the applied stator voltage is absorbed by the counter voltage generated by the revolving field. These relations are shown in the partial equivalent circuit of Figure 6-3 which is given on a per-phase basis — for one phase of an equivalent Y connection. The subscript 1 is used to describe stator quantities and the subscript ϕ is used to describe the counter voltage generated by the net revolving field. The action is like that of the transformer, where part of the applied voltage is absorbed in the winding resistance and leakage reactance, and the remainder is absorbed by the changing mutual flux of the core. It might at first seem that the revolving field of constant magnitude would have to be treated differently from the transformer flux, but it must be realized that, so far as one particular stator coil is concerned, the action of the revolving field is merely to cause a sinusoidally varying flux linkage with the coil.

The establishment of the mutual flux across the air gap requires ampere turns and part of the stator current is required for this purpose. Just as in the transformer equivalent, this can be shown by means of an exciting current in a fictitious shunt branch. There will be core losses as well as magnetizing reactive voltampere require-ments, and these are accounted for by the shunting G_c and B_c, respectively, in the figure. The ampere turns of the stator, which are not required to establish the flux by means of the exciting current branch, are cancelled in their magnetizing effect by the rotor currents, just as in the case of the transformer. Although the rotor currents are at a different frequency ($f_r = sf_s$) the stator coils "think" they see a current of the same frequency because of the motion of the rotor. If the turns ratio of the windings is one-to-one, the rotor current will be reflected in the stator, with equal amperes in order to balance the ampere turns. These ideas are shown in the equiva-lent circuit by designating the reflected rotor current as I_2 in the circuit diagram. If the turns ratio of stator to rotor is not one-to-one, the circuit must be interpreted as giving the current I_2, referred to the stator side, just as we do with the transformer.

The revolving field, which generates E_ϕ in the stator, would also generate E_ϕ in the rotor with unity turns ratio, but this is only true at standstill with the slip $s = 1$. Under motion the relative velocity of the revolving field with respect to the

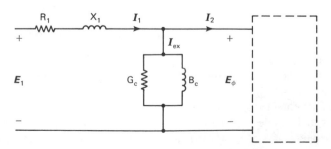

Figure 6-3   A partial equivalent circuit for the stator.

rotor conductors is only s per-unit, and therefore the generated voltage is $s\,E_\phi$. Since the rotor windings are closed, the voltage $s\,E_\phi$ is absorbed by the current I_2, passing through the rotor circuit resistance and leakage reactance.* We have then the equation expressing this relation

$$sE_\phi = R_2 I_2 + jx_{l2} I_2 \tag{6-6}$$

The reactance x_{l2} is a function of frequency. If we define the value of x_{l2} as X_2 at standstill ($s = 1$) then x_{l2} at any speed is given by $x_{l2} = sX_2$. Equation (6-6) then becomes

$$sE_\phi = R_2 I_2 + jsX_2 I_2 \tag{6-7}$$

An equivalent circuit for the rotor (also on a per-phase basis) is given in Figure 6-4.

To relate the stator quantities of Figure 6-3 and the rotor quantities of Figure 6-4 we divide both sides of Eq. (6-7) by the slip s and obtain

$$E_\phi = (R_2/s)I_2 + jX_2 I_2 \tag{6-8}$$

The equivalent circuit of Figure 6-4 is revised in Figure 6-5 to show the variables in terms of Eq. (6-8). It is said that this figure shows the rotor quantities as *viewed from the stator*.

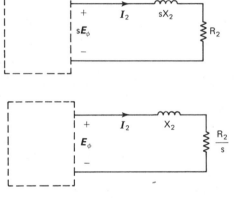

Figure 6-4   A partial equivalent circuit for the rotor.

Figure 6-5   The rotor equivalent as viewed from the stator.

As the culmination of this particular development, the stator and rotor equivalent circuits are joined in Figure 6-6, which is the complete equivalent circuit for an induction motor on a per-phase basis. The parameters of the circuit may be found for an actual motor in the laboratory subject to the usual approximations (representation of the exciting current branch as a linear circuit combination, etc.). The circuit may then be used to predict the performance of the motor under various conditions. The elements of the equivalent circuit may also be found by the designer from the dimensions and material properties of the design and the design altered to obtain a desired type of performance.

---

*In some types of motors, the rotor resistance is merely that of the rotor conductors themselves while in other types extra resistance may be inserted into the rotor circuit externally.

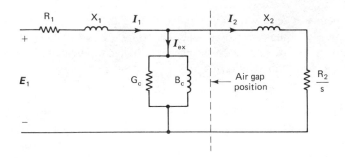

Figure 6-6   The equivalent circuit of the induction motor on a per-phase basis, viewed from the stator.

6.4 MACHINE PERFORMANCE FROM THE EQUIVALENT CIRCUIT

In the equivalent circuit of Figure 6-6, the power passing to the right through the dotted line is the power, P_{ag}, passing to the rotor across the air gap. From circuit theory we see that this is given by

$$P_{ag} = I_2^2 R_2 / s \qquad \text{watts per-phase} \qquad (6\text{-}9)$$

Part of this power is lost in the rotor windings; we will call this rotor copper loss, P_{rc} (although some rotor windings are actually aluminum).

$$P_{rc} = I_2^2 R_2 \qquad \text{watts per-phase} \qquad (6\text{-}10)$$

The difference between P_{ag} and P_{rc} is evidently the power converted to mechanical form, P_m*

$$P_m = P_{ag} - P_{rc} = I_2^2 R_2 \left(\frac{1}{s} - 1 \right)$$

$$P_m = I_2^2 R_2 \left(\frac{1-s}{s} \right) \qquad \text{watts per-phase} \qquad (6\text{-}11)$$

The separation of the air gap power into copper loss and mechanical power can be illustrated by splitting the quantity R_2/s as shown in Figure 6-7.

We note that the ratio of rotor copper loss to power across the air gap is particularly simple

$$P_{rc}/P_{ag} = I_2^2 R_2 / (I_2^2 R_2 / s) = s \qquad (6\text{-}12)$$

Equation (6-12) tells us that if, for example, we have a slip of 5 percent (0.05 per unit) then we lose 5 percent of the power across the air gap in the form of rotor copper losses, and the energy efficiency of the motor cannot possibly be more than 95 percent.[†] Most induction motors are designed with very low rotor resistance, and

---

*A small percentage of this power is lost as friction and does not appear as shaft output of the motor; we could call P_m *developed* mechanical power as contrasted with shaft *output* power but we will not clutter the page with this modifying adjective.

[†]There are other losses than the rotor copper losses; for example, stator copper losses in R_1, core losses in G_c, and mechanical friction losses. The quantity $(1 - s)$ would then be an upper bound for efficiency in per-unit.

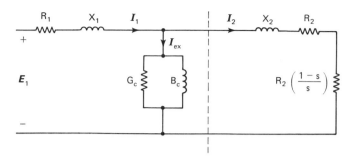

Figure 6-7 Equivalent circuit modified to show copper loss and mechanical equivalent separately.

therefore very low slip, but sometimes it is desirable to have a high slip for some special application reason. If such be the case we must be prepared to accept a high rotor copper loss and consequent low energy efficiency.

We can find the developed torque very easily from the expression for mechanical power, Eq. (6-11) since torque is given by

$$T = \frac{P_m}{\omega_m} = \frac{I_2^2 R_2}{\omega_m}\left(\frac{1-s}{s}\right)$$

but

$$\omega_m = (1-s)\omega_s \qquad \text{from Eq. (6-3)}$$

therefore

$$T = \frac{I_2^2 R_2}{\omega_s s} \qquad \text{newton-meters per-phase} \qquad (6\text{-}13)$$

In words, the developed torque is equal to the power across the air gap, divided by the synchronous speed. The MKS unit of torque is the newton-meter but we sometimes see the torque expressed as equal to the air gap power in units of "synchronous watts"; that is, the torque in synchronous watts is the power that *would* be developed by that torque *if* the rotor ran at synchronous speed.

When the load on the induction motor changes the current I_2 changes, but we normally consider the input voltage E_1 to remain fixed. It is desirable to analyze the motor performance in terms of the entire circuit, holding E_1 fixed as our electrical input. The circuit of Figure 6-7 involves several branches or nodes and the expression for torque, for example, becomes rather complex. It is common to simplify the circuit before solving for mechanical power or torque. One possible simplification would be to ignore the shunt-exciting branch just as we did for many purposes in the case of the transformer, but because of the air gap, the exciting current of the motor is much higher relative to the load current, and such an approximation may lead to large error. Another simplification quite commonly used is to move the exciting current branch to the left, directly across E_1 as in Figure 6-8(a). This leaves a single loop equation for finding I_2, but again we might wonder if this approximation is valid for the induction motor, even though it is the most precise representation usually encountered for the transformer. A more satisfying simplification used by several authors involves replacing the circuit to the left of the dotted line in Figure 6-7 by

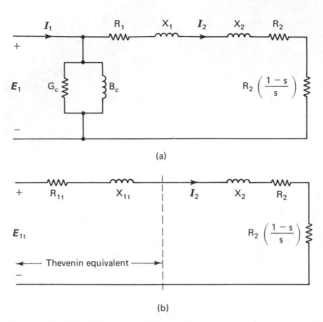

Figure 6-8   Simplifications of Figure 6-7 commonly used for analysis.

a Thevenin's theorem equivalent and this change is shown in Figure 6-8(b). The Thevenin circuit voltage is designated E_{1t} and is the open-circuit voltage that would exist across the exciting branch if the circuit were opened at the dotted line of Figure 6-7. This voltage is of course a specific fraction of the applied voltage within the limits of accuracy involved in considering G_c and B_c as linear constant circuit elements. The Thevenin equivalent impedance is marked R_{1t} and X_{1t} and is the parallel resultant of combining R_1, X_1, and the exciting current branch expressed as an impedance.

If we accept the circuit of Figure 6-8(b) as being reasonably valid as an equivalent to the motor, we find I_2 easily as

$$I_2 = \frac{E_{1t}}{\sqrt{(R_{1t} + R_2/s)^2 + (X_{1t} + X_2)^2}} \tag{6-14}$$

The mechanical power, P_m, follows easily from I_2

$$P_m = \frac{E_{1t}^2 R_2 \left(\dfrac{1-s}{s}\right)}{\left(R_{1t} + \dfrac{R_2}{s}\right)^2 + (X_{1t} + X_2)^2} \qquad \text{watts per-phase} \tag{6-15}$$

and, in turn, we find the torque, T, as

$$T = \frac{E_{1t}^2 \dfrac{R_2}{s}}{\omega_s \left[\left(R_{1t} + \dfrac{R_2}{s}\right)^2 + (X_{1t} + X_2)^2\right]} \qquad \text{newton-meters per-phase} \tag{6-16}$$

This last equation (6-16) is very important to users of induction motors, since it tells the relation between developed torque and speed and this is the information needed to apply the motor to the load with understanding of the resultant behavior. The value of R_2 is of great significance in this respect and this value is greatly affected by the construction of the rotor. We will digress to discuss some of the alternate types of rotor construction before investigating the application of Eq. (6-16) in detail.

6.5 ALTERNATE TYPES OF ROTOR CONSTRUCTION

In the beginning pages of this chapter, it was inferred that the rotor winding was more or less a facsimile of the stator winding, perhaps connected in Y, and with the terminals brought to slip rings on the shaft. Brushes riding on the slip rings, in turn, would connect the winding terminals to external resistors; the total resistance per-phase was called R_2 and involved the sum of the winding resistance itself plus the external resistance, if any. The one-to-one turn ratio resulting from such a winding leads to the ampere-turn balance of $(I_1 - I_{ex}) = I_2$, and to the statement that the voltage E_ϕ of the stator also appeared in the rotor modified by the relative velocity to become $s\,E_\phi$. This great similarity of the windings is convenient when developing the characteristics of the motor but is not necessary. We will discuss variations below.

It the number of turns per-phase of the rotor winding is not equal to that of the stator, our experience with transformers enables us to accept the idea that the turns ratio is merely a scale factor that can be removed by referring side 2 quantities to side 1.

It was inferred that, if the stator is a three-phase winding, the rotor winding would also be three-phase. This is not necessary. A two-phase rotor winding (but with the correct number of poles) would have two-phase balanced voltages and currents by induction, and result in a revolving field of constant magnitude, just as would a winding of any number of phases (greater than one) provided that it is balanced. If we viewed the rotor from the stator alone, we could not distinguish how many phases were involved in the creation of the revolving field of the rotor. A very common form of construction involves a large number of straight copper bars laid in slots or passed through holes in the rotor steel and connected at the end with rings to allow current circulation. Such a construction could be regarded as a winding of many phases with only one turn per-phase. This type of rotor is called a *squirrel cage* rotor because the bars and connecting end rings without the rotor steel would form a structure like the play wheel in a rodent's cage. A sketch of such construction is given in Figure 6-9.

The squirrel cage rotor (which gives its name to the motor using it) is very rugged and economical to build. It is virtually indestructible. The squirrel cage motor does suffer the disadvantage that external resistance cannot be inserted to change the value of R_2, but in many applications it is not necessary or desirable to do so. As a result, the majority of induction motors are of the squirrel cage type. Motors using the type of winding first discussed are called *wound rotor* motors. The

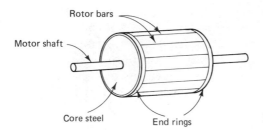

Figure 6-9   The squirrel cage rotor.

wound rotor has some features that will be brought out in the following pages, but nevertheless, the motor is not as often encountered as the squirrel cage version.

6.6 INDUCTION MOTOR SPEED—TORQUE CHARACTERISTICS

Equation (6-16) gives the developed torque per-phase as a function of the rotor slip, s. For given values of the motor constants, we could plot the torque versus slip over the range from zero (synchronous speed) to unity (standstill). It is instructive to consider the behavior of the machine at the extreme values of very low and very high slip by approximating Eq. (6-16) in these regions.

If we multiply numerator and denominator of Eq. (6-16) by s^2 and then consider very small values for s, we have

$$T \cong \frac{E_{1t}^2}{\omega_s R_2} s \qquad \text{(for small slip)} \qquad (6\text{-}17)$$

We note that at zero slip the torque is zero as we might expect, since there is no relative motion between the rotor and the stator revolving field and therefore no rotor currents. At small positive values of slip, the torque grows linearly with slip and the slope of the torque-slip curve is inversely proportional to the rotor resistance R_2. On a sketch of the torque vs. slip curve of Figure 6-10 this variation is shown as region 1 in the region of small slip.

If, on the other hand, we multiply numerator and denominator of Eq. (6-16) by s we have as an approximation for large s*

$$T \cong \frac{E_{1t}^2 R_2}{\omega_s (X_{1t} + X_2)^2} \frac{1}{s} \qquad \text{(for large slip)} \qquad (6\text{-}18)$$

and this equation says that, for large s the torque varies almost inversely with s as shown in region 2 of the figure.

It seems plausible that, somewhere between regions 1 and 2 of the sketch there must be a maximum of torque. The value of s for maximum torque may be found by differentiating Eq. (6-16) with respect to s and equating to zero. Alternately, we

*In this approximation, we exploit the fact that the reactance squared is greater than the resistance squared for typical motors, at least in the larger sizes. This approximation serves its purpose of showing the general shape of the torque-speed curve at high slip, but is not sufficiently accurate for actual calculations. Equation (6-16) should be used for that purpose.

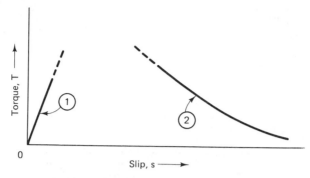

Figure 6-10   Variation of torque with slip.

might recognize that the torque is directly proportional to the power across the air gap (Eq. (6-13)), and it is therefore maximum when the power in R_2/s is a maximum. The maximum power transfer theorem tells us that the power in R_2/s (which is the sum of the resistances R_2 and $R_2(1 - s)/s$ in Figure 6-8(b)) is maximum when R_2/s is equal to the magnitude of the impedance

$$|Z| = |R_{1t} + j(X_{1t} + X_2)| = \sqrt{R_{1t}^2 + (X_{1t} + X_2)^2}$$

so the slip at maximum torque is found from

$$\frac{R_2}{s} = \sqrt{R_{1t}^2 + (X_{1t} + X_2)^2}$$

$$s = \frac{R_2}{\sqrt{R_{1t}^2 + (X_{1t} + X_2)^2}} \tag{6-19}$$

The value of slip at which maximum torque occurs is thus proportional to rotor resistance, R_2, and under control of the designer for a squirrel cage motor and by the operator for a wound-rotor motor. By inserting the value for slip from Eq. (6-19) into Eq. (6-16) we find the maximum value of the torque is

$$T_{max} = \frac{E_{1t}^2}{2\omega_s[R_{1t} + \sqrt{R_{1t}^2 + (X_{1t} + X_2)^2}]} \quad \text{N-m/phase} \tag{6-20}$$

It is at first surprising to see from Eqs. (6-19) and (6-20) that while the value of slip at which maximum torque occurs is strongly dependent upon R_2, the actual magnitude of the maximum torque is not affected by R_2.

The salient features of the torque-slip curve are outlined above; the initial steep slope—the ultimate decay—and the maximum in between, but application of this information is better done by recognizing that each value of slip corresponds to a particular value of the mechanical speed, ω_m, in accordance with the equation $\omega_m = (1 - s)\omega_s$. We can thus plot torque versus speed and note that where the slip is zero the speed is the synchronous speed, ω_s, and where the slip is unity the speed is zero. The curve of Figure 6-11 illustrates a typical curve of torque versus speed and amounts to the torque-slip curve shifted and folded. The torque-speed curve is very useful in studying the application of the motor to various loads, as will be seen in the pages following.

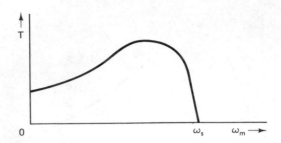

Figure 6-11 Typical torque-speed curve of an induction motor.

Let us consider a specific numerical example of the determination and use of the equivalent circuit(s) of Figure 6-8. We will phrase the discussion in terms of a test made on a motor rated 100 Hp, 2400 volts, three-phase, six-pole, 60 Hz. In testing an induction motor it is common to conduct two tests very analogous to the open- and short-circuit tests of a transformer.

The *open-circuit test* is performed by applying rated voltage to the stator with no mechanical load and observing the input with ammeters and wattmeters. This is called a *running light* test for a motor.

The *short-circuit test* is performed by holding the rotor fixed, with slip of unity, and applying voltage to the stator and observing current and power input. To limit the current to safe values, the voltage is reduced from normal. This is called the *locked rotor* test.

Suppose we observe the following data on our test motor:

Running Light	Locked Rotor
E = 2400 volts	E = 600 volts
I = 6 amperes	I = 40 amperes
P = 4 kilowatts	P = 14 kilowatts

These data are on a three-phase basis.

Under running light conditions, the power input is being consumed in the core losses of the steel and the mechanical friction losses, together with a very small $I_1^2 R_1$ loss in the stator copper. With these tests, it is usual to use the equivalent circuit of Figure 6-8(a) and lump all of these losses in the shunt conductance G_c because we really do not have enough data to separate the losses into components.* The slip is very small when running light, so the current and power going to the rest of the circuit is neglected in view of the high impedance of R_2/s. On this basis then:

$$G_c = \frac{4000}{3} \times \left(\frac{\sqrt{3}}{2400}\right)^2 = 6.94 \times 10^{-4} \quad \text{siemen/phase}$$

$$Y_c = \frac{6\sqrt{3}}{2400} = 4.33 \times 10^{-3} \quad \text{siemen/phase}$$

$$-B_c = \sqrt{Y_c^2 - G_c^2} = 4.27 \times 10^{-3} \quad \text{siemen/phase}$$

*See Problem 6 of the Homework Section for a more precise approach.

If we assume that the equivalent circuit is linear, the locked rotor current will be directly proportional to the voltage and the input power will be proportional to the square of the voltage, so we can ratio the test data in this proportion if we desire, although it would not really be necessary to do so in order to find the other elements of the equivalent circuit. At full voltage we would have with locked rotor:

$$E = 2400 \text{ volts}$$
$$I = 40 \times (2400/600) = 160 \text{ amperes}$$
$$P = 14 \times (2400/600)^2 = 224 \text{ kilowatts}$$

From these data we find that the net admittance looking into the total circuit with $s = 1$ is given by:

$$Y_0 = \frac{160\sqrt{3}}{2400} = 0.1155 \qquad \text{siemen/phase}$$

$$G_0 = \frac{224{,}000}{3} \times \left(\frac{\sqrt{3}}{2400}\right)^2 = 0.0389 \qquad \text{siemen/phase}$$

$$-B_0 = \sqrt{Y_0^2 - G_0^2} = 0.1087 \qquad \text{siemen/phase}$$

If we subtract the G_c and B_c from these values we have the admittance of the series branch

$$G_{\text{ser}} = 0.0389 - 6.94 \times 10^{-4} = 0.0382$$
$$-B_{\text{ser}} = 0.1087 - 4.27 \times 10^{-3} = 0.1045$$

and in turn if we convert to impedance we have

$$Z = \frac{1}{0.0382 - j0.1045} = 3.088 + j8.445 \qquad \text{ohms per-phase}$$

From this result we have

$$R_1 + R_2 = 3.088$$
$$X_1 + X_2 = 8.445$$

From the test data given we have no way of knowing how to split these values into the two components. We really don't care about $X_1 + X_2$, since these two appear as a sum in our work with the equivalent circuit. If we have no other information, we might assume that the stator and rotor copper losses are equal and divide the total resistance into two equal parts. We might make a more accurate division if we can measure the stator resistance separately, but we will use the simple division in the following.

We now have the elements of the equivalent circuit (Figure 6-12). With these elements known we can use the circuit to predict the performance of the motor under various circumstances. As an example, suppose we consider a slip of 3 percent or 0.03 per unit and wish to know the mechanical power, the torque, the power factor, and the input current all with rated voltage of $2400/\sqrt{3}$ volts applied per-phase. We proceed as follows:

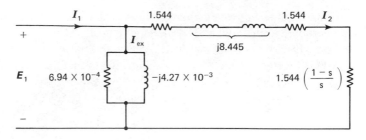

Figure 6-12 Per-phase equivalent circuit for example.

$$R_2\left(\frac{1-s}{s}\right) = 1.544\left(\frac{0.97}{0.03}\right) = 49.92$$

$$I_2 = \frac{(2400/\sqrt{3})\underline{/0}}{2 \times 1.544 + 49.92 + j8.445} = 25.81\underline{/-9.03}$$

$$P_{\text{mech}} = I_2^2 R_2\left(\frac{1-s}{s}\right) = 33.27 \text{ kW per-phase}$$

$$\text{Torque} = I_2^2\frac{R_2}{s} = 34.30 \text{ synchronous kW per-phase}$$

$$= \frac{34.30}{2\pi 60/3} \times 10^3 = 273 \text{ newton-meters/phase}$$

$$\omega_s = \frac{2\pi 60}{p/2} = 40\pi = 125.6 \text{ radians per second}$$

$$\text{Torque} = \frac{3 \times 34.30 \times 10^3}{125.6} = 819 \text{ newton-meters (total)}$$

$$I_{ex} = \frac{2400}{\sqrt{3}}\underline{/0}(6.94 \times 10^{-4} - j4.27 \times 10^{-3})$$

$$= 0.962 - j5.92$$

$$I_1 = I_2 + I_{ex}$$

$$= 26.45 - j9.97 = 28.2\underline{/-20.6°}$$

$$\text{power factor} = \cos 20.6 = 0.94$$

6.7 VARYING THE TORQUE—SPEED CURVE

The curve of Figure 6-11 is based on holding all of the parameters of Eq. (6-16) constant and considering varying values of slip, s. In practice, it may be that E_{1t} will vary owing to varying source voltage E_1 and R_2 may be varied, either as a matter of design of the squirrel cage rotor or by external resistance for the wound-rotor motor.

Since the Thevenin's equivalent voltage E_{1t} is a fixed fraction of the applied voltage E_1 (if G_c and B_c are constant), the developed torque will vary as the square of the applied voltage just as it varies as the square of E_{1t} in Eq. (6-16). For example, with 70 percent of rated voltage applied, the motor will develop only about 50 percent of rated torque at any particular value of slip or speed. At half voltage the torque would be only one-fourth of that available at full voltage! A family of curves is illustrated in Figure 6-13 for various values of applied voltage E_1.

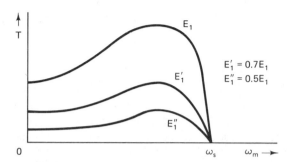

Figure 6-13   Effect of applied voltage on torque.

The other parameter to be discussed in terms of its influence on the motor torque-speed curve is the rotor resistance R_2. In the preceding section it was shown that the torque in the vicinity of zero slip varies with a slope inversely proportional to R_2 (Eq. (6-17)). The slip at which maximum torque occurs increases with increasing R_2 (Eq. (6-19)), but the maximum value itself is independent of R_2. When these factors are considered the variation of torque with R_2 illustrated in Figure 6-14 is easily understood.

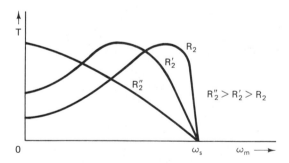

Figure 6-14   Effect of R_2 on motor torque.

Each of the variations shown in the above two figures has an influence on the performance of the motor under various loads, and must be considered in the application of the motor to a given task.

6.8 APPLICATION OF THE INDUCTION MOTOR TO LOADS

The point on the torque-speed curve at which the induction motor actually operates is determined by the load. The torque demanded by the load is a function of the speed, so we will operate at a point that simultaneously satisfies the speed-torque characteristics of the motor *and* the load. As an example, we might consider the torque-speed characteristic of a load that is sketched on the same plane as the motor characteristic in Figure 6-15. This type of load curve is typical of a number of loads like fans and centrifugal pumps. We see that at standstill, the starting condition, the motor produces more torque than the load demands. The excess torque then goes to accelerate the motor and load according to Newton's equation

$$T_a = J \, d\omega_m / dt \qquad\qquad (6\text{-}21)$$

where T_a is the accelerating torque ($T_{\text{motor}} - T_{\text{load}}$), J is the polar moment of inertia of the motor and load, and $d\omega_m/dt$ is the angular acceleration. Since T_a is a nonlinear function of ω_m, it is not possible to integrate Eq. (6-21) directly by usual methods, but we can integrate on the computer and get a numerical solution for ω_m as a function of time. In any case, we see that the motor picks up speed, and finally, when we have arrived at the intersection of the curves, the accelerating torque becomes zero and we operate at that speed.

Without going to more elaborate methods, we might obtain an approximate solution to the starting problem above by representing the motor and load torque-speed curves by straight line segments as in Figure 6-16. We call this a *piecewise-linear solution.* Values are marked on the sketch in per-unit on some convenient

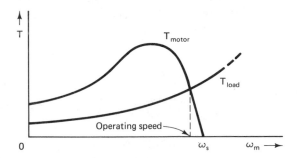

Figure 6-15   Motor and load characteristics.

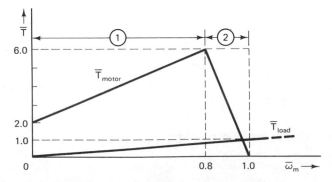

Figure 6-16   Linearized speed-torque approximation.

base. By representing the torques as shown, we may solve the resulting linear differential equation in region 1 and use that result to find the initial conditions to solve a second linear differential equation in region 2. Suppose we know that the polar moment of inertia of the motor and load is given by $\bar{J} = 8$ per-unit and we wish to find the time required for the motor to accelerate to 0.95 times the final velocity. We proceed as follows:

In region 1

$$\bar{T}_M = 2 + 5\bar{\omega}_m; \qquad \bar{T}_L = 1 \times \bar{\omega}_m$$

$$\bar{T}_a = 2 + 5\bar{\omega}_m - \bar{\omega}_m = 2 + 4\bar{\omega}_m = 8\frac{d\bar{\omega}_m}{dt}$$

from which we find

$$\bar{\omega}_m = 0.5\varepsilon^{0.5t} - 0.5$$

and at

$$\bar{\omega}_m = 0.8$$
$$t = 1.91 \text{ seconds}$$

In region 2

$$\bar{T}_M = 30 - 30\bar{\omega}_m; \qquad \bar{T}_L = 1 \times \bar{\omega}_m$$

$$\bar{T}_a = 30 - 30\bar{\omega}_m - \bar{\omega}_m = 30 - 31\bar{\omega}_m = 8\frac{d\bar{\omega}_m}{dt}$$

If we solve this differential equation with $\bar{\omega}_m = 0.8$ at $t = 0$, we have

$$\bar{\omega}_m = \frac{30}{31} - 0.1677\varepsilon^{-31t/8}$$

from which we find at $\bar{\omega}_m = 0.95(30/31)$

$$t = 0.32 \text{ seconds}$$

So total time to reach 95 percent of the final velocity is

$$t = 1.91 + 0.32 = 2.23 \text{ seconds}$$

Many motors operate very satisfactorily if they are chosen so that rated horsepower matches the demand of the load at operating speed, but some types of loads may give trouble in getting the motor started, even though the motor may be perfectly capable of handling the load demand when once up to speed. An example might be a conveyor belt or ski lift, which tends to have a fairly constant torque requirement with speed, or at any rate one that rises more slowly than the fan or pump. Worse yet, the load may have a static friction or "breakaway" torque requirement. Such a load is sketched in Figure 6-17, together with two motor characteristics corresponding to two different values of rotor resistance R_2. It will be noted that, with the motor characteristic marked R_2, the motor is unable to develop sufficient torque to start the load moving. By going to a higher rotor resistance, as with the curve marked R_2', we can get the load started, because the motor develops an excess of torque at zero speed. Unfortunately, though, we note that the ultimate speed of operation is less for the high rotor resistance machine; that is, the slip is high. Remember that the rotor copper loss is equal to the slip in per-unit times the power

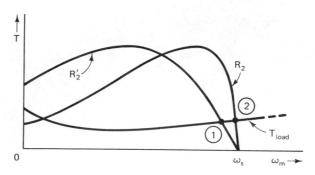

Figure 6-17 Solution of a starting problem.

across the air gap, so, even though we might not object to the somewhat lower speed, we may not wish to tolerate the high losses associated with this type of operation. An application such as this often calls for the use of a wound rotor type of motor, since we can add external rotor resistance for starting, and then when we are up to speed we can cut out the resistance and run at high efficiency. The wound rotor motor is more expensive and requires more elaborate controls, but it may be the answer to a problem load.*

In some cases the wound rotor motor is used where it is desired to vary the speed, as between points 1 and 2 in Figure 6-17. We do so at the price of wasted energy, however, when operating in the high slip condition.

Another problem in motor starting concerns the effect of the motor on the system to which it is connected. Examination of the equivalent circuit of Figure 6-6 shows that when starting, the slip is unity and the motor presents a relatively low impedance to the source connected at E_1, and therefore the current is high compared to what it will be later when R_2/s is higher. Nevertheless, many medium-size, and sometimes even fairly large, motors are started by simply connecting them directly "across the line." If the voltage regulation of the source is poor (a high source impedance), the voltage may dip severely on motor starting, and other apparatus on the same circuit may suffer from the low voltage. If necessary, the starting current may be limited by inserting extra impedance in series with the motor during the starting period, or by reducing the applied voltage for starting with a transformer or an autotransformer. References to Figure 6-13 reveal that a lowered voltage will also result in greatly reduced starting torque (the starting torque is reduced more than the starting current — why?), so the engineer must give careful attention to all these factors in the design of an electric motor drive for a load.

Before leaving this topic, we will once again illustrate these matters with a numerical example. Consider again the motor of Section 6.6 and let us ask what the starting torque and the starting current are with rated voltage applied and the motor data as given. Then, if perhaps these quantities are not satisfactory, how would they

---

*It is also possible to design the squirrel cage rotor in such a way as to exploit the variation of resistance with frequency in ac circuits, an effect analogous to skin effect. The rotor frequency is high on starting and we have a higher R_2 than we do later, when the rotor frequency becomes very low. One way of accenting this effect is to make the rotor bars thin and deep.

be changed by inserting an external 2 ohms per-phase in the rotor circuit? This presumes a wound rotor type of construction for this alteration.

Starting conditions are $\omega_m = 0$, and therefore the slip $s = 1.0$. With $s = 1.0$ we find:

$$I_2 = \frac{(2400/\sqrt{3})\underline{/0}}{3.088 + j8.445} = 154.10\underline{/-69.91}$$

$$\text{Torque} = \frac{I_2^2 R_2}{\omega_s} = \frac{154.1^2 \times 1.544}{125.6}$$

$$= 291.92 \text{ newton-meters per-phase}$$

$$I_1 = I_2 + I_{ex} = 160\underline{/-70.3}$$

With addition of 2 ohms to R_2:

$$I_2 = \frac{(2400/\sqrt{3})\underline{/0}}{5.088 + j8.445} = 140.54\underline{/-58.93}$$

$$\text{Torque} = \frac{I_2^2 R_2}{\omega_s} = \frac{140.5^2 \times 3.544}{125.6}$$

$$= 557.33 \text{ newton-meters per-phase}$$

$$I_1 = I_2 + I_{ex} = 146\underline{/-59.8}$$

By comparing the results with and without the external resistance, we see that the starting torque has been greatly increased, but the starting current has not been greatly reduced. If this level of starting current is more than the supply system can stand, we might consider going to a reduced voltage starter, but further consideration of this will be reserved for the problem section.

6.9 THE SINGLE-PHASE INDUCTION MOTOR

The discussion thus far in this chapter has dealt entirely with induction motors fed from a polyphase source (usually three-phase), and indeed most large motors are polyphase types, but small shops and homes do not have polyphase power service and it is necessary to operate the motors from single-phase power.

The construction of a single-phase motor is shown in Figure 6-18, which shows a single stator coil that is connected to the single-phase source. The rotor is ordinarily built as the squirrel cage type. Suppose that the current in the stator coil varies sinusoidally as

$$i(t) = \sqrt{2}I \sin \omega t$$

The magnitude of the B vector at the top of the air gap where $\theta = \pi/2$ will vary sinusoidally with time as

$$B^{\max}(t) = \frac{\mu_o N i(t)}{2l_g}$$

where it is assumed that all the ampere turns of the stator winding are consumed in the air gap of length l_g. If the turns of the stator winding are distributed, the magnitude of B will vary around the periphery as θ varies as in

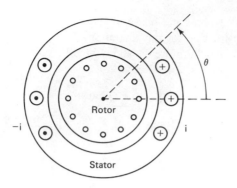

Figure 6-18 Construction of a single-phase induction motor.

$$B(\theta, t) = \frac{\mu_o N}{2l_g} \sqrt{2}I \sin \theta \sin \omega t$$

or

$$B(\theta, t) = KI \sin \theta \sin \omega t \tag{6-22}$$

where the constants are gathered together in the single symbol K in Eq. (6-22) for convenience. If we plot B versus θ for various instants of time, t, we have a picture just like Figure 5-6 (page 182), which illustrates the flux density distribution with θ and time for one coil of a three-phase machine. This type of variation with θ and t was recognized as that of a *standing wave,* which students may have met in courses in physics or in transmission lines.

The field set up by a single coil is seen to be a pulsating field and not a revolving field, and it would at first seem that the single-phase induction motor would not run. This would indeed be the case if it were not for two things: first the value of $B(\theta, t)$ is also affected by the rotor currents, and second, the rotor is initially set in motion by other forces. We proceed to discuss these things in the paragraphs to follow.

If Eq. (6-22) is changed by the trigonometric identity

$$\sin x \sin y = \frac{1}{2}[\cos(x - y) - \cos(x + y)]$$

we may obtain the form

$$B(\theta, t) = \frac{KI}{2} \cos(\theta - \omega t) + \frac{KI}{2} \cos(\theta + \omega t + \pi) \tag{6-23}$$

Equation (6-23) will be recognized as the sum of two *traveling waves* traveling in opposite directions. The interference pattern of two equal traveling waves going in opposite directions is a standing wave. The mechanism of the production of the standing wave is illustrated in Figure 6-19, where the two components, labeled A and B are shown in the left-hand column at various instants of time, and their sum is shown in the right-hand column. The time variation shown extends over one-quarter cycle but the pattern persists cyclically after this time. If we take the point of view of Eq. (6-23) we seem to have the contradiction that one traveling wave is urging the rotor in one direction and the other is working in the opposite direction.

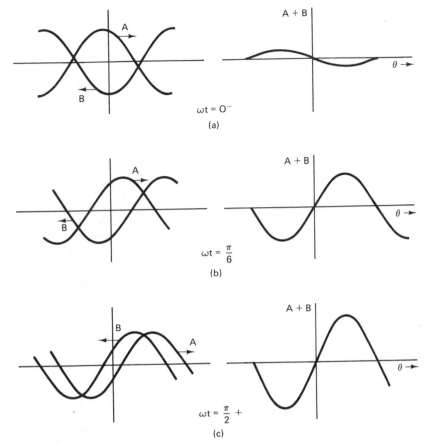

Figure 6-19   Illustration of combination of forward and backward traveling wave to form a standing wave.

One way of explaining that there can still be a net torque is known as the *double revolving field theory*.

If we consider the effect of one of the components of Eq. (6-23) acting alone we have the usual speed-torque characteristic of a motor with a revolving field but we must consider the region of negative ω_m —a slip greater than unity. From Eq. (6-18) we see that the torque curve continues to decay with slip as slip becomes greater and greater, and therefore the torque-speed curve assumes the form of curve *a* in Figure 6-20. The revolving field component in the opposite direction gives rise to a speed-torque relation that is the reverse of *a* and this is shown as curve *b* in the figure. The result of the two fields acting concurrently is the curve *c* of the figure. From curve *c* we see that the net torque is indeed zero at zero rotor velocity, but if somehow the rotor is set in motion there is a net positive (or negative) torque to keep the rotor moving as long as the developed torque is sufficient to overcome the resisting torques of the load and/or friction torques.

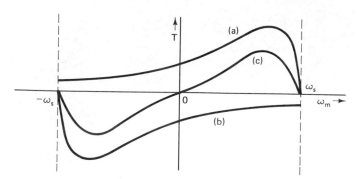

Figure 6-20 Illustration of the double revolving field concept.

We may gain further insight into the operation of the motor by considering the role of the rotor currents in the motor performance. The rotor bars form short-circuited paths. If the rotor is in motion, one of the traveling waves travels in opposition to the direction of rotor travel, whereas the other component travels along with the rotor. In the vicinity of synchronous speed ω_s (small slip) the forward-traveling wave is almost in step with the rotor, and the ampere turns of the stator winding are able to establish a substantial flux linkage with the rotor through the closed paths of the rotor circuit. On the other hand, the backward-traveling wave passes the rotor conductors at almost $2\omega_s$, and the induced voltages produce currents that fight the tendency for the flux to penetrate the rotor circuits according to Lenz's law. There is very little flux set up by the backward wave as a result, and we really have a net revolving field in the forward direction as a result of the combination of the stator and the rotor currents.* In the light of these comments, we observe that the B wave of Eq. (6-22) really never develops because of the influence of the rotor currents, but it is convenient conceptually to visualize two revolving flux waves.

6.10 STARTING THE SINGLE-PHASE MOTOR

We see from the preceding section that the single-phase motor will develop a running torque if the angular velocity is not zero, but the problem is starting the motor from a standstill. Auxiliary windings are provided for this purpose. There are various forms that the auxiliary starting windings take, and often the motor takes its name from the particular method used. Figure 6-21 illustrates two schemes that are popular for general-purpose fractional horsepower single-phase motors. In addition to the main stator winding, a second winding is provided as though for a two-phase motor,

---

*The predominance of the revolving field component in the forward direction could be seen by a probe in the air gap but the effect may be more easily observed in the case of a three-phase motor with two phases open. If started, the motor goes on running — so-called single-phasing. A voltmeter on the other coils will still detect polyphase voltage owing to the net forward revolving field. This effect forms the basis of phase changers, which amount to a polyphase motor running with single-phase power input, but with polyphase windings used to supply a polyphase load.

oriented with the magnetic axis at 90° with respect to the main winding. This winding is sketched in (a) of the figure. The winding is provided with a current that is out of phase with the main winding current by either (b) using a high-resistance winding as the starting winding or (c) placing a capacitor in series with the starting winding.

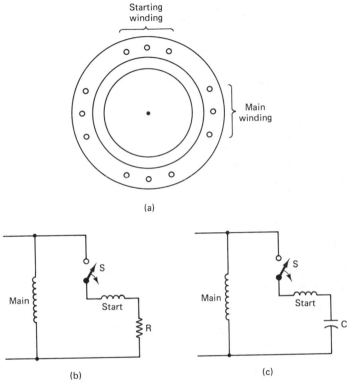

(a)

(b)   (c)

Figure 6-21   Starting windings for the single phase motor.

The combination of main and starting winding thus has a component of a revolving field owing to the phase difference between the currents and the motor starts (if the load torque is not too great). In many cases a centrifugal switch, S, is used to disconnect the starting winding after coming up to speed, since the winding may not be rated thermally to operate continuously. Other starting schemes are also in common use. One form is the *shaded-pole* motor, where a portion of the main pole is encircled by a copper band to delay the buildup of flux through the portion of the pole encircled and thus establish a component of a revolving field to start the motor. Such motors are common in small electric fans. Another scheme involves a commutator type of construction (see Chapter 7) and starts with brushes riding on the commutator which are later lifted by a centrifugal force–acted device. This is the so-called repulsion-start induction motor, and finds use where a large starting torque is desired. The many other varieties of single-phase induction motors must be left to texts more detailed than this introductory treatment.

ADDITIONAL READING MATERIAL

1. Brown, D., and E. P. Hamilton III, *Electromechanical Energy Conversion,* New York: MacMillan Publishing Company, 1984.

2. Chapman, S. J., *Electric Machinery Fundamentals,* New York: McGraw-Hill Book Company, 1985.

3. Chaston, A. N., *Electric Machinery,* Reston, Virginia: Reston Publications Company, Inc., 1986.

4. Del Toro, V., *Electric Machines and Power Systems,* Englewood Cliffs, New Jersey: Prentice-Hall, Inc., 1985.

5. Elgerd, O. I., *Basic Electric Power Engineering,* Reading, Massachusetts: Addision-Wesley Publishing Company, 1977.

6. Fitzgerald, A. E., C. Kingsley, Jr., and S. D. Umans, *Electric Machinery,* New York: McGraw-Hill Book Company, 1983.

7. Lindsay, J. F., and M. H. Rashid, *Electromechanics and Electric Machinery,* Englewood Cliffs, New Jersey: Prentice-Hall, Inc., 1986.

8. Nasar, S. A., *Electric Machines and Electromechanics,* Schaum's Outline Series in Engineering, New York: McGraw Hill Book Company, 1981.

9. Nasar, S. A., *Electric Energy Conversion and Transmission,* New York: MacMillan Publishing Company, 1985.

10. Shultz, R. D., and R. A. Smith, *Introduction to Electric Power Engineering,* New York: Harper & Row, Publishers, 1985.

11. Slemon, G. R., and A. Straughen, *Electric Machines,* Reading, Massachusetts: Addison-Wesley Publishing Company, 1980.

STUDY EXERCISES

1. Consider Figure 6-8(b) and Eq. (6-16) of the text.
   a. For typical values of induction machine parameters, how does starting torque vary with applied voltage? With R_2?
   b. If we consider operation in the normal range of very low slip (R_2/s dominates), how does the slip vary with applied voltage for a *constant torque* load?
   c. What is the expression for mechanical *power* developed in terms of the applied voltage and the machine parameters given in the equation?

2. Sometimes very small induction motors are speed controlled by varying the applied voltage as from a transformer of variable ratio. If it is desired to vary the speed over a wide range without stalling the motor, should the rotor resistance, R_2, be relatively small or large with respect to usual normal values? Explain.

3. Suppose that we want to use a three-phase, 4-pole, 60-Hz, wound rotor induction motor to change the power frequency for a certain load. We apply normal 60-Hz power to the stator and drive the rotor from a mechanical source with a speed controlled anywhere from 200 to 1000 rpm. What range of frequency is available at the slip rings of the rotor?

4. Given: a 100-horsepower, 480-volts, 60-Hz, 6-pole, three-phase induction motor. The motor runs at full load with a slip of 3 percent with rated voltage applied. Under conditions of stress on the power supply system the voltage drops to 420 volts. If the load is of the constant torque type, then with the lower voltage:

a. What is the slip (Use small slip approximation.)?

b. What is the shaft speed in revolutions per minute?

c. What is the horsepower output?

d. What is the rotor copper loss, $I_2^2 R_2$, in terms of the original rotor copper loss with full rated voltage?

5. The properties of a series R-L circuit with constant applied voltage and variable R are sometimes illustrated by a locus diagram showing the locations of the current phasor as the resistance, R, is varied. The figures below are examples.

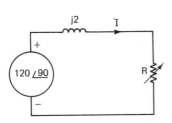
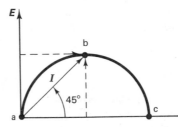

It may be shown by trigonometry, or by cut-and-try, that the tip of the current phasor will always lie on the semicircle shown, as R varies from zero to infinity. One sample phasor is shown. The dotted arrows show the in-phase and quadrature components of the current. We are reminded that, with constant voltage, the in-phase and quadrature components of current are proportional to P and Q, respectively.

a. Three points on the circle are identified as a, b, and c. What are the values of R at each of these points?

The circuit at the left might be regarded as the highly simplified equivalent circuit of an induction motor where $R_1 = 0$, $G_c = B_c = 0$, and $R = R_2/s = R_2 + R_2(1 - s)/s$.

b. If R_2 of the specific example $= 0.5$ ohm, show on a sketch, the region of the circle where I will lie as slip, s, varies from 0 to 1.0.

c. Show on the same sketch the position of I at starting and at maximum torque.

d. If the motor is a three-phase, four-pole, 60-Hz machine, what is the maximum torque and at what speed does it occur?

   Hint: Remember that torque is proportional to power across the air gap!

6. Suppose that the motor of the example of Sections 6.6 and 6.7 is driven by a prime mover so that the slip is -3%; that is, the shaft is driven *faster* than synchronous speed. With the terminal voltage maintained at the rated value by an infinite bus (perfect potential source):

a. What is the speed of the shaft in revolutions per minute?

b. How much power must the driving source *supply* to the shaft?

c. How much complex power, S, is *drawn from* the electrical source at the machine terminals?

d. Viewed as a *generator,* what is the efficiency of this machine?

7. When the motor of the example of Sections 6.6 and 6.7 was started across the line from standstill ($s = 1.0$) the starting current was 160 amperes. Suppose that it is desired to reduce the current taken *from the source* to one-half this value (80 amperes):

a. What should be the turns ratio of a step-down transformer to feed the motor on starting such that only 80 amperes will be drawn at first connection?

b. If the starting transformer is used, what will be the starting torque of the motor?

8. Consider yet again the motor in the example of Sections 6.6 and 6.7. In the example, an increment of two ohms was added to R_2 and the starting torque computed. What should be the increment added to R_2 if we wish to develop the maximum possible starting torque for a given voltage?

SOLUTIONS TO STUDY EXERCISES

1. a. On starting from standstill $s = 1.0$ and Eq. (6-16) becomes:

$$T = \frac{E_{1t}^2 R_2}{\omega_s[(R_{1t} + R_2)^2 + (X_{1t} + X_2)^2]}$$

So we see that starting torque varies as the square of E_{1t} and hence, since E_{1t} varies directly with the applied voltage, the starting torque varies as the square of the applied voltage.

If R_2 is *small,* relative to the X_1 and X_2, we see that the starting torque varies directly with R_2. (See also Figures 6-12 and 6-13)

b. If s is very small, R_2/s is very large and Eq. (6-16) becomes approximately:

$$T = E_{1t}^2 s/\omega_s R_2$$

or

$$s = T\omega_s R_2/E_{1t}^2$$

and therefore slip varies inversely as the square of the voltage.

c. Since $P = T\omega_m$ and $\omega_m = \omega_s(1 - s)$

$$P/\phi = \frac{E_{1t}^2 R_2(1 - s)/s}{(R_{1t} + R_2/s)^2 + (X_{1t} + X_2)^2}$$

2. Considering typical speed-torque curves with varied voltage, as we see in the first figure:

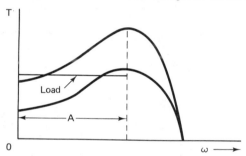

Note that over the speed range shown as A there is danger of instability, as, for example, with a constant torque load. On the other hand with a high value of R_2, the curves may appear as

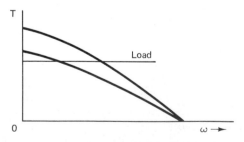

and we see that stable operation with constant torque is possible as the applied voltage varies. Therefore, a *high* value of R_2 is desirable for such a motor.

A relatively *high* value of R_2 does imply high rotor copper losses, however, so this method of speed control would normally be used only when small amounts of energy are involved, as, for example, in the case of instrument drives or servomotors.

3. Note that

$$f_r = sf_s$$ Eq. (6-2)

and

$$s = (\omega_s - \omega_m)/\omega_s$$ Eq. (6-1)

or

$$s = (n_s - n_m)/n_s \quad \text{where } n \text{ is speed in r/min}$$
$$f_r = (n_s - n_m)f_s/n_s$$

where

$$n_s = 120f_s/P$$ from Eq. (5-1)
$$= 120 \times 60/4 = 1800 \text{ r/min}$$

therefore at

$$n_m = 200$$
$$f_r = (1800 - 200)60/1800 = \underline{53.33 \text{ Hz}} \quad \longleftarrow$$

and at

$$n_m = 1000$$
$$f_r = (1800 - 1000)60/1800 = \underline{26/67 \text{ Hz}} \quad \longleftarrow$$

Note: The positive sign on n_m infers rotation in the same direction as the revolving field. If the rotor were driven oppositely, the values of n_m would be -200 and -1000 r/min and the rotor frequencies become 66.67 Hz and 93.33 Hz, respectively.

4. a. $s = T\omega_s R_2/E_{1t}^2$ as in Problem 1(b)
        $= 0.03(480/420)^2 = \underline{0.0392} \quad \longleftarrow$ a
   b. $n_s = 120f_s/P = 120 \times 60/6 = 1200$ r/min
        $n_m = (1 - s)n_s = (1 - 0.0392)1200 = \underline{1153}$ r/min $\longleftarrow$ b
   c.   With constant torque, $P\alpha n_m$, therefore
        $P = 100 \times 1153/1200(1 - 0.03) = \underline{99.1}$ horsepower $\longleftarrow$ c
   d. $P_{rc} = I_2^2 R_2 = s(P_{ag})$ Eq. (6-12)
        But power across the air gap, P_{ag}, is constant if torque is constant.   Eq. (6-13)
        Hence if we consider: Case 1, $E_1 = 480$ and
                              Case 2, $E_1 = 420$

$$\frac{P_{rc2}}{P_{rc1}} = \frac{s_2}{s_1} = \frac{0.0397}{0.03} = 1.31$$

Rotor copper loss is increased by 31 percent and the rotor might overheat!

5. a.   When at a, $I = 0$, so $R \to \infty$
        When at b, $\underline{/I} = 45°$, so $R = X = 2$
        When at c, $\underline{/I} = 0°$, so $R = 0$
   b., c.
        When $R_2 = 0.5$ and slip, $s = 0$, $R_2/s \to \infty$ and $I = 0$ and we are at the origin.
        When $R_2 = 0.5$ and slip, $s = 1.0$, $R_2/s = 0.5$ and

$$I = \frac{120\underline{/90°}}{0.5 + j2.0} = \underline{58.2\underline{/14°}}$$

Therefore, the tip of the current phasor ranges over the heavy portion of the locus, as is shown in the following figure.

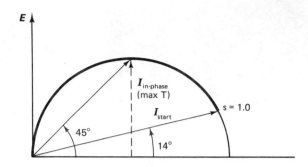

Note that P_{ag} and hence the developed torque, are maximum when the in-phase component of I is maximum; that is, when I is at an angle of 45°.

d. At $I = 45°$, $R_2/s = X = 2$, and $s = 0.5/2 = 0.25$

   $\omega_m = (2\pi 60/2)(1 - 0.25) = 141.4$ rad/sec, or 1350 r/min

   Diameter of circle $= 120/2 = 60$ A, $I_{\text{in-phase}} = 60/2 = 30$ A (max)

   $P_{ag} = 3EI_{\text{in-phase}} = 3 \times 120 \times 30 = 10,800$ W

   $$T = P_{ag}/\omega_s = \frac{10,800}{(2\pi 60/2)} = 57.3 \text{ N-m} \quad \longleftarrow \quad \text{d}$$

6. a. $n_s = 120 \times 60/6 = 1200$, $n_m = [1 - (-0.03)]1200 = \underline{1236 \text{ r/min}} \quad \longleftarrow \quad \text{a}$

b.

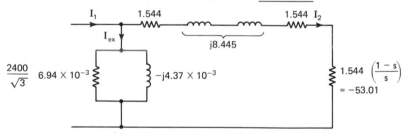

$2400/\sqrt{3} = 1385$, $R_2(1 - s)/s = 1.544(1 + 0.03)/(-0.03)$
$$= -53.01 \text{ ohms}$$

$$I_2 = \frac{1385\underline{/0°}}{2 \times 1.544 - 53.1 + j8.445} = 27.37\underline{/-170.4°}$$

$P_m = I_2^2 R_2(1 - s)/s = 27.37^2(-53.01) = -39,700 \text{ W}/\phi$

$P_{m(\text{tot})} = 3(-39.7) = -119.1 \text{ kW}$

We note that the circuit references are in terms of motor action (P_m *out* of shaft) and hence the negative sign infers that positive $P_m = 119.1$ kW must be *supplied* to the shaft by the prime mover and the machine is acting as a *generator*.

c. $I_{ex} = 1385\underline{/0°}(6.94 \times 10^{-3} - j4.27 \times 10^{-3}) = 11.2\underline{/-31.6°}$

   $I_1 = I_{ex} + I_2 = 11.2\underline{/-31.6°} + 27.37\underline{/-170.4°}$
   $$= 20.28\underline{/-148.8°}$$

   $S/\phi = 1385\underline{/0°} \times 20.28\underline{/+148.8} = (-24.06 + j14.52) \times 10^3$ VA

   $S_{\text{tot}} = 3S/\phi = \underline{-72.2 + j43.6 \text{ kVa}} \quad \longleftarrow \quad \text{c}$

d. $\eta = P_{\text{out}}/P_{\text{in}} = 72.2/119.1 = \underline{0.61} \quad \longleftarrow \quad \text{d}$

7. a.

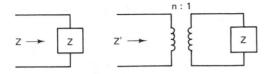

The motor appears as a certain fixed impedance, Z, at slip $s = 1.0$. We want the apparent impedance, Z' to be twice this value, $Z' = 2Z$, so
$$n^2 Z = 2Z$$
$$n = \sqrt{2} \longleftarrow a$$

b. $E_{motor} = E_{source}/\sqrt{2}$
   but $T \alpha E^2$, so $T = (1/\sqrt{2})^2 T_{full\ voltage} = (1/2)T_{full\ voltage}$
   $$= 291.2/2 = \underline{145.9}\ \text{N-m/}$$

8. Maximum torque occurs when $R_2/s = \sqrt{R_1^2 + (X_1 + X_2)^2}$
   When starting $s = 1.0$ so $\quad R_2 = \sqrt{1.544^2 + 9.445^2}$
   $$= 8.58\ \text{ohms}$$
   therefore $\Delta R_2 = 8.58 - 1.544 = \underline{7.04}\ \text{ohms} \longleftarrow$

HOMEWORK PROBLEMS

1. A 100-horsepower, 480-volt, three-phase, 60-Hz wound rotor induction motor is run without load on the shaft and found to run at 897 revolutions per minute.
   a. What is the probable number of poles in the motor?
   b. If we observe the rotor currents, what would be their frequency at no load?
   c. If, at full load, the motor slows to 860 r/min, what is the frequency of the rotor currents?
   d. What would be the approximate frequency of the rotor currents at one-half rated load?

2. Suppose that we arrange an electronic frequency changer to provide three-phase power to the slip rings of a 4-pole, 60-Hz, wound rotor induction motor as in the figure below. If regular 60-Hz power is provided to the stator, and rotor power ranges over the frequencies from 20 to 40 Hz, at what steady-state speeds could the machine run?

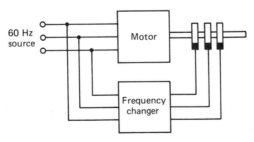

3. Given: a 300-horsepower, 6-pole, 60-Hz, 600-volt, three-phase squirrel cage induction motor with the following parameters of the circuit of Figure 6-6 as obtained from test. All values are per-phase quantities.

$$G_c = 0.015\ S, \quad B_c = -0.068\ S, \quad R_1 = 0.05\ \text{ohm}, \quad R_2 = 0.06\ \text{ohm}$$
$$X_1 = X_2 = 0.12\ \text{ohm}$$

a. Find the values of the parameters in the Thevenin's theorem type of equivalent circuit of Figure 6-8(b).
b. What is the total full voltage starting torque of the motor in n-m?
c. If the motor is loaded to a torque of 1780 n-m, what will be the slip in per-unit?
d. Suppose that the voltage applied to the motor is only 0.70 times normal and the same constant torque load of 1780 n-m is applied to the shaft:
     i. Will the motor stall; that is, will maximum torque capability be exceeded?
     ii. *If* the motor runs without stalling, what is the percent increase in the rotor copper loss, $I_2^2 R_2$?

4. Given: a 10-horsepower, 240-volt, 60-Hz, 4-pole, three-phase induction motor of the squirrel cage type. The motor runs at a slip of 3.5 percent at full load and has a starting torque with full voltage applied of 150 percent of full load torque.

What is the maximum torque possible and at what slip does it develop?

Note: To make the problem tractable, we will assume that the parameters G_c, B_c, and R_1 of the equivalent circuit are negligible.

5. Given: a 150-horsepower, 6-pole, 60-Hz, 2400-volt, three-phase induction motor. The motor is tested in the laboratory using the two-wattmeter method of measuring total input power in each of the tests. The wattmeter current coils are placed in phases b and c, and the common point of the potential coils is phase a. The following data are obtained:

Running light test

Impressed voltage:	2400 volts
Line current:	8.5 amperes
Wattmeter b	13.4 kW
Wattmeter c	-7.2 kW

Locked rotor test

Impressed voltage:	580 volts
Line current:	50 amperes
Wattmeter b	23.0 kW
Wattmeter c	-2.4 kW

Measured $R_1 = 1.52$ ohms per phase.
a. What was the phase sequence of the applied voltage?
b. Find the parameters of the equivalent circuit of Figure 6-8(a).
c. What is the maximum torque developed by this motor with full rated voltage applied?
d. What are the torque and power output with slip $s = 0.032$?
e. What is the full load efficiency of this motor?

6. The method used in Section 6.6 of arriving at the equivalent circuit of Figure 6-12 is expedient, but it results in an approximation that might be unacceptable for certain purposes. The fault lies in the assignment of the mechanical friction losses to the shunt element, G_c, rather than subtracting them from the mechanical developed power. An alternate test procedure that might be used is as follows.

Let us *drive* the motor from a 6-pole, 60-Hz, *synchronous* motor while we measure the stator input from a rated voltage source. In this way the mechanical friction losses are supplied by the synchronous motor while the stator input goes to the stator copper loss and the stator core loss. The new data for stator input running light is:

$$E = 2400 \text{ V}$$
$$I = 5.95 \text{ A}$$
$$P = 2.5 \text{ kW}$$

The locked rotor data is unchanged from that given in the example.

a. What is the mechanical friction loss of this machine at rated speed? *Note:* We assume this to be a constant at all speeds near synchronous.

b. What are the new values for the elements of the equivalent circuit of Figure 6-12?

c. Now assuming a slip of 0.03 as in the example, what are the mechanical power and torque output using the new values in the equivalent circuit?

7. An industrial plant has a 150-horsepower, 600-volt, 6-pole, 60-Hz, three-phase squirrel cage induction motor which draws 140 A at 0.9 power factor at full load. When started with full voltage it draws five times rated current at 0.4 power factor.

The motor and some incidental lighting are supplied by a three-phase 200-kVA, 4160-600–volt transformer with a series impedance of $0.025 + j0.12$ ohm per phase referred to the 600-volt side. The transformer is fed from a 4160-volt utility bus with negligible source impedance — an "infinite bus."

a. When the motor is started "across-the-line" (without any limiter) by what percentage does the voltage dip? By "dip" we mean a momentary voltage change at first connection.

b. If the lighting load cannot tolerate a dip of more than 5 percent, what must be the ratio of a starting transformer used to feed the motor when starting?

8. A certain motor drive involves two 10-horsepower, 480-volt, 4-pole, 60-Hz induction motors connected to a common shaft as shown below. One motor is a wound rotor type with a slip of 0.05 at full load and the other motor is a squirrel cage type with a full load slip of 0.04.

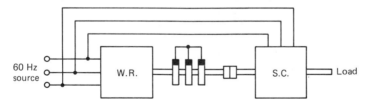

a. When supplying a total load of 20 horsepower by the common shaft, how much load does each motor carry?

On some occasions we see two such motors connected, as shown below where power from the slip rings of the wound rotor motor is used to supply the stator of the squirrel cage motor. The connection is known as *concatenation*.

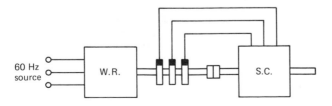

b. What will be the approximate no-load speed of the shaft with losses neglected?

9. A three-phase, 2400-V, 60-Hz, 6-pole induction motor delivers 250 horsepower to a mechanical load. The mechanical losses are 2 percent of the delivered power and the slip is 0.05.

   If tests show that $R_1 = R_2$, what is the efficiency of the motor at the specified load?

10.

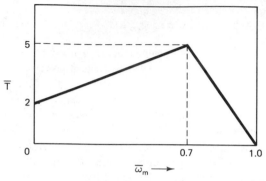

For ease in computation, the speed-torque curve of a certain induction motor is idealized, as shown above where speed and torque are plotted in per unit on some suitable bases such that 1.0 per unit speed corresponds to ω_s, the synchronous speed of the machine. The load is a constant torque of 1.0 per unit. The curve is shown for full rated (1 per unit) voltage.

a. What is the slip, s, under the conditions given?

b. What is the per-unit mechanical power developed?

c. If it is desired to reduce voltage to limit the starting current, how low can the voltage be and still start the motor?

Chapter 7

The

Commutator

Machine

- Another type of rotating machine.
- The basic equations describing performance.
- Illustrations of application of the machine to various loads.

7.1 INTRODUCTION

In addition to the family of synchronous machines and induction machines (which are ac machines primarily) there is another family of machines that use a switching mechanism called a *commutator* to realize characteristics different from the first two classes. While the commutator machine is perhaps primarily regarded as a type used with dc systems, there are ac versions of this type as well. The following discussion will concentrate mainly on the dc power versions of this versatile machine.

7.2 MACHINE CONSTRUCTION

In most versions of the commutator type of machine the stationary member, or stator, is comprised of a cylindrical steel ring supporting pole pieces that supply a magnetic field that is relatively constant over short intervals of time at least. The field may be the result of permanent magnets, or it may be the result of dc currents flowing in windings around the pole pieces. An end view of the stator of such a machine is shown in Figure 7-1(a).

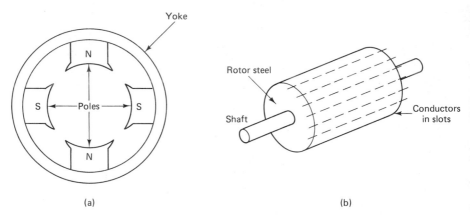

(a)   (b)

Figure 7-1   (a)   Stator construction of a dc machine and (b) rotor construction of the machine.

The rotating portion of the machine is a steel cylinder supported by the shaft. The rotor cylinder is made of stacked laminations — the purpose of lamination being to reduce the eddy currents induced in the steel as the magnetic field reverses during rotation. The surface of the rotor (like other smooth rotor types previously discussed) is provided with axial slots in which conductors are embedded (Figure 7-1(b)). The conductors, in turn, are interconnected in coils to form the armature *winding* and this winding is connected to the switching mechanism known as the *commutator*. The commutator is comprised of axially oriented strips of copper, insulated from each other, and forming a cylinder concentric with the shaft. Electrical contact with the rotor windings is made through graphite brushes that ride on the commutator surface to control the flow of current in the armature conductors, as will be described in the following paragraphs.

The individual conductors are connected first as *turns,* where one side of a given turn is under a pole of one polarity and the other side is under an opposite pole. There may be just one turn in one pair of slots connecting to the commutator, or more often, several turns are placed in the same slot thus forming a *coil.* A developed view showing how the conductors are interconnected to each other and to the commutator is shown in Figure 7-2. Only three coils are shown, but the armature surface is actually uniformly covered by similar coils. For simplicity, single-turn coils are assumed. The "developed view" is a picture of the surface of the rotor and commutator rolled out to a flat surface. The two sides of the picture connect to each other at the ends.

The actual details of armature windings are quite involved and of interest primarily to designers of these machines. From the standpoint of understanding the general relationships of the armature coils and their interconnection, the simplified drawing of Figure 7-3 may help. The coils form series paths between the segments, and hence form a path for current between the brushes. In the diagram of Figure 7-3, the coils are shown taking current, I, from the left-hand brush and passing $I/2$ through each half of the winding group to the right-hand brush, where the current is removed from the armature circuit. It is to be understood that the diagram uses circuit diagram symbols to show the coils, and the actual physical layout of the coils may be similar to Figure 7-2 or quite different, depending upon the type of armature winding devised by the designer. One feature worth comment even in this symbolic diagram is that armature windings are characterized by the number of parallel paths into which the total armature current is divided. In the case of Figure 7-3, we see that there are *two* such paths, but for other numbers of poles and brushes and other types of windings there may be many such parallel paths. In the development to follow, the number of parallel paths will be symbolized by the letter a.

In the operation of this device, two electromagnetic actions are of significance: (a) If current passes through the conductors a force is experienced on the active portion of the conductors lying under the magnetic field poles and (b) if there is

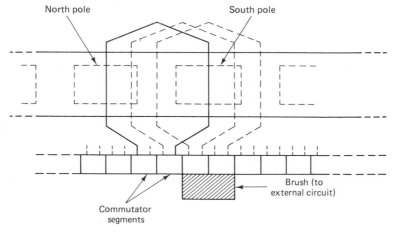

Figure 7-2 Developed view of armature winding connections.

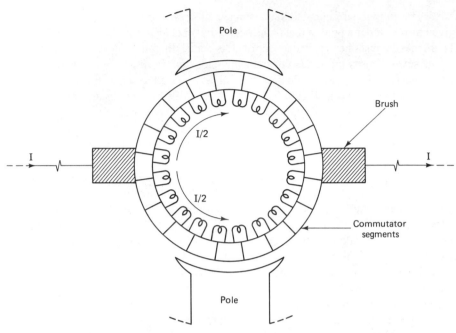

Figure 7-3 Symbolic representation (circuit diagram type) of the relationship between armature coils, brushes and commutator segments.

motion a voltage will be generated in the coil. For computation of these effects Eqs. (4-16) and (4-17) are perhaps the most useful forms of the electromechanical relations and these are reproduced as Eqs. (7-1) and (7-2), respectively, for ready reference here. A review of Section 4.9 of the text would be in order here.

$$f = B\ell i \qquad (7-1)$$

$$e = B\ell v \qquad (7-2)$$

To illustrate the application of Eqs. (7-1) and (7-2) we show an "end-on" view of a machine in Figure 7-4 where a machine with four poles has been chosen for illustration. The flow lines of the B vector are shown as dotted lines, oriented as given for the particular polarity of each pole. The armature conductors, which lie in the slots, are shown in cross section, with assumed direction of currents indicated by the dots and crosses. These currents connect to the world outside the armature through the commutator segments and the brushes riding upon them. It will be noted that the direction of the currents is opposite under poles of opposite polarity, and this is the function of the commutator and brushes — to switch the connection of the coils as the rotor changes position under the poles. A reference to the symbolic diagram of Figure 7-3 will make it easy to understand how the direction of the current in any given coil is reversed when the commutator segment to which that coil is connected passes under the brush.

If we consider the current in any individual armature conductor, we see that a force will be developed in the tangential direction according to Eq. (7-1). The

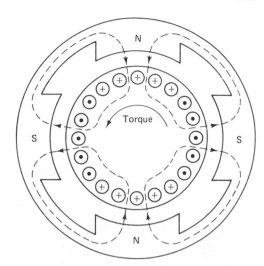

Figure 7-4   Illustration of machine operation.

direction of the force may be determined by any of the memory aids of first physics courses or perhaps by application of the vector relation

$$\vec{f} = i\vec{l} \times \vec{B}$$

where $\vec{l}$ is the length of a straight conductor and $\vec{B}$ is the magnetic flux density. By whatever means, we see that *for the assumed polarity relations* a torque will be developed on the rotor *tending* to move the rotor in a counter-clockwise direction. Whether motion actually ensues depends upon the mechanical constraints on the rotor.

If indeed there *is* motion, either owing to the developed force or owing to external torque applied to the shaft of the rotor, then according to Eq. (7-2) a voltage will be induced. The polarity of the voltage may be seen by considering energy conversion matters. If, for example, the rotor is allowed to rotate in the direction shown above, then the device is a *motor* and the voltage induced will be of such a polarity as to oppose the current flow direction; that is, a *drop*. If, on the other hand, the rotor is forced in a direction opposed to the electromagnetic forces, the machine is a generator and the voltage polarity will be such as to push the current in the direction shown—a *rise*. These conclusions are based upon the necessity for electric power to be *absorbed* when developing mechanical power, and *generated* when absorbing mechanical power. The same polarity conclusions may be reached by any of the elementary physics memory aids (the right-hand rule, for example) or from the electromagnetic field relation showing the force on a charge moving through a magnetic field

$$\vec{f} = q\vec{u} \times \vec{B}$$

where q is the charge, u is the velocity of the charge, and B is the magnetic flux density.

If we again assume that the exact details of the armature winding are of interest only to design specialists, a simplified representation of the armature portion of a

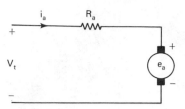

Figure 7-5   Electric circuit equivalent of a dc machine armature circuit.

machine for use in circuit analysis may be used, as, for example, that of Figure 7-5. In Figure 7-5 the circle with dark squares touching it represents a commutator with brushes connecting. Actually, in multipole machines there is a multitude of brushes, but all positive brushes are connected together electrically as are all of the negative brushes, thus only two symbolic brushes suffice for the electrical equivalent of the figure. The armature conductors have resistance, which is actually distributed throughout the length of the conductors, but as usual in circuit diagrams, this resistance is symbolized by an external lumped value, R_a, in the figure.* Likewise, the voltage generated owing to motion of the conductors in the field is shown as a net equivalent voltage, e_a, in the diagram.

The voltage V_t in the diagram is a source voltage and may be a constant in some applications, or it may be variable to control the action of the device.

It will be noted that the particular references chosen for the variables are those of *motor* action — electrical power *into* the armature and hence mechanical power *out* of the shaft. Other choices may be made for other applications.

7.3 THE TORQUE—FLUX—CURRENT RELATION

For a machine like that of Figure 7-4, with a number of conductors uniformly distrubuted around the surface of the rotor, we may easily derive an expression for the torque developed by the machine with a given armature current. The flux density vector, B, under each of the P poles is typically distributed as in Figure 7-6 where the magnitude of B is plotted against the angle, θ, around the periphery of the rotor. It may be assumed that the direction of B is normal to the surface of the steel on each side of the air gap, since the permeability of the steel is far greater than that of the air. The angular distance subtended by one pole is $2\pi/P$, and the B vector varies from zero at the neutral zone midway between the poles to some relatively constant value directly under the poles on a contour sometimes described as "hat shaped."

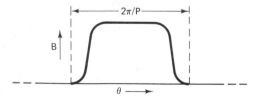

Figure 7-6   Distribution of B under a pole.

*More detailed representations sometimes include the voltage drop between the graphite brushes and the commutator surface as a separate entity. We will assume that this drop is approximated by appropriate values assigned to R_a.

Consider a sample conductor under a pole. If the total armature current is i_a and the winding pattern is such that there are a parallel paths carrying this current, then the current in a given conductor is i_a/a. The force on the sample conductor is then a function of the angle θ as

$$f(\theta) = B(\theta)li_a/a$$

where l is the active length of the conductor under the pole face. The average force on a conductor may then be evaluated by

$$f_{ave} = \frac{P}{2\pi} \int_0^{2\pi/P} f(\theta)\, d\theta$$

$$= \frac{Pi_a}{2\pi ra} \int_0^{2\pi/P} B(\theta)lr\, d\theta$$

but this latter integral is simply the flux, ϕ, per pole, hence

$$f_{ave} = \frac{Pi_a\phi}{2\pi ra}$$

It is interesting to note that the average force is a function of the total flux, ϕ, per pole and does not depend on the actual distribution of B with angle, θ. The force on each conductor is tangential to the surface of the rotor because of the normal direction of the B vector. If there is a total of Z conductors, all lying at the radius r, the total torque developed is given by*

$$T = Zrf_{ave}$$

$$= \frac{ZPi_a\phi}{2\pi a}$$

or

$$T = K_a\phi i_a \tag{7-3}$$

where

$$K_a = \frac{ZP}{2\pi a}$$

It will be noted that K_a is fixed for a given machine.

The positive direction of the torque may be obtained by application of the right-hand rule. In the case of Figure 7-4, we see that the torque is positive in a counterclockwise direction.

7.4 THE VOLTAGE—FLUX—SPEED RELATION

If a machine such as that symbolized in Figure 7-4 rotates, either because of the torque developed as in the preceding section, or because the rotor is driven, a voltage

---

*The validity of this approach depends on a smooth, continuous distribution of current-carrying conductors (as would be the case with a *current sheet*). The conductors are distributed discretely in a finite number of slots and the total torque has a slight pulsating component, owing to this distribution and the effects of the commutator as it switches the current. This slight pulsation is of no consequence in operation because of the flywheel effect on the rotor, and will be ignored.

will be developed between the brushes. Application of right-hand rules will reveal that, for counterclockwise rotation, the voltage, e_a, will be positive in opposition to the current flow. This voltage is variously called the *generated voltage* or the *counter emf*. The circuit diagram representation of Figure 7-5 has the references chosen consistent with this case.

If we write a Kirchhoff's voltage law equation for the circuit of Figure 7-5, we obtain

$$V_t = i_a R_a + e_a \quad \text{(volts)}$$

If, in turn, we multiply both sides by i_a, the voltage equation becomes a power equation

$$V_t i_a = i_a^2 R_a + e_a i_a \quad \text{(watts)}$$

The last equation says that the source supplies power in the amount $V_t i_a$ and some of this power goes into $i_a^2 R_a$, the armature copper loss. The remainder of the power, $e_a i_a$, is the power converted to mechanical form. On the other hand, the mechanical power developed must be equal to the torque, T, times the angular velocity, ω_m, (newton-meters times radians per second) therefore, we have the equation

$$e_a i_a = T \omega_m$$

But, from Eq. (7-3) we have

$$T = K_a \phi i_a$$

hence

$$e_a i_a = K_a \phi \omega_m i_a$$

and

$$e_a = K_a \phi \omega_m \tag{7-4}$$

where K_a is the same constant of the machine as in Eq. (7-3), ϕ is the flux per pole, and ω_m is the mechanical rotational velocity in radians per second.*

7.5 APPLICATIONS—THE DC MOTOR

The commutator type machine, while expensive to construct and to maintain, as contrasted with the induction machine, for example, still has properties that make it the preferred choice for certain applications. In the most common form, the dc motor, the controllability of the motor under varying speed and torque requirements is unsurpassed. It is also a logical choice, of course, where the primary power source is dc in nature, as in automobiles or small boats or airplanes.

The equations developed thus far, together with Kirchhoff's laws, serve quite well to analyze the performance of the dc motor. Consider the application shown in Figure 7-7, where a commutator machine is supplied from a source voltage, V_t, and the shaft is connected to a mechanical load that demands a certain torque at a certain

---

*Eq. (7-4) could have been derived starting with Eq. (7-2) by a means analogous to that used in deriving Eq. (7-3). By using this method, and observing that the constant, K, is the same in both Eqs. (7-3) and (7-4) we may verify that the product $e_a i_a$, is indeed the electrical power converted to mechanical form. This alternate derivation is recommended to the student as a study exercise!

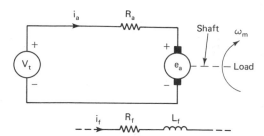

Figure 7-7   An illustration of motor usage.

speed. The magnetic flux, ϕ, of the motor may be supplied either by a permanent magnet (common in smaller motors) or by an electromagnet. The electromagnet is shown as R_f and L_f in the circuit diagram and is supplied with a current i_f. A desired flux, ϕ, may be created by means of a small current in a large number of turns on a field winding (often placed in parallel or *shunt* with the main supply voltage) or by a large current in a small number of turns (by passing the armature current, i_a, through the field winding in *series*).* The influence of the flux on motor performance as derived in various ways will be discussed below.

Equations (7-3) and (7-4) and Kirchhoff's voltage law are repeated

$$T = K_a \phi i_a$$

$$e_a = K_a \phi \omega_m$$

$$V_t = i_a R_a + e_a = i_a R_a + K_a \phi \omega_m$$

By algebraic manipulation the following relations may be found.

$$\omega_m = \frac{V_t - i_a R_a}{K_a \phi} \tag{7-5}$$

$$i_a = \frac{V_t - K_a \phi \omega_m}{R_a} \tag{7-6}$$

$$T = \frac{K_a \phi V_t - K_a^2 \phi^2 \omega_m}{R_a} \tag{7-7}$$

The performance of the motor follows the above equations, but what happens depends on what is held fixed and what varies. As an example in Eq. (7-5), suppose that the terminal voltage is held fixed and also the field flux, ϕ. If a mechanical load is applied to the motor shaft, mechanical power is delivered to the load and the armature current must increase to provide this power from the electric source. As a result, the speed will drop with increasing load. This effect is shown in Figure 7-8. The slope of the curve is usually small, since R_a is made small to conserve energy, thus the motor approximates a constant speed device in this mode of operation. The field flux for this mode of operation of the motor is often supplied by a winding connected in shunt with the source voltage, V_t, and by a current controlled by the field resistance, R_f. If the field flux is reduced by — for example — increasing the field resistance, R_f, the motor will speed up under no-load conditions where i_a

---

*The matter of the creation of the field will be discussed further in the pages to follow. For the moment, we assume that we know the flux per pole ϕ.

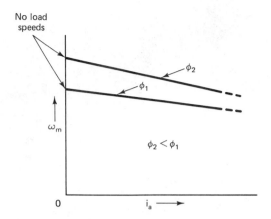

Figure 7-8 Speed variation with armature current with fixed source voltage, V_t, and for two different values of flux per pole.

approaches zero.* The speed then falls off with increasing i_a with a steeper slope as load picks up, and this is shown by a second curve marked ϕ_2 on Figure 7-7.

Equation (7-6) shows us that the motor will draw a very high current on starting when the speed, ω_m, is zero. Except for very small motors it is necessary to reduce the applied voltage and current under starting conditions to avoid damage. A simple way to reduce the voltage is to add resistance to the armature circuit. When starting, ω_m is zero and Eq. (7-6) becomes $i_a = V_t/(R_a + R_{ext})$. If the voltage V_t is obtained by rectifying an ac source, it may be possible to reduce the average value of V_t by phase angle control of thyristors.

Equation (7-7) shows the relation between torque and speed; this is a very important relation in considering motor application (see the analogous curves for the induction motor in Chapter 6). If the applied voltage and the field flux are held constant, the torque varies with speed according to the straight line plot of Figure 7-9. Part (a) of the figure shows the variation over the entire range, and the student is encouraged to solve for the intercepts on the two axes and the slope of the curve. The curve tends to give the wrong impression of speed variation, however. It will be noted from Eq. (7-3) that the armature current i_a is a parameter along the curve, and increases as the torque increases (or vice versa). The usual range of rated operation for this type of motor is in the lower region of the speed-torque curve for thermal reasons, and in part (b) of the figure an enlarged view is given of this portion to reinforce the idea that this type of motor is relatively constant in speed.†

By changing the applied source voltage or the field flux or both, the curve of Figure 7-9(a) may be caused to have different intercepts on the axes and different slopes. This versatility speaks to the greater controllability mentioned earlier, and equipment drives with critical control problems are often based on the use of dc

---

*That the motor speeds up with decreasing flux, ϕ, at first seems a contradiction but Eq. (7-5) clearly tells us that this is so and this is sometimes verified in the laboratory even to the extent of having the motor "run away" to a self destructive speed when the field current is inadvertently interrupted!

†Think back—what would the torque vs. speed curve of a synchronous motor look like?

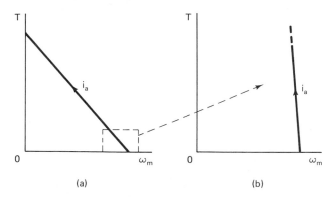

Figure 7-9   (a) Torque versus speed for a dc motor with constant flux and terminal voltage. (b) An enlarged view of the curve over the usual range of operation.

motors for these reasons. As a specific example, consider a motor operated with constant field flux, ϕ, and a variable applied armature voltage V_t. The motor speed will now be almost directly proportional to the applied voltage but relatively constant, with varying torque load, as illustrated in Figure 7-10, where torque-speed curves are given for several different values of applied armature voltage. Sometimes the voltage V_t may be obtained from a dc generator (see Section 7.12), or, again, one convenient way of obtaining variable voltage supply is to use a transformer with silicon-controlled rectifiers (SCR) with phase angle control. The *average* value of the variable width pulses from such a source is the value used for V_t in Eq. (7-7).

The significance of curves such as those of Figure 7-10 may be seen by superposing the speed-torque curves of a given load upon the speed-torque curves of the motor.* An example is shown in Figure 7-11.

It will be seen from Figure 7-11 that the speed of the motor and load may be set to any intersection point, such as 1, 2, or 3 by varying the applied voltage V_t. It should be realized that the curves illustrate steady-state conditions, and in changing from one operating point to another, there will be dynamics involved.

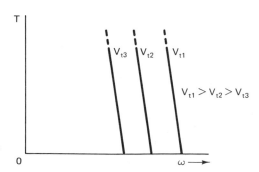

Figure 7-10   Speed control by armature voltage.

---

*The torque-speed curves show *developed* torque, not all of which is available to the load because of mechanical (rotational) losses. If we are concerned with a slight inaccuracy, we might compensate by raising the load torque curve a tiny bit.

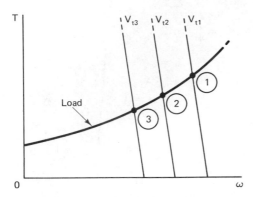

Figure 7-11 Relation of load and motor torques to the motor operating point.

7.6 MAGNETIZATION CURVES

Unless we have a permanent magnet device, the field flux, ϕ, is created by currents flowing in coils wound around the field pole structure. The field passes over the air gap between the field poles and the armature surface and H is far greater in the air gap than in the steel. When we come to integrating the total magnetomotive force, $\mathcal{F} = Ni = \oint \vec{H} \cdot \vec{dl}$ many of the ampere turns are absorbed by the air gap, and sometimes we approximate the magnetic field relations by a linear relation between flux and field current. An economical design, however, requires that we work the material harder than the B values that would give rise to negligible ampere turns in the steel portions of the magnetic circuit, so we usually find departure from a strictly linear relation (air gap line) in the actual ϕ versus i plot. At this stage it is well to note the similarity of Figures 7-4 and 4-7. Magnetization curves for a magnetic circuit with an air gap are shown in Figure 4-6.

Since, from Eq. (7-4) $(e_a = K_a \phi \omega_m)$ the generated voltage, e_a, is proportional to the field flux per pole, ϕ, at any given speed, ω_m, a curve of e_a versus Ni differs from those of Figure 4-6 only by scale constants on the axes. Such curves are shown in Figure 7-12.

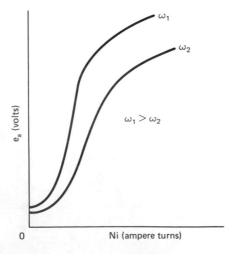

Figure 7-12 Typical magnetization curves for a dc machine.

Some comments should be made in regard to the curves of Figure 7-12. First note that two curves are given, for two different speeds designated ω_1 and ω_2. Since, at a given value of flux, ϕ, the generated voltage, e_a, is directly proportional to speed, any one speed could be chosen to illustrate and the values for other speeds found by direct proportion. Second, note that the axis of absissae is marked in Ni or ampere turns. If there is only one coil on each pole then the ampere turns and the current are directly proportional, and we could as well plot against field current, i_f. If, on the other hand, there is more than one coil per pole, it is the total of the ampere turns which determines the flux and therefore the voltage at any given speed. More specifically, if we have a *compound* motor with one field coil per pole connected in series with i_a and a second field coil per pole connected in shunt with the armature and carrying i_f, the total ampere turns are given by $Ni = N_{series}i_a + N_{shunt}i_f$.

Another detail — since part of the magnetic circuit is through steel, we have magnetic hysteresis, and the value of flux obtained at any given value of ampere turns depends on the history of how we arrive at that point on the curve. For an all-steel path we find *hysteresis loops* as we pass through a cyclic variation (see Figure 3-14 for an illustration). If an air gap is present, the ampere turns consumed by the steel become of minimal importance and any loop becomes very slim — so we approximate by means of a single curve. One concession toward recognizing hysteresis effects is the inclusion of a residual flux at field current zero. Note that a very small voltage is induced even at current zero, owing to previous operations of the machine and the residual magnetism left in the core.

When the field flux is established by an increasing current, starting from zero, the initial permeability is low as seen from the slope of the curve. After the current becomes greater the curve becomes steeper until saturation effects result in the knee of the curve. The initial permeability region is exaggerated in the figure for clarity.

7.7 ARMATURE REACTION

Thus far in the discussion, the field flux, ϕ, has been attributed to the ampere turns of the field coils only. Actually, the currents in the armature conductors affect the distribution of the field flux, and even the magnitude of the flux — in some cases enough to have a significant effect on the machine performance. To illustrate, consider Figure 7-13. A two-pole machine is illustrated in the figure with the armature conductors and their currents shown by the circles with crosses and dots to indicate a sample current direction. Sample flow lines for the B field set up by the armature currents are shown, where it will be noted that the axis of symmetry for the resultant field pattern is *across* (or in quadrature with) the main axis of the field poles. If we superpose this action on that of the main field we see that the main field flux distribution is altered from that of Figure 7-6 and would look more like that of Figure 7-14.*

---

*The word *superpose* must be used guardedly, since we are dealing with a nonlinear material in the steel. We may not *add* the armature and field effects quantitatively, but we may certainly use the term qualitatively to describe the action in this case.

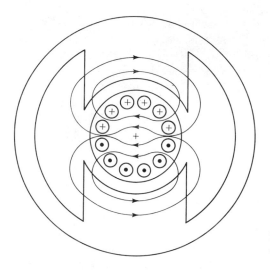

Figure 7-13 Magnetic effects of the armature current.

It will be noted from the figure that the B vector magnitude is *increased* under one pole tip and *decreased* under the other.* Also, it will be noticed that the point of zero B field is shifted from the center line between the poles. This latter effect is said to be *neutral shift,* and discussion of the effects of neutral shift will be delayed until the next section under the topic of *commutation.*

It might at first be thought that the distortion of the flux distribution would be of little effect since, for example, in the derivation of Eq. (7-3) (Section 7.3) the value

$$\phi = \int_0^{2\pi/P} B(\theta)lr\,d\theta$$

was simply identified as the total flux per pole and the details of distribution of the B vector turned out to be unimportant in the derivations for torque and generated voltage. The distortion of flux distribution, as in Figure 7-14, results in a decrease

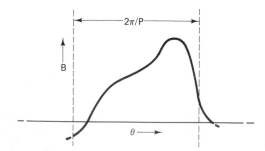

Figure 7-14 The distortion of the B vector distribution, owing to the armature reaction.

*The question of *which* pole tip is strengthened and which is weakened is often discussed in terms of *leading* and *trailing* pole tips, expressions that relate to the direction of rotation. For a study exercise, the student is encouraged to analyze this situation by assuming a polarity for the field poles and then determining the direction of rotation for the motor action and the generator action case. The direction of shift of the field will be found to be opposite in the motor action and the generator action cases.

in ϕ, the total flux per pole, and hence affects machine performance. To work the material economically, it is necessary to design the magnetic circuit so that some portions of the circuit, such as the teeth between the armature slots, operate at B values above the knee of the B-H curve — thus approaching saturation. As a result, the incremental ampere turns on one pole tip tending to increase the B have less effect than those ampere turns on the other tip tending to decrease the B vector, and the total flux is reduced under load current conditions. Many of the curves in the preceding sections describing motor performance were based on constant field flux, ϕ, and this assumption serves as a reasonable first approximation and simplifies the understanding of motor performance. For more precise analysis we must take account of changes in ϕ with variation of armature current, i_a, and this we proceed to do in the next section.

7.8 EFFECT OF ARMATURE CURRENT ON SPEED-TORQUE CHARACTERISTICS

We have seen that the armature current can have a demagnetizing effect on the field, owing to the distortion of the field and the nonlinear properties of the steel. Also, in some motors, we may provide a field coil in series with the armature current, in addition to a field coil in *shunt* with the machine terminals. Such a series coil may either strengthen or weaken the effects of the shunt field coil, depending upon the relative winding direction. The circuit diagram, Figure 7-15, shows such a connection. A motor built in such a way, with both series and shunt fields, is called a *compound* motor. The dots on the coils of Figure 7-15 indicate that positive field current, i_f, and positive armature current, i_a, reinforce each other, and this is said to be *cumulative compounding*. The opposite relative direction of the coils is called differential compounding, and is not very common for reasons discussed below.

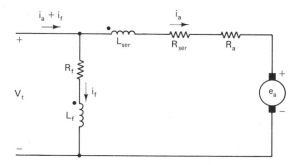

Figure 7-15   A *compound* connected dc motor.

Analysis of machine performance, including the effects of armature current (either through a series winding or because of demagnetizing effect of armature reaction), may be made in terms of the basic equations (7-5) through (7-7), but quantitative calculations are difficult because of the nonlinear relation of the flux,

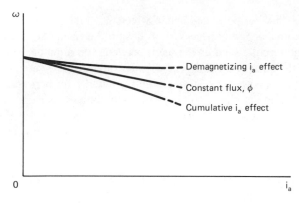

Figure 7-16 The effect of armature current on speed.

ϕ, and the armature current, i_a. We may note qualitatively, however, how curves such as ω versus i_a in Figure 7-8 will change by examining Eq. (7-5). In Figure 7-16 we sketch variation of ω with i_a for cases where i_a reinforces the shunt field flux or where i_a opposes the shunt field flux as from a differential compound connection or the effect of armature reaction.

It will be noted from Eq. (7-5) and Figure 7-16 that the effect of armature current may be to cause the speed to droop more than with the constant flux case, or it may cause the speed to droop less or even rise, depending upon whether the armature current increases the field flux or decreases the field flux.

To analyze machine performance with various loads the speed-torque curves like Figure 7-9(b) are desirable. We might, as a first approximation, describe the flux, ϕ, by a linear relation such as

$$\phi = \phi_0 + Ki_a \qquad (7\text{-}8)$$

where ϕ_0 is the flux with zero armature current and the various cases are considered in terms of different values for the constant, K. If K is positive we have the cumulative case, if K is zero we have the constant flux case, and if K is negative we have the differential case, or the case of demagnetizing armature reaction.

A numerical example of speed-torque computation using the linear approximation will be helpful in understanding these matters — and to show the typical mode of variation of speed with torque for the various armature current effects exemplified by Eq. (7-8).

Suppose we have a hypothetical machine built for 120 volts input and a rated armature current of 60 amperes. Suppose that the armature circuit resistance is given as 0.1 ohm.* We will ignore rotational losses, and thus assume that, with no load on the shaft, zero current is drawn. We will set the current in the shunt field so that

---

*A more precise representation of armature dissipative elements would involve a separate allowance for brush drop (typically 2 volts), and resistive allowances for not only the armature conductors but brush rigging and any series windings. To simplify the example, we will take a specific value for R_a (0.1 ohm) and use that value for all cases in illustration.

at no load we run at 1200 r/min or $\omega = 40\pi$ rad/sec, and leave this current fixed. We thus establish the base value for flux as by

$$e_a = 120 - 0.1 \times 0 = 120 = K_a\phi_0 40\pi$$
$$K_a\phi_0 = 0.955$$

By rearranging the constants of Eq. (7-8) we have

$$K_a\phi = K_a\phi_0(1 + K_m i_a/60)$$
$$= 0.955(1 + K_m i_a/60)$$

where K_m is the fractional change in ϕ_0 caused by the rated armature current of 60 amperes. From Eq. (7-5) we find the speed at any given armature current to be

$$\omega = \frac{120 - 0.1i_a}{0.955(1 + K_m i_a/60)}$$

and the torque from Eq. (7-3) as

$$T = 0.955(1 + K_m i_a/60)i_a$$

We will check three cases, $K_m = 0.1$ for a cumulatively connected series winding, $K_m = 0$ for a constant field flux case with no armature current effect, and $K_m = -0.1$ for a demagnetizing effect of armature current owing to either a differentially connected series winding, or armature reaction, or both. We evaluate speed and torque for values of $i_a = 30$ amperes and $i_a = 60$ amperes.

i_a	$K_m = 0.1$		$K_m = 0$		$K_m = -0.1$	
	ω	T	ω	T	ω	T
0	125.7	0	125.7	0	125.7	0
30	116.7	33.0	122.5	31.4	129.0	29.8
60	108.5	69.1	119.4	62.8	132.6	56.5

A plot of the speed-torque curves for the three cases is given in Figure 7-17. A few remarks about the curves of Figure 7-17: It will be noted that, the stronger the field the more torque will be developed for a given current, as seen from Eq. (7-3). It is seen that, at constant flux, the speed drops with increasing torque, owing to the armature resistance drop. With a stronger field the speed drops even more at a given armature current, owing to the term in the denominator of Eq. (7-5). Last when armature current has a demagnetizing effect, the speed may actually rise under increasing torque load.

A rising speed-torque curve is undesirable as may be seen by superposing a load speed-torque curve on the figure; for example, a horizontal line represents a constant torque load. The equilibrium points for the three cases are seen to be points A, B, and C. It will be seen that point C is an unstable equilibrium point similar to examples treated in Chapter 4. For example, a small departure to the right of point C will result in a developed torque greater than that demanded by the load and the excess will go into acceleration of the load ($T_{dev} - T_{load} = J\,d^2\theta/dt^2$). A runaway condition is thus established, usually resulting in excess current and clearing of the fuse or breaker supplying the motor.

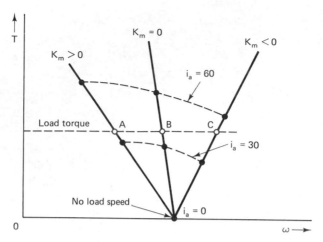

Figure 7-17   Speed-torque curves showing effects of armature current.

Since, even without a differentially connected series winding, armature reaction alone may result in a rising speed-torque curve, we sometimes see shunt motors supplied with a weak cumulatively connected series winding and then called a "stabilized shunt" motor.

7.9 THE SERIES MOTOR

For some applications, it may be desirable to have motor speed-torque characteristic, where the speed drops with increasing torque load even more than does the cumulatively compounded motor. Such applications are epitomized by traction service, where it is desired to have a high torque to accelerate when starting up at low speed. A circuit diagram for a series motor connection and typical speed-torque curve are shown in Figure 7-18. If we have a magnetization curve giving the field flux versus armature current through the series field, then by assuming a number of values for armature current, we may use Eqs. (7-3) and (7-5) to calculate points on the speed torque curve. This approach is involved in one of the problems at the end of this chapter.

7.10 COMMUTATION

The commutator type of machine is valuable because of its excellent controllability of speed and torque, and it is much used for critical drives in spite of the fact that primary energy supply is often in the ac form. Unfortunately, the commutator, which makes these excellent characteristics possible, also presents a difficult problem in operation and maintenance. Consider the diagram of Figure 7-19, which might be regarded as a developed or "rolled-out" view of the armature winding and commutator of Figure 7-3. As the commutator sweeps past the brushes, any given coil connected to one of the segments has current in a particular direction as the brush

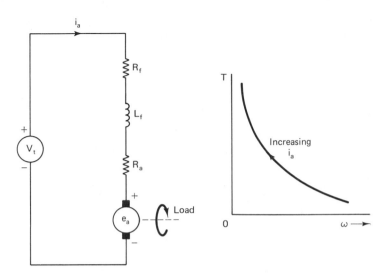

Figure 7-18   The series motor connection and its torque-speed characteristic.

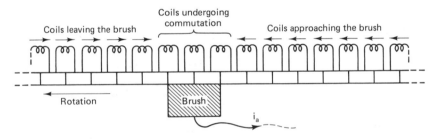

Figure 7-19   The need for sudden current reversal in the commutated coil.

is approached and, in the time that the segment passes the brush, the current in the coil must be reversed. The reversal is difficult because of the property of an inductor in keeping current going over short periods of time. The reversal is further hindered by the distortion of the field flux by armature reaction, as shown in Figure 7-14. Because of the shift in the zero B, or neutral, position a component of voltage is induced in the coil under commutation, also tending to keep the current going in the direction that it had in approaching the brush. The result of these problems may be excessive sparking and erosion of the commutator and/or electromagnetic interference with other devices in the vicinity.

The effects of commutator sparking may limit the maximum current to values less than would be permissible on a short-time thermal overload basis for fear of damage as extensive as the development of an arc completely around the commutator surface, or *flashover*!

Small machines tend to be relatively high in resistance in the armature circuit, and commutation problems may not be serious. Larger machines (integral horsepower, for example) are often fitted with special devices to aide the commutation

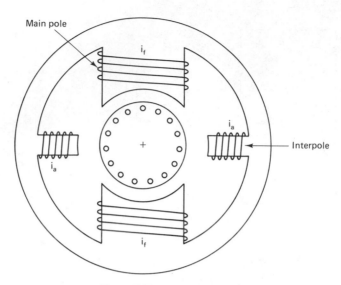

Figure 7-20 Use of interpoles.

process. Such devices are illustrated in Figure 7-20, where auxiliary poles called *interpoles* or *commutating poles* are placed within the machine frame and located between the main poles. Armature current is passed through series windings around the interpoles, polarized in such a direction as to (a) reverse the portion of main flux shifted into the neutral zone by armature reaction and (b) generate a small counter EMF in the coil undergoing commutation in order to assist in current reversal. Such auxiliary poles are almost universal in larger machines, and they are highly successful in allowing good commutation.

7.11 THE AC COMMUTATOR MOTOR

The armature current of the commutator machine has been described as i_a, inferring the possibility of variation with time, whereas in the dc mode of operation we might as well have used the symbol I_a associated with a constant current. It is interesting to consider an armature current of a sinusoidal, ac, nature. If the field flux, ϕ, were a constant, then an ac armature current would produce a sinusoidally varying torque in accordance with Eq. (7-3). Such a torque would produce no perceptible motion because the mass of the rotor would not permit the rotor to follow a torque variation at ordinary power frequencies. If, on the other hand, the field flux is made to reverse each time the armature current reverses, the torque will be unidirectional with an average magnitude proportional to the product of the rms magnitudes of the sinusoidally varying flux and armature current. This mode of operation may be obtained by using a series field arrangement like that of Figure 7-18, except that the source voltage, the armature current, and the generated voltage, or counter EMF, would all be sinusoidal time functions.* A motor such as this—the ac series motor—has excellent speed-torque characteristics and is much used where a difficult starting torque requirement must be met. Many of the small hand tool and kitchen appliance motors are of this type.

The ac series motor will operate from a dc source equally well (sometimes even better) and such motors are called *universal motors*. On the other hand, a motor designed for series dc operation might not operate well or at all on ac, because the ac series motor must have a stator constructed of laminations of quality electrical steel, or otherwise the eddy currents in the solid steel yoke used for some dc machines would prevent the establishment of significant field flux.

7.12 THE DC GENERATOR

In some applications the dc machine may be driven mechanically, as by a prime mover. With field flux, ϕ, supplied the voltage, $e_a = K_a \phi \omega_m$, will be generated, and if an electrical load is connected to the brushes electric power will be supplied. Figure 7-21 illustrates such an application. The field flux may be provided in various

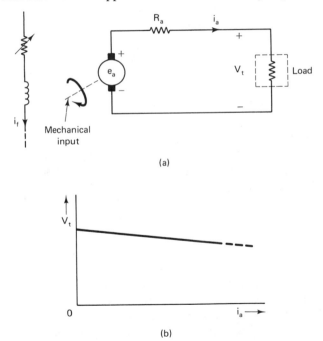

Figure 7-21   (a) dc machine used as a generator, and (b) voltampere characteristic under load.

*A shunt field, with ac voltage applied, would not serve in this case since the current, and hence the flux, would lag the applied voltage by almost 90 degrees. The student will recall that the average value of the product of two sinusoids differing in phase by 90 degrees is zero!

ways. A permanent magnet field is used, for example, in a tachometer application, where the no-load voltage of the armature is directly proportional to speed, ω_m. For applications where more power is supplied, it is usual to use a field coil winding with field current, i_f, supplied from either the machine's own terminal voltage or from a separate source. In such case the field current (and hence the flux) is controllable by varying either the field circuit resistance or the magnitude of the separate voltage source. It will be noted that the reference direction for current is reversed in this diagram from that used for the motor action case, which was the basic case used for most of the preceding discussion. Use of the right-hand rule (as on Figure 7-4) will reveal that the torque developed by positive current in this case will oppose the motion of the rotor. In other words, the mechanical source will supply energy to work against the opposition of the electromagnetic torque of the rotor, as this is the energy that is converted to electrical form. The Kirchhoff's law equation now becomes

$$V_t = e_a - i_a R_a \tag{7-8}$$

and the product $e_a i_a$ is the electrical power generated while the product $T\omega_m$ is the mechanical power absorbed by the machine.

As an example of generator usage of the machine, suppose that the driving speed of a prime mover is constant and the field flux is maintained as a constant, then the output (terminal) voltage, V_t, of the generator will fall under load as seen from Eq. (7-8) and illustrated in part (b) of Figure 7-21. Since R_a is relatively small, the drop in voltage under load is small under these conditions. The speed of most prime movers drops as load is applied, and in many cases the field flux of the machine is reduced with increasing load; hence the voltage may fall more rapidly under load than is indicated by the above. If constant voltage is a requirement, automatic control devices may be employed to maintain the voltage. A speed governor may be used on the prime mover and a voltage sensitive pickup on the generator to increase the field excitation if the voltage tends to fall (a voltage regulator). These devices fall under the scope of automatic *feedback* control systems, which form a discipline all their own.

Other forms of operation of dc generators are seen in practice as, for example, the use of series field windings or combinations of series and shunt field windings (compound generators), but further discussion of these matters will be reserved for more detailed courses on dc machines. In many cases, the basic source of energy in a given location is the ac power supply system and in such cases it is more economical and less troublesome to supply a dc load requirement by use of silicon rectifier or SCR circuits rather than employ a commutator machine as a dc generator. The dc generator has been rendered less important with the advances in power electronics.

ADDITIONAL READING MATERIAL

1. Brown, D., and E. P. Hamilton III, *Electromechanical Energy Conversion,* New York: MacMillan Publishing Company, 1984.

2. Chapman, S. J., *Electric Machinery Fundamentals,* New York: McGraw-Hill Book Company, 1985.

3. Chaston, A. N., *Electric Machinery,* Reston, Virginia: Reston Publications Company, Inc., 1986.

4. Del Toro, V., *Electric Machines and Power Systems,* Englewood Cliffs, New Jersey: Prentice-Hall, Inc., 1985.

5. Elgerd, O. I., *Basic Electric Power Engineering,* Reading, Massachusetts: Addison-Wesley Publishing Company, 1977.

6. Fitzgerald, A. E., C. Kingsley, Jr., and S. D. Umans, *Electric Machinery,* New York: McGraw-Hill Book Company, 1983.

7. Lindsay, J. F., and M. H. Rashid, *Electromechanics and Electric Machinery,* Englewood Cliffs, New Jersey: Prentice-Hall, Inc., 1986.

8. Nasar, S. A., *Electric Machines and Electromechanics,* Schaum's Outline Series in Engineering, New York: McGraw Hill Book Company, 1981.

9. Nasar, S. A., *Electric Energy Conversion and Transmission,* New York: MacMillan Publishing Company, 1985.

10. Shultz, R. D., and R. A. Smith, *Introduction to Electric Power Engineering,* New York: Harper & Row, Publishers, 1985.

11. Slemon, G. R., and A. Straughen, *Electric Machines,* Reading, Massachusetts: Addison-Wesley Publishing Company, 1980.

STUDY EXERCISES

1. A certain dc motor operates with a constant terminal voltage of 250 volts and has a rated armature current of 60 amperes at full load. The armature circuit resistance is 0.15 ohm.

When loaded such that 30 amperes of armature current are drawn, the motor runs at 1250 r/min (revolutions per minute).

a. What is the magnitude of the product $K_a\phi$ of Eq. (7-4)?

b. What is the torque developed under the above conditions?

c. What is the developed horsepower? (One hp = 746 W)

If we assume that the flux, ϕ, per pole and the applied voltage, V_t, are constant, but the total torque load (external load plus the torque consumed by the mechanical losses) is doubled:

d. At what speed will the motor run?

e. If, on the other hand, the effect of armature reaction is to reduce the original flux (at 30 A) by 10 percent when the load increases, then at what speed does the motor run with the same current as in (d)?

2. As an approximate index to the power handling capabilities of a given amount of material, consider a machine like that of Figure 7-4 with the following data given:

Radius of rotor steel,	$r = 10$ cm
Length of rotor steel,	$l = 12$ cm
Number of poles,	$P = 4$
Average B under poles,	$B = 1\ T$
Fraction of periphery that is subtended by the pole surface	$= 0.7$
Depth of slots in rotor steel	$= 1$ cm

Fraction of rotor periphery occupied by rotor slots	= 0.5
Fraction of slot cross section occupied by copper	= 0.75
Allowable current density in rotor copper	$J = 200$ A/cm^2
Rated speed	$n = 1200$ r/m

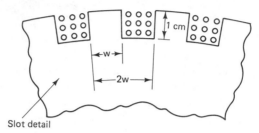

Slot detail

a. What is the approximate power conversion capability of this machine in watts and in horsepower?

b. At the rated speed, what is the centrifugal force tending to throw a single rotor conductor of 15 grams out of the slot? (Give in newtons and pounds)

3. Given a dc commutator type of motor with a shunt (parallel connected) field and the following properties:

$$V_t = 250 \text{ volts}, \quad I_a \text{ (rated)} = 125 \text{ amperes}, \quad R_a = 0.08 \text{ ohm}$$
$$4 \text{ poles, with } \phi \text{ per pole} = 0.02 \text{ weber}, \quad K_a = 95$$
$$\text{Rotational (mechanical) loss} = 1000 \text{ watts}; \quad \text{field current} = 3.5 \text{ amperes}.$$

a. What is the horsepower output at rated armature current? *Note:* 1 horsepower = 746 watts.

b. What is the speed in revolutions per minute (r/min) at rated armature current?

c. What is the efficiency at the above load?

d. If the initial starting current must be limited to 225 amperes, how much additional resistance must be inserted in the armature circuit for starting?

e. Under the conditions of (d) what is the starting torque in newton-meters?

4. A certain dc motor is tested in the laboratory with a constant applied terminal voltage of 240 volts and a constant shunt field excitation resulting in constant flux, ϕ, per pole.

At no load the motor runs at 850 rev/min and when a torque load of 900 N-m is applied, the speed drops to 800 rev/min. It is assumed that the torque lost to supply the mechanical losses is small and constant at 30 N-m.

The motor drives a load of the viscous friction type wherein the load torque is directly proportional to the speed at which the load is driven. A calibration test finds one point on the speed-torque curve of the load where driving at 900 rev/min requires 600 N-m of torque.

a. At what speed will the motor drive this load?

b. What is the horsepower output of the motor at this load?

c. In part (a), above, we find the speed at a given torque load and we know that the developed torque is 30 N-m more than the load torque. We also know that with no external load and a developed torque of 30 N-m the speed is 850 rev/min. What, then, is the armature resistance drop, $I_a R_a$, at the specified load?

d. What is the electrical input to the armature at the specified load?

e. If the shunt field winding requires 8 amperes to establish the flux in this application, what is the overall efficiency, η, of the motor at this load?

5. Suppose that the machine of Problem 3 is driven by an external prime mover (perhaps a small diesel engine) at a speed of 1250 revolutions per minute. The same field current is supplied and it is assumed that the flux is constant.

As a *generator:*

a. What will be the no-load terminal voltage?

b. What will be the full-load terminal voltage ($I_a = 125$ A) assuming that the prime mover speed is controlled by a speed governor and does not change?

c. What is the percent voltage regulation of this machine?

SOLUTIONS TO STUDY EXERCISES

1. $e_a \quad = V_t - i_a R_a = 250 - 30 \times 0.15 = 245.5$ V

$e_a \quad = K_a \phi \omega_m$

$$\omega_m \quad = \frac{1250 \times 2\pi}{60} = 130.9 \text{ rad/sec}$$

$$K_a \phi = \frac{245.5}{130.9} = \underline{1.88} \quad \longleftarrow \quad a$$

$T \quad = K_a \phi i_a = 1.88 \times 30 = \underline{56.3 \text{ N-m}} \quad \longleftarrow \quad b$

$P \quad = T\omega_m = 56.3 \times 130.9 = 7365$ W

$$\text{or} = \frac{7365}{746} = \underline{9.87 \text{ horsepower}} \quad \longleftarrow \quad c$$

$$i_a \quad = \frac{T}{K_a \phi} = \frac{2 \times 56.3}{1.88} = 60 \text{ A}$$

$$\omega_m \quad = \frac{V_t - i_a R_a}{K_a \phi} = \frac{250 - 60 \times .15}{1.88} = 128.5 \text{ rad/sec}$$

$$\text{or} = \frac{128.5 \times 60}{2\pi} = \underline{1227 \text{ rev/min}} \quad \longleftarrow \quad d$$

$K_a \phi = 1.88 \times 0.9 = 1.69$

$$i_a \quad = \frac{2 \times 56.3}{1.69} = 66.7 \text{ A}$$

$$\omega_m \quad = \frac{250 - 66.7 \times 0.15}{1.69} = \frac{240}{1.69} = 142.2 \text{ rad.sec}$$

$$\text{or} = \frac{142.2 \times 60}{2\pi} = \underline{1358 \text{ rev/min}} \quad \longleftarrow \quad e$$

2. $\phi/\text{pole} = BA/\text{pole} = 1.0 \times \dfrac{10}{100} \times 2\pi \times 0.7 \times \dfrac{12}{100} \times \dfrac{1}{4} = 0.0132$ Wb/pole

Consider Z conductors, connected in a paths:

Cross section of conductors

$$= 0.75 \times (\text{slot area}) = \frac{0.75 \times 2\pi r \times 1 \times 0.5}{Z}$$

$$= \frac{23.5}{Z} \text{ cm}^2$$

$$\text{Current per conductor} = 200 \times \frac{23.5}{Z}$$

$$i_a = a(\text{current/conductor}) = \frac{a\,200 \times 23.5}{Z}$$

$$K_a = \frac{ZP}{2\pi a} = \frac{4Z}{2\pi a}$$

$$e_a = K_a\phi\omega_m = \frac{4Z0.0132 \times 1200 \times 2\pi}{2\pi a}$$

$$= 1.06\frac{Z}{a}$$

$$p = e_a i_a = 1.06\frac{Z}{a} \times \frac{200 \times 23.5a}{Z}$$

$$= 4961 \text{ W or } \frac{4961}{746} = \underline{6.7 \text{ horsepower}} \quad \longleftarrow \quad a$$

$$f = M\omega^2 r = \frac{15}{1000} \times \left[\frac{1200 \times 2\pi}{60}\right]^2 \times \frac{10}{100}$$

$$= \underline{23.7N \text{ or } 5.33 \text{ lbs (force)}} \quad \longleftarrow \quad b$$

3. $e_a = V_t - i_a R_a = 250 - 125 \times 0.08 = 240 \text{ V}$

$P_{\text{dev}} = e_a i_a = 240 \times 125 = 30\,000 \text{ W}$

$P_{\text{out}} = P_{\text{dev}} - P_{\text{loss}} = 30\,000 - 1000 = 29\,000 \text{ W}$

$$\text{or} = \frac{29\,000}{746} = \underline{38.9 \text{ horsepower}} \quad \longleftarrow \quad a$$

$$\omega = \frac{E_a}{K_a\phi} = \frac{240}{95 \times 0.02} = 126 \text{ rad/sec}$$

$$\text{or} = \frac{126 \times 2\pi}{60} = \underline{1206 \text{ rev/min}} \quad \longleftarrow \quad b$$

$$\eta = \frac{P_{\text{out}}}{P_{\text{in}}} = \frac{29\,000}{250 \times 125 + 250 \times 3.5} = \underline{0.90} \quad \longleftarrow \quad c$$

$$225 = \frac{250}{R + \Delta R}, \quad \Delta R = \underline{1.03 \text{ ohm}} \quad \longleftarrow \quad d$$

$$T = K_a\phi i_a = 95 \times 0.02 \times 225 = \underline{427.5 \text{ N-m}} \quad \longleftarrow \quad e$$

4. a. Since speed vs. developed torque is linear, Eq. (7-7), then output torque (dev. torque − loss torque) vs. speed is linear if the loss is a constant, hence:

$$n = n_0 - KT_{\text{out}}$$
$$800 = 850 - K900$$
$$K = 0.0556$$
$$n = 850 - 0.0556T \quad (1)$$

For the load:

$$n = 950\frac{T}{600} = 1.583T \quad (2)$$

Subst. Eq. (2) into Eq. (1)

$$n = 850 - 0.0556 \frac{n}{1.583}$$

$$n = \underline{821.2 \text{ rev/min}} \quad \longleftarrow \quad a$$

b. $P = T\omega = \left(600\dfrac{821.2}{950}\right)\left(821.2\dfrac{2\pi}{60}\right) = 44\,600 \text{ W}$

$$\text{or} = \frac{44\,600}{746} = \underline{59.8 \text{ horsepower}} \quad \longleftarrow \quad b$$

c. From Eqs. (7-5) and (7-3)

   $n \propto (V_t - i_a R_a)$

   $T \propto i_a R_a$

$$\frac{n_1}{n_2} = \frac{240 - i_{a1} R_a}{240 - i_{a2} R_a} = \frac{850}{821.2}$$

$$\frac{T_1}{T_2} = \frac{30}{(821.2/1.583) + 30} = 0.0547 = \frac{i_{a1} R_a}{i_{a2} R_a}$$

hence: $\dfrac{850}{821.2} = \dfrac{240 - 0.0547 i_{a2} R_a}{240 - i_{a2} R_a}$

$$i_{a2} R_a = \underline{8.59 \text{ V}} \quad \longleftarrow \quad c$$

d. Mech $P_{\text{out}} = 44\,600 \text{ W}$

$$\text{Mech } P_{\text{dev}} = \frac{T_{\text{out}} + 30}{T_{\text{out}}} \times 44\,600 = 47\,180 = e_a i_a$$

   But $e_a = 240 - 8.49 = 231.4 \text{ V}$

   $i_a = 47\,180/231.4 = 203.9 \text{ A}$

   $P_{\text{arm}} = 240 \times 203.9 = \underline{48\,931 \text{ W}} \quad \longleftarrow \quad d$

e. $\eta = \dfrac{P_{\text{out}}}{P_{\text{in}}} = \dfrac{44\,600}{48\,931 + 240 \times 8.0} = \underline{0.88} \quad \longleftarrow \quad e$

5. a. $e_a = K_a \phi \omega$

$$= 95 \times 0.02 \times 1250 \times \frac{2\pi}{60} = \underline{248.7 \text{ V}} \quad \longleftarrow \quad a$$

b. $V_t = e_a - i_a R_a$

   $= 248.7 - 125 \times 0.08 = \underline{238.7 \text{ V}} \quad \longleftarrow \quad b$

c. $\%\text{Reg} = \dfrac{248.7 - 238.7}{238.7} \times 100 = \underline{4.2\%} \quad \longleftarrow \quad c$

HOMEWORK PROBLEMS

1. A four-pole motor has poles with an area of 0.01 m^2 on the face of each pole next to the armature, and operates with a constant flux density of 0.9 tesla between the pole face and the armature surface. For simplicity, we ignore the fringing of the flux at the pole tips and the distortion of the field caused by armature reaction. There are 600 conductors (Z) on the armature surface, so connected that there are four parallel paths (a) between the brushes. The effective radius of the armature is 0.08 m. The armature resistance is 0.06 ohm, including the effect of brush drop and other series elements.

If 120 volts is applied to the armature and the machine is loaded to the rated armature current of 90 amperes:

  a. At what speed does the motor run?

  b. What torque (in newton-meters) is developed?

  c. If 5% of the developed torque is lost in rotational (mechanical) losses, what is the output of the motor in horsepower?

  d. If, in addition to the power input to the armature, another increment of 3% of that power is required to excite the field, then what is the efficiency of the motor at the above load of 90 amperes into the armature?

  e. If the mechanical losses are the same as in (c) above, what is the no-load speed of the motor?

  f. What is the speed regulation of the motor in percent?

2. a. If the armature reaction effect (previously ignored) is to reduce the main field flux by 15 percent under the 90-ampere load, what will be the speed and horsepower output of the motor? All other data remain the same as in Problem 1.

  b. Under the above conditions, will the motor be stable with a constant torque load? Explain.

3. A 250-volt dc motor has an armature resistance of 0.08 ohm and a rated full load current, $I_a = 120$ amperes. It is operated with a constant field flux, ϕ, in what follows:

  a. If started "across the line" with full rated voltage suddenly applied, as is often the method with induction motors, a very large current will flow. How much would the initial current be with 250 volts applied to the motor at standstill?

  b. When the motor accelerates from standstill and the current falls to 200 A, at what speed does the motor run in percent of full load rated speed?

The currents involved in a starting operation as above are usually excessive, even to the point of destroying the motor, except for very small motors. Instead of applying full voltage, most motors of this dc class are started by applying a reduced voltage. If the motor is supplied through electronic means, the average voltage may be reduced by phase angle control of SCRs (see Chapter 9), or the voltage may be reduced by means of series resistance inserted during the starting operation.

If the armature current, I_a, is to be limited to twice rated (240 A in this case):

  c. To what value should the average applied voltage be reduced when first starting?

  d. What extra resistance, R, in series with the armature would accomplish the desired reduction?

  e. With the extra resistance of (d) above in place and the motor allowed to accelerate under no-load, at what percent of normal full load speed will the motor run when the armature current has dropped to 120 A?

4. We have a 100 kW dc generator which is rated at 250 V, 400 A, and which has an armature circuit resistance of 0.02 ohm. When the field is excited at a constant current of 15 A and the machine is driven at 1500 rev/min the no-load voltage is 250 V.

When loaded to rated current of 400 A the terminal voltage drops to 230 volts, in part because the driving prime mover speed drops. We assume that armature reaction effects are negligible because of the construction of the machine.

  a. What is the driving speed at the 400 A load?

  b. If we assume a linear relation between flux and field current, how much must the field current be increased by voltage regulator equipment in order to maintain the terminal voltage at 250 V under load?

  c. If the driving speed of the prime mover is maintained constant at the no-load value by means of a speed governor then what is the terminal voltage under full load and what is the answer to (b) above?

  d. A more accurate representation of the flux-field current relation takes account of the nonlinear magnetization curve. One way sometimes used to represent the nonlinear relation is an equation such as

$$K_a\phi = \frac{0.117I_f}{1 + 0.007I_f}$$

If we use this relation (known as Frölich's formula) what is the answer to (c) above?

5. A certain commutator machine has a rated (full load) armature current of 40 amperes. The armature resistance (with an approximation for brush drop) is 0.25 ohm. The machine is built with a field intended for series connection, and this winding has an additional resistance of 0.10 ohm.

Information on the magnetization characteristics was obtained from a laboratory test, where the machine was driven at a constant speed of 1500 revolutions per minute and various currents were passed through the series-connected armature and field. The generated voltage, E_a, was then computed from the equation

$$E_a = V_t - I_a(R_a + R_{ser})$$

and the results plotted as below.

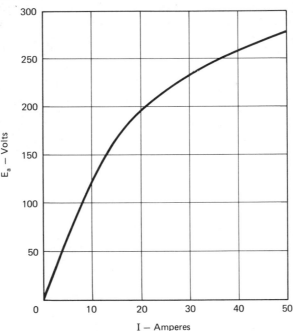

I — Amperes

Now consider that the machine is to be run as a series motor with an applied terminal voltage, V_t, of 230 volts.

a. Calculate the developed torque and the speed at armature currents of 10, 20, 30, and 40 amperes.

b. Assuming that 2 percent of the developed torque is consumed in mechanical losses, plot the output torque versus speed for this motor over the range of current loads as specified above.

c. In starting a heavy inertia load, the current is to be limited to 25 percent overcurrent (50 amperes). How much external resistance must be placed in the armature circuit with the same 230 volts applied?

d. With the resistance of part (c) in place, how much torque is developed at standstill to start accelerating the heavy inertia load?

e. What is the energy conversion efficiency (mechanical output/electrical input) at each of the load currents of (a) above?

6. Consider a dc motor with rated voltage, $V_t = 240$ V, rated current, $I_a = 100$ A, armature resistance $R_a = 0.06$ ohm, and $K_a\phi = 2.0$. For simplicity it will be assumed that the load on the motor is purely a heavy inertia with the combined polar moment of inertia of load and motor, $J = 12.0$ Kg-m^2. Mechanical (rotational) losses will be ignored.

   a. If external R is used in starting to limit the current to two times rated with full voltage applied, how long will it take to accelerate the motor and load from standstill to 95 percent of the final velocity? (External starting resistance is left in place for whole time.)

   b. How many joules of energy will be dissipated in the armature resistance, R_a, during the entire starting sequence?

   c. Repeat parts (a) and (b) with $K_a\phi$ increased to 3.0 by increasing the field current. Do we reduce the energy dissipated in R_a (and hence armature heating) by this procedure?

   Note: Before working on this problem it may be well to review Section 6.8 of Chapter 6, which deals with somewhat similar matters concerning an induction motor.

7. It is sometimes noted that a motor such as that of Problem 6 is analogous to a capacitive R-C circuit. What are the parameters of a series R-C circuit which behaves the same as the motor above with starting resistance in place?

Chapter 8

Energy

Transmission

- The characteristics of overhead lines and underground cables.
- The equations applying to distributed constant lines.
- Models for use in analyzing the performance of the lines.

8.1 INTRODUCTION

The conductors that connect an electric energy source to a load might be very short, as in the case of a plant with both generation and load. On the other hand, the distance from generators to load may be hundreds of miles, as in the typical large electric utility. In this latter case, the conductors are known as a *line*. A long, high-voltage line carrying bulk power (often 100 MW and above) is known as a *transmission line*. Shorter lines leading from substations to load areas and carrying less power (perhaps 10 MW and below) are known as *distribution lines*. The majority of transmission lines are polyphase ac in nature, whereas distribution lines are almost invariably ac and may be either three-phase or single-phase. The transmission lines that are not ac are high-voltage dc, with power electronic conversion apparatus at the ends to interconnect with otherwise totally ac systems. Although dc lines are the oldest form historically, dating back to Thomas Edison, it is only recently that the electronic conversion apparatus has developed to the point of making dc transmission at high power levels a feasible alternative.

Lines may be of the form of overhead conductors supported by poles or steel towers, or they may be in the form of cables, perhaps installed underground or even under water. In either case, the conductors have resistance; they have magnetic fields represented by inductive reactance and electric fields represented by capacitive susceptance. Only if the connecting conductors are very short may these parameters be ignored in analyzing the performance of a system. The evaluation of the line parameters and the use of line models accounting for these parameters form the largest part of this chapter.

8.2 TRANSMISSION AND DISTRIBUTION

The classification of lines as either transmission or distribution is not always clear cut. As rather extreme examples, consider a 765-kV, three-phase, 300-mile line carrying 1500 MW versus a 7.2-kV, single-phase, 5-mile line carrying one MW. The high-voltage line would clearly be called a *transmission line,* whereas the low voltage line is more apt to be called a *distribution line*. The dividing point between the two classifications is not fixed, although some things may be said to characterize the two cases. Lines of high voltage carrying large amounts of power between major substations or generating plants and forming a high-voltage network or *grid* would be called transmission lines. Medium- to low-voltage lines, often radiating out from a substation to supply local load would be called distribution lines. Some utilities identify a medium class of line as a *subtransmission* line (perhaps 69 kV carrying 100 MW for 30 miles as an example). Transmission lines are usually polyphase, whereas distribution lines may be polyphase, or they may be made up of a single phase tapped into a three-phase system. Small loads such as residences are normally fed from these single-phase lines through distribution transformers. As mentioned in Chapter 2, care is taken to allocate equal loads between each of the three phases at the substation, thus insuring *balanced* operation of the major system components, including the transmission lines. The relative unbalance between phases is usually less than 5 percent.

An analogy to illustrate the difference between transmission and distribution is the difference between a large network of interstate highways and the small country roads that might radiate out from the major junction points. Symbolic one-line diagrams in Figures 8-1 and 8-2 illustrate two typical cases.

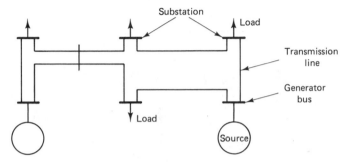

Figure 8-1   One-line diagram of a transmission network. Each line represents one phase of a three-phase component.

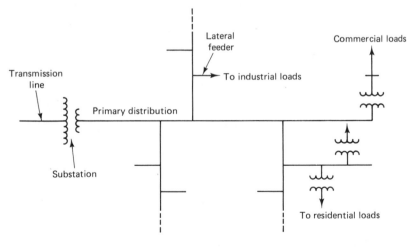

Figure 8-2   One-line diagram of a distribution system.

8.3 LINE CONSTRUCTION

The choice of construction forms for lines, open wire lines overhead or cables underground, is made on the basis of many factors such as right-of-way availability, safety, environmental effects, and economy. Power transmission by cables is very expensive — it may cost as much as five times the cost of overhead construction.

Very high-voltage transmission by cables over long distances becomes technologically unfeasible because of the high charging current (discussed later), but underground cable distribution in residential areas is often very attractive in spite of the high cost. Some typical constructions are illustrated in Figures 8-3 through 8-5.

The selection of voltage level for transmission lines or distribution lines is dominated by the amount of power and the distance of transmission, as will be

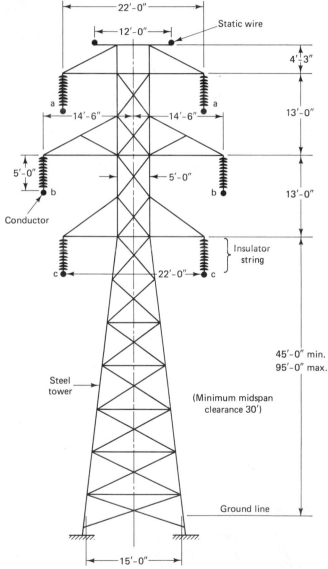

Figure 8-3 Typical construction of a 138-kV, (120-kV class) three-phase, double-circuit line. There are two three-phase lines supported by the tower. The two conductors at the tower top are ground wires used to shield the lines from lightning strokes.

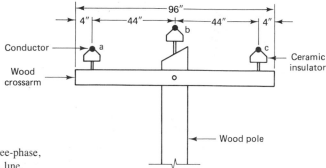

Figure 8-4 Typical 12.47-kV, three-phase, single-circuit overhead distribution line.

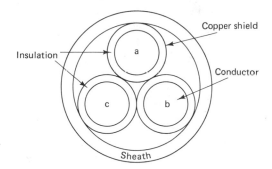

Figure 8-5 12.47 kV, three-phase single-circuit underground cable configuration.

illustrated in the examples to follow. The voltage used will determine the insulation and clearances required, as may be seen by comparing Figures 8-3 and 8-4. The thickness and construction of the cable insulation is determined by the voltage, as in the construction of Figure 8-5. The conductor cross section is determined by the current to be carried, although the outer diameter of conductors is also influenced by the dielectric stress of the electric field at the conductor surface.

As an example of voltage and current relations of a line, consider a sample case where we desire to transmit 100 MW of power over a suitable three-phase transmission line of 90 miles (144 kilometers) in length.

As a first trial, suppose we choose 12.47 kV as the line voltage. The current then will be given by

$$I = 100 \times 10^6 / \sqrt{3} \times 12.47 \times 10^3 = 4630 \text{ A}$$

if we assume unity power factor. This amount of current would require a very large conductor, which would be expensive, but perhaps more important is the question of the power loss in the line. If we take the resistance of the conductor typically as 0.02 ohm per mile, the power loss in the three phases would be

$$P = 3 \times 4630^2 \times 0.02 \times 90 = 115 \times 10^6 \text{ W} = 115 \text{ MW}$$

which is more than the power to be transmitted!

As a second trial consider a line voltage of 115 kV. The current will then be given by

$$I = 100 \times 10^6 / \sqrt{3} \times 115 \times 10^3 = 502 \text{ A}$$

which is a value requiring considerably less conductor cross section. If we take a typical resistance for such a conductor as 0.2 ohm per mile, the total loss is given by

$$P = 3 \times 502^2 \times 0.2 \times 90 = 13 \times 10^6 \text{ W}$$
$$= 13 \text{ MW}$$

which corresponds to a loss of 13 percent; this value is a little high, but perhaps we may accept it. More likely, we will consider a yet higher voltage for the line, but we will drop the example at this point.

Transmission voltages commonly encountered range from 115 kV to 765 kV three-phase. The range from 115 to 345 kV is called *High Voltage* (HV), the range from 345 to 765 kV is called *Extra High Voltage* (EHV), and the range greater than 765 kV is known as *Ultra High Voltage* (UHV). Voltages in this latter range, such as 1200 kV, are currently under development. The voltages mentioned are given in terms of ac lines. High-voltage dc lines are in existence or planned in the range of ±500 kV, that is 500 kV on each side of ground potential.

8.4 TYPES OF CONDUCTORS

Copper has been the most common conducting material in electrical work, but other metals and combinations have been much used for transmission and distribution lines and cables. Aluminum is perhaps the most common material used for transmission lines — often reinforced with steel for mechanical strength. Aluminum conductors have often replaced copper because the aluminum is lighter in weight and lower in cost for conductors for the same resistance. The following types of conductors are used for the transmission and distribution of electric energy:

1. AAC: All-Aluminum Conductor
2. AAAC: All-Aluminum-Alloy Conductor
3. ACSR: Aluminum Conductor, Steel-Reinforced
4. ACAR: Aluminum Conductor, Alloy-Reinforced

The AAC type is predominantly used for distribution lines and the ACSR type for transmission lines. The steel reinforcement in the ACSR provides the desired mechanical strength, whereas the aluminum portions carry the largest part of the current. Figure 8-6 shows the cross section of an ACSR conductor with seven steel strands and twenty-four aluminum strands. (24Al/7St)

Conductor manufacturers provide the characteristics of the standard conductors used for transmission and distribution. Code words (bird names) have been assigned to each conductor for easy reference. These code names are uniform throughout the industry. An example of some of the data for the conductor of Figure 8-6 is given in Table 8-1.

The manufacturers provide, in addition to the ac resistance, other parameters to be discussed later. In the above table, the conductor size in kcmil gives the

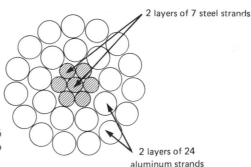

2 layers of 7 steel strands

2 layers of 24
aluminum strands

Figure 8-6 Cross-sectional view of 556.6 kcmil ACSR "Parakeet" conductor (not to scale).

TABLE 8-1   Electrical Characteristics of Parakeet ACSR Conductor

Code	mega circular mils? Conductor Size (kcmil or MCM)	Stranding (Al/St)	Radius (cm)	Approximate Current Rating (amperes)	ac Resistance at 75 °C (Ω/km)
Parakeet	556.5	24/7	1.161	730	0.1240 (= 0.1995 Ω/mile)

conductor cross-sectional area in kilo circular mils. The mil is one one-thousandth of an inch (0.001″) and the circular mil is defined as the area of a circle of one mil in diameter. The area of a circle varies as the square of the diameter and hence the area of a circle of d mils in diameter is d^2 times that of a circle of one mil, or simply d^2 circular mils. As an example, one million circular mils would be the area of a solid wire with circular cross section and one inch in diameter. Although this text uses mainly the metric (SI) system of units, exceptions are made in this chapter because of the very general practice of using feet, inches, mils, and circular mils in transmission line computations.

8.5 TYPES OF INSULATION

For overhead lines, the insulation between conductors is simply air (the conductors are bare metal for transmission lines). Air has a dielectric constant the same as that of a vacuum ($\varepsilon_0 = 8.85 \times 10^{-12}$ farad/meter) and a breakdown strength which varies between 10 and 30 kV/cm depending upon the weather conditions. For a 138-kV line like that of Figure 8-3, an analysis of the electric field around the conductors would show that a spacing of only a few inches would be sufficient to avoid arc-over between conductors, but the actual spacing is far more for several reasons. Reasons of safety, vibration effects, and corona (an ionized air discharge at the conductor surface) dictate spacings of at least three to four feet. Thus it will be seen that the spacings shown in Figure 8-3 are very conservative. Some newer lines are built with reduced spacings and are called *compact* lines.

Underground cables commonly use either oil-impregnated paper or plastic (polymer) as insulation. The thickness of the insulation is determined by the voltage (line-to-ground) the cable must withstand. The relative permittivity of the materials used is about three or four. Development is under way to use compressed gas such as SF_6 for insulation of cables at transmission line voltage levels.

8.6 LINE PARAMETERS — RESISTANCE

In analyzing transmission line performance, we need to know the circuit parameters of the line, such as the line resistance. Transmission line conductors are of sufficient cross section that skin effect becomes noticeable, even at the power frequency of 60 Hz, hence it is the ac resistance, R_{ac}, which is needed, and tabular data on the common conductors is available from the manufacturers, usually in terms of resistance at dc, 50 Hz, and 60 Hz. The skin effect factor, which is the ratio R_{ac}/R_{dc}, increases with frequency and is typically 1.02 to 1.04 at 60 Hz.

The dc resistance of a conductor is given by:

$$R_{dc} = \frac{\rho l}{A} \tag{8-1}$$

where ρ is the conductor resistivity in ohm-meters (Ω-m)

$$= 2.83 \times 10^{-8} \quad \text{for aluminum at 20 °C}$$

and,

$$= 1.77 \times 10^{-8} \quad \text{for copper at 20 °C}$$
$$l = \text{length of the conductor in meters}$$
$$A = \text{conductor cross-sectional area in } m^2$$

Values for the resistivity, ρ, are also available for Eq. (8-1) when the length is given in feet and the cross-sectional area is given in circular mils (or kilocircular mils)

The ac resistance of a given conductor differs from the dc resistance of Eq. (8-1) because of three factors: skin effect (mentioned above), temperature, and spiraling of the strands forming a stranded conductor like that shown in Figure 8-6.

Effect of temperature

The resistivity of common conductor materials fortunately varies almost linearly with temperature, and hence we may write an equation such as (8-2)

$$\rho_2 = \rho_1 \frac{t_2 + T}{t_1 + T} \tag{8-2}$$

where T is a constant for a given material, and ρ_1 and ρ_2 are the resistivities of the material at temperature of t_1 and t_2 respectively. It follows that:

$$\frac{R_2}{R_1} = \frac{t_2 + T}{t_1 + T} \tag{8-3}$$

Sample values for the constant T are:

T_{al} = 228 for hard-drawn aluminum of 60% conductivity.*
T_{cu} = 241 for hard-drawn copper of 97% conductivity.

*Conductivity (the reciprocal of resistivity) is commonly given in terms of the percent of the conductivity of pure, soft, annealed copper.

In summary, to find the ac resistance of a conductor from the dc resistance given as Eq. (8-1) requires the consideration of three effects: skin effect, temperature effect, and spiraling the strands, as in the example following.

Example

a. Find the dc resistance of "Parakeet" conductor at 20 °C in ohms per mile from the information of Table 8-1.

Solution

$$l = 1609 \text{ meters}$$

$$A = 556.5 \text{ kcmil} = 556.5 \times 10^{-3} \frac{\pi}{4} \times 0.0254^2 = 2.82 \times 10^{-4} \text{ m}^2$$

$$\rho = 2.83 \times 10^{-8} \ \Omega - m$$

$$R_{dc} = \frac{\rho l}{A} = \frac{2.83 \times 10^{-8} \times 1609}{2.82 \times 10^{-4}} = 0.1615 \ \Omega/\text{mile} \tag{8-4}$$

b. If the dc resistance of Parakeet conductor is given as 0.1626 Ω/mile at 20 °C by a manufacturer, what is the percent spiraling effect?

Solution

$$\text{percent spiraling effect} = \frac{0.1626 - 0.1615}{0.1615} \times 100$$

$$= 0.7\%$$

c. If the manufacturer provides R_{ac} as 0.1669 Ω/mile at 20 °C, what is the skin effect factor?

Solution

$$\text{skin effect factor} = \frac{0.1669}{0.1626} = 1.0264$$

$$\text{or} \quad 2.64\% \text{ increase}$$

d. Find the ac resistance of the same conductor at 75 °C.

Solution

$$R_{75} = R_{20} \frac{228 + 75}{228 + 20} = 0.1669 \times \frac{303}{248} = 0.2039 \ \Omega/\text{mile}$$

The small difference from the value of Table 8-1 may be accounted for by small differences in the manufacturer's method of computation.

e. What is the overall percent increase in the ac resistance of Parakeet conductor at 75 °C compared to its dc resistance at 20 °C?

Solution

$$\frac{R_{ac} @ 75 \text{ °C}}{R_{dc} @ 20 \text{ °C}} = \frac{0.2039}{0.1615} = 1.2625$$

or a 26.25% increase, owing to the three factors discussed above.

8.7 LINE PARAMETERS—INDUCTANCE

An overhead line or a cable is surrounded by a magnetic field when carrying current. When the current varies, the magnetic field varies, and a voltage is introduced according to Faraday's law. The voltage is conveniently accounted for by defining the parameter, *inductance,* and/or for the sinusoidal case of current variation, *inductive reactance.*

The evaluation of the inductance parameter from the line geometry may be approached from the flux linkage point of view since flux linkages, λ, are given by an expression

$$\lambda = LI \tag{8-5}$$

Alternately, we might approach the evaluation from an energy point of view, using the relation

$$\text{Magnetic field energy} = LI^2/2 \tag{8-6}$$

The magnetic field of a long, straight wire carrying a current is easily analyzed to obtain inductance constants. For the benefit of those who may not be familiar with the process from field courses, Appendix C outlines the procedure.

Inductance of a single-phase line

As an example of finding the inductance of an overhead line, consider the single-phase line shown in Figure 8-7. The magnetic field picture of the flow lines of $\vec{H}$ and $\vec{B}$ surrounding this structure is rather more complicated than the single-wire case of Appendix C, where the flow lines form concentric circles around the conductor. Fortunately, however, we are dealing here with a linear medium (μ is a constant) and the principle of superposition can be used. We may consider one wire at a time and assemble the inductance expressions from the building blocks of Appendix C.

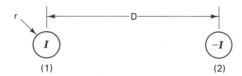

Figure 8-7   Two long, straight parallel conductors, forming a single-phase line.

We denote the left-hand wire as 1. There is a current, I, in this wire. The conductor has a radius, r, and is separated from a return conductor, 2, which carries a current, $-I$, in terms of the same reference direction (into the paper for example). Conductor 1 then has two components of inductance, corresponding to the internal and the external flux linkages. From Eq. (C-12)

$$L_{1(\text{int})} = \frac{1}{2} \times 10^{-7} \qquad H/m$$

and from Eq. (C-19)

$$L_{1(\text{ext})} = 2 \times 10^{-7} \ln(D/r) \qquad H/m$$

We note that the numerator in the term x_2/x_1 of Eq. (C-19) becomes simply D, since, if we were integrating the flux linkages as in the equation derivation, we would stop the summation at D—paths exterior to D would encircle no net current.*

*It might seem that this approach would be only approximate unless r were very small with respect to D. It may be shown, though, that the results are *exact* if D is the center-to-center distance.

The total inductance of conductor 1 is then

$$L_1 = \frac{1}{2} \times 10^{-7} + 2 \times 10^7 \ln(D/r) \qquad H/m \tag{8-7}$$

Eq. (8-7) is often rearranged as follows

$$L_1 = 2 \times 10^{-7}\left[\frac{1}{4} + \ln(D/r)\right]$$

$$= 2 \times 10^{-7}[\ln \varepsilon^{1/4} + \ln(D/r)]$$

$$= 2 \times 10^{-7} \ln \frac{D}{r\varepsilon^{-1/4}} \tag{8-8}$$

The term $r\varepsilon^{-1/4}$ is known mathematically as the *geometric mean radius* of a circle with radius r. The term *geometric mean radius* is abbreviated *GMR*. For this conductor of circular cross section the $GMR = r\varepsilon^{-1/4} = 0.7788r$ and Eq. (8-8) becomes

$$L_1 = 2 \times 10^{-7} \ln \frac{D}{GMR} \qquad H/m \tag{8-9}$$

Equation (8-9) can be given an interesting physical interpretation. If the original conductor were replaced by a hollow tube with a thin wall and a radius equal to the GMR then Eq. (C-19) would lead directly to (8-9) since there would be no internal flux and no internal component of inductance!

For conductor cross sections other than circular we may find the *GMR* by mathematical means but it is more common to find that the conductor manufacturer gives the *GMR* in tables of conductor properties. In many cases the *GMR* in tables will be given in *feet*. If such is the case, it is not necessary to convert to metric units, but simply remember to express D in feet as well. Note that Eq. (8-9) still gives the inductance of one *meter* of conductor in such cases.

Equation (8-9) gives the inductance of one conductor. The inductance L_2 of the second conductor would be identical if the conductor has the same radius. In this case of single-phase, we often define the total inductance of the loop formed by the two conductors as

$$L = L_1 + L_2 = 4 \times 10^{-7} \ln \frac{D}{GMR} \qquad H/m \tag{8-10}$$

We may also choose to reform this last equation using logarithms to the base 10 and using different lengths for our unit.

$$L = 1.482 \log \frac{D}{GMR} \qquad mH/\text{mile} \tag{8-11}$$

or

$$L = 0.921 \log \frac{D}{GMR} \qquad mH/\text{km} \tag{8-12}$$

Any of the three forms of equations (8-10) to (8-12) may be used as suits the convenience of the specific problem at hand.

Three-phase inductance—Delta spacing

Next we consider a three-phase line with three conductors, each with radius, r, or with known GMR, and placed at the corners of an equilateral triangle as shown in Figure 8-8. Such an arrangement is known as *delta* spacing.

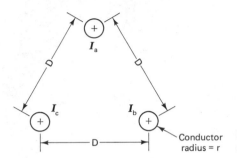

Conductor
radius = r

Figure 8-8 A three-phase line with delta spacing.

We will assume that the currents form a balanced three-phase set, $I_a + I_b + I_c = 0$, either in terms of phasor representation or in terms of instantaneous time functions. This is consistent with the earlier chapters, where primarily the balanced case was considered.

If the distance, center-to-center between the conductors is D, then Eqs. (C-12) and (C-19) give as the total inductance (internal plus external) of any one of the conductors as

$$L = \frac{1}{2} \times 10^{-7} + 2 \times 10^{-7} \ln(D/r)$$

which may be condensed as in the single phase case to

$$L = 2 \times 10^{-7} \ln(D/GMR) \qquad H/m \qquad (8\text{-}13)$$

Again it might be noted that, if we were evaluating the external component of inductance by integrating the flux linkages from the conductor surface to a distance x_2, we would take x_2 only to D, since beyond D the $\vec{H}$ and $\vec{B}$ vectors must be zero when no net current is encircled by the line integral used to evaluate $\vec{H}$.

From symmetry, the inductance, L, of Eq. (8-13) applies to any one of the conductors. The inductance may be referred to as a *per-phase* quantity. The equation may be modified as was the single-phase case to use logarithms to base ten and/or to express the distance along the line in miles or kilometers.

Three-phase inductance—Asymmetrical spacing

The delta configuration of conductors discussed above is convenient for analysis purposes, but many lines are built with an asymmetrical spacing between the conductors. Such a spacing is shown in Figure 8-9. With such a spacing, even with balanced currents, the voltage drops along the line will be unbalanced. To avoid this unbalanced voltage condition, it is common to *transpose* the conductors. The position of each conductor is moved periodically with distance down the line so that

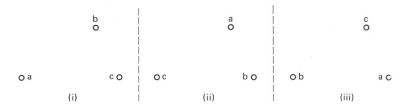

Figure 8-9 A three-phase line with asymmetrical spacing.

each conductor occupies each position for an equal distance of the total line length. The three elements of the figure show such a pattern of transposition.

In the evaluation of inductance, there are three different distances, D_{ab}, D_{bc}, and D_{ca}, involved. The equation for the inductance per phase becomes the same as Eq. (8-13), except that the distance D is replaced by the geometric mean of the three distances where*

$$GMD = \sqrt[3]{D_{ab} D_{bc} D_{ca}}$$

The inductance per phase in these terms becomes

$$L = 2 \times 10^{-7} \ln \frac{GMD}{GMR} \qquad H/m \qquad (8\text{-}14)$$

It is important to note that the meaning of any given distance symbol is given in terms of a reference or base configuration of the three conductors such as (i) of the above figure. We do not change the position of measurement of distance D_{ab}, for example, when the conductors move to a new position in the transposition scheme.

Alternate forms for the inductance per-phase follow as in the single-phase case

$$L = 0.7411 \log \frac{GMD}{GMR} \qquad mH/mile \qquad (8\text{-}15)$$

$$L = 0.4605 \log \frac{GMD}{GMR} \qquad mH/km \qquad (8\text{-}16)$$

For steady-state sinusoidal analysis we will want the inductive *reactance*, $X_L = \omega L = 2\pi f L$. For the particular case of $f = 60$ Hz the expressions for X_L become

$$X_L = 0.2794 \log \frac{GMD}{GMR} \qquad \Omega/mile/phase \qquad (8\text{-}17)$$

$$X_L = 0.1736 \log \frac{GMD}{GMR} \qquad \Omega/km/phase \qquad (8\text{-}18)$$

For typical high-voltage transmission lines at 60 Hz, the ratio of GMD/GMR is such that most lines have a reactance of approximately 0.8 ohm per phase per

---

*Verification of this assertion will not be pursued in these pages. For the interested student, it might be said that the process involves averaging the flux linkages of a sample conductor over each of the three transposition positions. We then use the property of logarithms that gives

$$\frac{1}{3}(\ln x_1 + \ln x_2 + \ln x_3) = \ln\sqrt[3]{x_1 x_2 x_3}$$

mile. If large "bundle" conductors are used as in EHV applications, the reactance may be more nearly 0.5 ohm per phase per mile. Distribution lines, on the other hand, may be as low as 0.25 ohm per phase per mile. Some examples follow which develop typical values.

Example
A 138 kV-transmission line has the configuration shown in Figure 8-10. Each phase conductor is "Drake" ACSR with 795 kcmil of cross-section. The *GMR* of the conductor is 0.0373 feet. Find the inductive reactance per mile of the line assuming balanced currents and complete transposition. Frequency is 60 Hz.

15'

28'

15'

Figure 8-10 Example configuration of transmission line.

Solution   From Eq. (8-17)

$$X_L = 0.2794 \log \frac{GMD}{GMR} \quad \Omega/\text{mile/phase}$$

$$GMD = \sqrt[3]{D_{ab} D_{bc} D_{ca}} = \sqrt[3]{15 \times 15 \times 28} = 18.47 \text{ feet}$$

Therefore

$$X_L = 0.2794 \log(18.47/0.0373) = 0.7529 \ \Omega/\text{phase/mile}$$

As a matter of practical interest, it might be mentioned that a manufacturer lists the resistance of the conductor as 0.1283 Ω/mile at 50 °C. The resulting X/R ratio is typical of high-voltage overhead lines.

Example
Figure 8-11 shows a typical configuration of a 12.47-kV underground distribution cable line. Find the inductive reactance in ohms/mile at 60 Hz, assuming balanced currents, complete transposition, and no metallic cable sheaths with circulating currents. Each conductor has a *GMR* of 0.03211 feet.

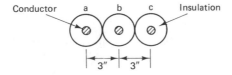

Conductor   a   b   c   Insulation

3"   3"

Figure 8-11   Configuration of underground cable line.

Solution

$$GMD = \sqrt[3]{0.25 \times 0.25 \times 0.50} = 0.3153 \text{ feet}$$
$$X_L = 0.2794 \log(0.3153/0.03211)$$
$$= 0.2772 \text{ ohm/phase/mile}$$

Also it may be noted that the manufacturer gives the resistance of this conductor as 0.1684, thus showing a ratio of X/R much smaller than that of a typical overhead transmission line.

8.8 LINE PARAMETERS—CAPACITANCE

If a transmission line is of appreciable length, the capacitance between conductors and/or the capacitance from conductor to earth must be considered in analyzing the performance of the line. In the case of cables, the higher permittivity of the insulating medium and the closer spacing of conductors makes the capacitance even more significant.

The physical constant, capacitance, comes from the statement that in a linear medium, charge and voltage between conductors are proportional according to the equation

$$Q = CV \tag{8-19}$$

In the case of an overhead line with a multitude of conductors, an analysis of the field to find the relation between charges and potential differences could be rather complex. Fortunately, we may use the principle of superposition and compute the effects of each conductor and its charge one at a time and then superpose. This is analogous to the methods used in the computation of inductance in the preceding section. In some of the cases of capacitance calculation, involving two or more parallel wires, there is a minor difficulty in this approach. The presence of a second wire (as an equipotential region) disturbs the field of a first wire, even if the second wire has no charge. We normally ignore this effect if the distance of separation of the wires is great with respect to the radii of the conductors. There is no analogous problem in the magnetic field case, since the permeability of copper and aluminum conductors is μ_0, the same as the rest of the region, and the field is not distorted by the presence of parallel conductors. We begin with what is perhaps the simplest example of capacitance calculation. Appendix D reviews the field theory background.

Capacitance of a coaxial cable

As a first example of capacitance calculations, consider the coaxial cable shown in Figure 8-12. There is a circular conductor in the center with radius r_1 and a concentric metal sheath with an inner radius r_2. The space between is filled with an insulating medium with a relative permittivity ε_r.

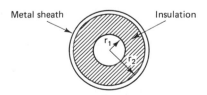

Figure 8-12   A coaxial cable illustration.

Equation (D-5) may be used except that we must recognize that the net permittivity is the product $\varepsilon_r \varepsilon_o$. We may then write

$$V_{12} = \frac{Q}{2\pi\varepsilon_r\varepsilon_o} \ln(r_2/r_1)$$

where Q is the charge per unit length on the inner conductor. The outer sheath will normally have a charge of $-Q$ per unit length but this charge does not contribute to the field *inside* the sheath. From Eq. (8-19) we recognize that capacitance is given by

$$C = Q/V$$

so we solve for the capacitance, which turns out to be

$$C_{12} = \frac{2\pi\varepsilon_r\varepsilon_o}{\ln(r_2/r_1)} \qquad \text{farads/m} \tag{8-20}$$

While we might perhaps think of r_1 and r_2 in meters, any units may be used since the two quantities appear as a ratio. Other forms of the equation may be given for other length units such as the mile, and examples of these alternates will be given in some of the following cases.

Capacitance of a two-wire line

Next we consider a single-phase circuit of two wires of radius r and separation D as shown in Figure 8-13. We will assume a charge per-unit length on conductor a of $+Q$ coulombs per meter and an opposite charge of $-Q$ coulombs per meter on conductor b. The potential difference from a to b may be found by using Eq. (D-5) to compute the contribution from each of the charges and then adding to get the total effect.

$$V_{ab} = \frac{Q}{2\pi\varepsilon_o} \ln(D/r) + \frac{-Q}{2\pi\varepsilon_o} \ln(r/D)$$

$$= \frac{Q}{2\pi\varepsilon_o} \ln(D^2/r^2)$$

and, from Eq. (8-19), $C = Q/V$

$$C_{ab} = \frac{2\pi\varepsilon_o}{\ln(D^2/r^2)} = \frac{\pi\varepsilon_o}{\ln(D/r)} \qquad \text{farads/m} \tag{8-21}$$

Recalling that $\varepsilon_o = 8.85 \times 10^{-12}$ we may put Eq. (8-21) in alternate forms for convenience in application:

$$C_{ab} = \frac{0.0194}{\log(D/r)} \qquad \mu\text{F/mile} \tag{8-22}$$

$$C_{ab} = \frac{0.0121}{\log(D/r)} \qquad \mu\text{F/km} \tag{8-23}$$

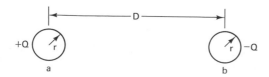

Figure 8-13   Cross section of a single-phase line.

In the preceeding equations, D and r can be in any convenient units, but both must be in the *same* units whatever is used. The radius, r, was taken to be that of a solid smooth conductor of circular cross section. For stranded conductors, we use the outer radius as illustrated in Figure 8-14. The slight distortion of the field in the "corners" of the strands is ignored in such a case.

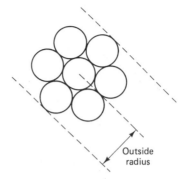

Outside radius

Figure 8-14 Cross section of a stranded conductor.

Capacitance to neutral

For some purposes, we find it convenient to define a capacitance to neutral as illustrated in Figure 8-15. In the figure we show two capacitances, C_{an} and C_{bn} to a neutral position (perhaps a point of zero or ground potential). From symmetry, we assume that $C_{an} = C_{bn}$. If the combination is to be equivalent to C_{ab}, then it must be true that

$$\frac{1}{C_{ab}} = \frac{1}{C_{an}} + \frac{1}{C_{bn}} = \frac{2}{C_{an}}$$

and

$$C_{an} = C_{bn} = 2C_{ab} \tag{8-24}$$

Eqs. (8-21) through (8-23) may then be modified to give capacitance to neutral as:

$$C_{an} = C_{bn} = \frac{2\pi\varepsilon_o}{\ln(D/r)} \qquad \text{F/m} \tag{8-25}$$

$$= \frac{0.0388}{\log(D/r)} \qquad \mu\text{F/mile} \tag{8-26}$$

$$= \frac{0.0241}{\log(D/r)} \qquad \mu\text{F/km} \tag{8-27}$$

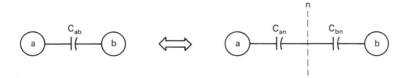

Figure 8-15 Illustration of capacitance to neutral.

For circuit computation purposes, we may wish to know the capacitive reactance, $X_C = -(1/\omega C)$, or capacitive susceptance $B_C = +\omega C$. Sample equations for these quantities for a frequency of 60 Hz appear below:

$$X_C = -0.0684 \log(D/r) \qquad \text{M}\Omega\text{-mile} \tag{8-28}$$

$$B_C = +\frac{14.63}{\log(D/r)} \qquad \mu\text{S/mile} \tag{8-29}$$

Capacitance of a three-phase line (delta spacing)

The simplest case of a three-phase line is the symmetrically spaced (delta) configuration illustrated in Figure 8-16.

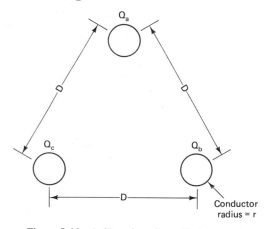

Figure 8-16   A three-phase line with delta spacing.

We assume that each of the conductors has a charge and that the three charges form a balanced set such that $Q_a + Q_b + Q_c = 0$. This equation may be interpreted as a relation between phasors representing the sinusoidal time function or as a relation between the time functions themselves. We assume a region of zero potential S meters from the array of conductors, where S is a distance so great that it is the same from any one of the three conductors. We may then find the potential drop from a sample conductor, perhaps a, by superposing the contributions of all the charges and using Eq. (D-5). We designate this voltage as V_{an} where

$$V_{an} = \frac{Q_a}{2\pi\varepsilon_o} \ln(S/r) + \frac{Q_b}{2\pi\varepsilon_o} \ln(S/D) + \frac{Q_c}{2\pi\varepsilon_o} \ln(S/D)$$

since

$$Q_b + Q_c = -Q_a$$

$$V_{an} = \frac{1}{2\pi\varepsilon_o} [Q_a \ln(S/r) - Q_a \ln(S/D)]$$

$$= \frac{Q_a}{2\pi\varepsilon_o} \ln(D/r)$$

and since $C = Q/V$

$$C_{an} = \frac{2\pi\varepsilon_o}{\ln(D/r)} \qquad \text{farad/m} \qquad\qquad (8\text{-}30)$$

As with the single-phase case Eq. (8-30) may be altered to use logarithms to base ten and use other units for length of conductor.

$$C_{an} = \frac{0.0388}{\log(D/r)} \qquad \mu\text{F/mile} \qquad\qquad (8\text{-}31)$$

$$C_{an} = \frac{0.0241}{\log(D/r)} \qquad \mu\text{F/km} \qquad\qquad (8\text{-}32)$$

It will be noted that these expressions happen to be the same as those for the capacitance of the single-phase line to neutral. Likewise, Eqs. (8-28) and (8-29) for reactance and susceptance at 60 Hz are valid for the three-phase delta spaced line.

Asymmetrically spaced three-phase line

If a three-phase line is asymmetrically spaced as in Figure 8-9 but transposed for balance, the capacitance may be found from equations the same as those for the delta spaced line except that the distance, D, must be replaced by the geometric mean distance (GMD) of the three conductors as for example

$$C_{an} = \frac{2\pi\varepsilon_o}{\ln \dfrac{GMD}{r}} \qquad \text{F/m} \qquad\qquad (8\text{-}33)$$

where

$$GMD = \sqrt[3]{D_{ab}D_{bc}D_{ca}}$$

As in the case of the inductance of the asymmetrical three-phase line, the details of derivation will be omitted. Expressions for reactance, X_C, and susceptance, B_c, will likewise be the same as Eqs. (8-28) and (8-29) but with the GMD used instead of D. Examples of capacitance calculation follow.

Example

Consider the asymmetrically spaced overhead line shown in Figure 8-10. The line is completely transposed. The conductor has an outer diameter of 1.108 inch.

a. What is the capacitance per phase of one meter of this line?
b. What is the capacitive susceptance, B_C, of ten miles of this line at 60 Hz?

Solution

a. $\qquad GMD = \sqrt[3]{15 \times 15 \times 28} = 18.47$ feet

$$C_{an} = \frac{2\pi 8.85 \times 10^{-12}}{\ln \dfrac{18.47}{1.108 \times 1/2 \times (1/12)}} = 9.28 \times 10^{-12} \text{ farad/m}$$

$$= 9.28 \text{ picofarad/m}$$

b.   Modifying Eq. (8-31)

$$C_{an} = \frac{0.0388}{\log \dfrac{GMD}{r}} \quad \mu F/mile$$

$$B_C \text{(ten miles)} = 2\pi 60 \frac{0.0388 \times 10^{-6}}{\log \dfrac{18.47}{1.108 \times 1/2 \times (1/12)}} \times 10$$

$$= 56.2 \times 10^{-6} \ S$$
$$= 56.2 \ \mu S$$

Example

Given a coaxial cable like that sketched in Figure 8-12: The central conductor has an outside diameter of one inch. There is 0.5 inch of insulating material around the conductor. The insulating material has a relative permittivity of 3.5. There is an outer lead sheath.

a. What is the capacitance per meter of this cable?
b. If a line-to-ground (sheath) voltage of 34 kV is applied to a five-mile length of this cable, how much charging current flows?

Solution

a.   From Eq. (8-20)

$$C = \frac{2\pi 3.5 \times 8.85 \times 10^{-12}}{\ln(1.0/0.5)} = 280 \times 10^{-12} \ farad/m$$

$$= 280 \ picofarad/m$$

b. $$B_C = 2\pi 60(280 \times 10^{-12})5 \times 1609 = 8.49 \times 10^{-4} \ S$$
$$I = 34 \times 10^3 \times 8.49 \times 10^{-4} = 28.9 \ amperes$$

8.9 LINE PARAMETERS—SHUNT CONDUCTANCE

The fourth parameter which characterizes lines and cable is the shunt conductance from line to ground or neutral. In the case of overhead lines, the major contribution to shunt conductance is the leakage over insulators. An appreciable leakage current can affect the transmission line performance, and it represents a power loss, but fortunately this effect is seldom large enough to be significant. Occasional exceptions occur in regions where insulators may be contaminated by salt spray, dust, or chemical deposits. In such cases the insulators may be cleaned periodically to reduce the extent of the problem.

In the case of cables, shunt conductance represents leakage currents through the cable insulation. This effect is very difficult to evaluate quantitatively. The leakage is sensitive to temperature and especially to any moisture that might penetrate the cable. For cable in good condition, the effect is of little importance in relation to line performance, but leakage resistance might be measured as an index to cable condition.

In line analysis we sometimes represent the effect by a shunting element from line to neutral of G siemens conductance per-unit length of line, but most often this

element is ignored, partly because of the unimportance and partly because of igno-
rance of the actual magnitude of the quantity. The net shunting admittance,
$Y = G + jB_C$, may be included in the mathematical analysis but in numerical
solution we will assume G to be zero.

8.10 THE DIFFERENTIAL EQUATIONS

A transmission line or cable represents a *distributed* system wherein the constants
such as resistance, inductance, and capacitance are not lumped in discrete packages
but are smoothly distributed along the length of the line. Such a system does not
yield directly to usual circuit solution methods, where the elements are represented
in discrete packages or *lumps*. Because of the distributed nature of the constants, the
voltage and current variables are functions of both time and distance along the line.
Thus the system equations are *partial* differential equations in the general case. The
partial differential equations present intolerable difficulties in obtaining an explicit
solution, except when simplifying assumptions are made. Fortunately for our
present purposes we are primarily concerned with steady-state sinusoidal behavior
of the lines. We can thus use the *phasor* representation of voltage and current and
our variables become functions of distance alone. The differential equations become
ordinary differential equations for which solution is much simpler. We now proceed
to find the equations describing a line's behavior by means of phasor representation.

   Consider Figure 8-17 which represents one phase to neutral of a line between
a source voltage at one end and a load at the other. Variables are designated on the
diagram for purposes of mathematical treatment. A key element in the description
is a small segment of line dx long. We assume that the line parameters have been
evaluated and we know the series impedance elements per-unit of length,
$Z = R + jX_L$, as well as the shunt admittance elements per unit of length,
$Y = G + jB_C$. In Figure 8-18 we show an enlarged view of the small segment for
greater clarity in what follows.

   In terms of the references of the figures, we may write the Kirchhoff's law
equations

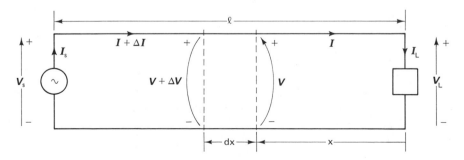

Figure 8-17   Representation of a transmission line.

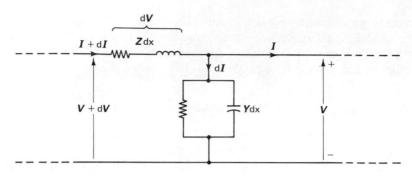

Figure 8-18 An infinitesimal increment of the line.

$$dV = IZ\,dx$$
$$dI = VY\,dx$$

or

$$\frac{dV}{dx} = IZ \tag{8-34}$$

$$\frac{dI}{dx} = VY \tag{8-35}$$

which are the simultaneous first-order differential equations describing the line behavior. To solve we also need boundary values on the variables which are $V = V_L$, and $I = I_L$ at $x = 0$.

Subject to the boundary values, we find the solutions to be

$$V(x) = V_L\left[\frac{\varepsilon^{\gamma x} + \varepsilon^{-\gamma x}}{2}\right] + I_L Z_0\left[\frac{\varepsilon^{\gamma x} - \varepsilon^{-\gamma x}}{2}\right] \tag{8-36}$$

$$I(x) = \frac{V_L}{Z_0}\left[\frac{\varepsilon^{\gamma x} - \varepsilon^{-\gamma x}}{2}\right] + I_L\left[\frac{\varepsilon^{\gamma x} + \varepsilon^{-\gamma x}}{2}\right] \tag{8-37}$$

where

$$Z_0 \triangleq \sqrt{\frac{Z}{Y}} = \sqrt{\frac{R + j\omega L}{G + j\omega C}} \tag{8-38}$$

and

$$\gamma \triangleq \sqrt{ZY} = (\alpha + j\beta) \tag{8-39}$$

The quantities Z_0 and γ are known as the *characteristic impedance* and *propagation constant* respectively.

The details of solution to arrive at Eqs. (8-36) and (8-37) will not be given here. The curious student may verify the solution by substituting (8-36) and (8-37) into (8-34) and (8-35) to show that the solutions do indeed satisfy the differential equations and the boundary values. This is a time-honored method of solution of differential equations anyway—assume an answer and verify.

Equations (8-36) and (8-37) are often put in an alternate form as

$$V(x) = V_L \cosh(\gamma x) + I_L Z_0 \sinh(\gamma x) \tag{8-40}$$

$$I(x) = \frac{V_L}{Z_0} \sinh(\gamma x) + I_L \cosh(\gamma x) \tag{8-41}$$

The equations (8-36) through (8-41) describe the voltage and current as a function of the distance, x. In this work we are primarily interested in the overall performance of the line for the total length, l, from the load end (often known as the *receiving* end) to the source end (often known as the *sending* end). Thus we insert the value $x = l$ and the equations become

$$V_S = V_L \cosh(\gamma l) + I_L Z_0 \sinh(\gamma l) \tag{8-42}$$

$$I_S = \frac{V_L}{Z_0} \sinh(\gamma l) + I_L \cosh(\gamma l) \tag{8-43}$$

We now proceed to use these equations to develop a model for the transmission line.

8.11 THE MODEL OF A TRANSMISSION LINE

Equations (8-42) and (8-43) give us an "exact" solution for the performance of a transmission line with distributed constants between two busses, designated here as the source and load busses.* For some circuit solution purposes it is convenient to have a lumped parameter circuit model to use in circuit computations, either by pencil, paper, and a calculator, or perhaps by a digital computer program that is capable of handling combinations of lumped circuit elements but not the distributed parameter equations. Circuit theory tells us that it is always possible to find an equivalent pi section to replace any given two-terminal pair network at a given frequency. Such a pi section is shown in Figure 8-19. We may find the elements of the pi such that the circuit will satisfy Eqs. (8-42) and (8-43) by suitable algebraic manipulation. It turns out that

$$Z_\pi = Z_0 \sinh(\gamma l) \tag{8-44}$$

$$\frac{Y_\pi}{2} = \frac{1}{Z_0} \frac{\cosh(\gamma l) - 1}{\sinh(\gamma l)} \tag{8-45}$$

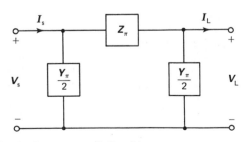

Figure 8-19 Equivalent pi section for a line.

*At any rate as "exact" as we know the circuit parameters Z, Y and l.

If these equations are used to evaluate the elements of the pi then we do indeed have an *equivalent pi* which represents the line section exactly at the given frequency.

The hyperbolic sine and cosine function have complex number arguments and are themselves complex numbers. Sometimes the functions are approximated by using their series expansions

$$\cosh(\gamma l) = 1 + \frac{(\gamma l)^2}{2!} + \frac{(\gamma l)^4}{4!} + \cdots \tag{8-46}$$

$$\sinh(\gamma l) = (\gamma l) + \frac{(\gamma l)^3}{3!} + \frac{(\gamma l)^5}{5!} + \cdots \tag{8-47}$$

The series converges rapidly for small values of l, the line length. Various approximations are used based on using just a few terms of the series.

The short line model

If a line is short (l is small), we might use just the first term of each series, from which we find

$$Z_\pi = Zl \tag{8-48}$$

$$\frac{Y_\pi}{2} = 0 \tag{8-49}$$

In other words, for a short line, we simply use a single series impedance equal to the total series impedance of the line, Zl. Also, for the short line, we ignore the shunt admittance (which is largely capacitance). What constitutes a "short" line deserves discussion and depends on the relative values of Z and Y vis-a-vis the line length. Treatment of this question will be presented in example problems to follow. Figure 8-20 shows the model for this case.

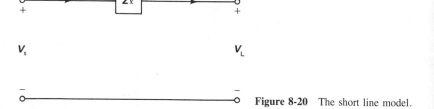

Figure 8-20 The short line model.

Medium line model—the nominal pi

If we desire more precision we may use just the first term of the sinh series but the first *two* terms of the cosh series. After doing this we find

$$Z_\pi = Zl$$

$$\frac{Y_\pi}{2} = \frac{Yl}{2}$$

which amounts to using the total series impedance for Z_π and dividing the total shunt

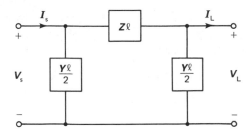

Figure 8-21 The nominal pi representation.

admittance into two pieces, one at each end of the line. This is known as the *nominal pi* approximation.

Figure 8-21 shows the model for this case.

In summary, three models have been presented: the short line model, the nominal pi or medium line model, and the equivalent pi or "exact" model of Figures 8-20, 8-21, and 8-19, respectively. The question of which model should be used depends upon the length of the line, the relative values of the line parameters, the precision desired, and the particular problem to be solved. General answers to the question of model selection cannot be given, but some numerical examples of typical situations will help to give an insight into the situation.

Examples and discussion

As an example case consider the line shown in Figure 8-10 for which the parameters were found to be:

$$X_L = 0.7529 \ \Omega/\text{mile}$$
$$R = 0.1283 \ \Omega/\text{mile} \qquad \qquad Z = 0.7638\underline{/80.33°} \ \Omega/\text{mile}$$
$$B_C = 5.62 \times 10^{-6} \ S/\text{mile}$$
$$G = 0 \qquad \qquad Y = 5.62 \times 10^{-6}\underline{/90°} \ S/\text{mile}$$

From these line constants we find

$$Z_0 = \sqrt{\frac{0.1283 + j0.7529}{0 + j5.62 \times 10^{-6}}} = 368.65\underline{/-4.84°} \ \Omega$$

$$= 367.33 - j31.07 \ \Omega$$

$$\gamma = \sqrt{(0.1283 + j0.7529)(0 + j5.62 \times 10^{-6})}$$

$$= 0.00207\underline{/85.16°} = 0.000175 + j0.00206$$

We now assume various value for l, the line length and compute Z_π and $Y_\pi/2$ for the equivalent pi from Eqs. (8-43) and (8-44). The values for the nominal pi are simply obtained from $Z_\pi = Zl$ and $Y_\pi/2 = Yl/2$. The impedance of the short line model is of course the same as the Z_π of the nominal pi model.

The results of such computation are tabulated in Table 8-2.

A comparison of the values in the table reveals very little difference between the nominal (approximate) and the equivalent (exact) pi models for lines of 150 miles and less. For lines longer than 150 miles, there is a noticeable difference. For the line used in the example of 138 kV, a line longer than 150 miles would be very

TABLE 8-2   Examples of Constants of Line Models

l	Z_π (Ω)		$Y_\pi/2$ (μS)	
	Nom.	Equiv.	Nom.	Equiv.
10	7.638	7.637	28.10	28.10
30	22.914	22.898	84.30	84.33
50	38.190	38.120	140.50	140.62
100	76.380	75.838	281.00	281.99
150	114.570	112.754	421.50	424.87
300	229.140	214.863	843.00	870.79
500	381.900	318.094	1405.00	1543.21

unusual, but for lines of EHV (Extra High Voltage) construction we do find lines of 500 miles in length and so it may be necessary or desirable to use the *equivalent* pi model for such lines.

The choice of model is also affected by the particular problem at hand. For example, for short-circuit computations, where the short-circuit current is much greater than the shunt capacitive current, the short line model ignoring shunt admittance is usually used. For voltage regulation problems, the short line approximation even for lines of say 30 miles in length may not be adequate. This fact will be illustrated in examples to follow in the next section.

8.12 PERFORMANCE MEASURES—VOLTAGE REGULATION

An index to the performance of power apparatus, which applies to lines as well as other apparatus is *voltage regulation*. Voltage regulation was discussed in relation to transformers in Section 3.14 of Chapter 3 and illustrated by Figure 3-35, which the reader may wish to review at this time. The matter was further pursued in Problem 7 of the homework problems of Chapter 3.

If we use the short line model for a line, ignoring the shunt admittance, the computations and diagrams for the line are the same as those for other apparatus, except for the particular designation of the parameters and variables. To review in terms of line descriptions, we consider the change in the voltage magnitude, $|V_L|$, at the load end when a specified load is reduced to zero with the source voltage magnitude, $|V_S|$, maintained constant. *Voltage regulation* is then the *change* expressed as a percentage of the normal rated load voltage, $|V_L|$, under the specified load.

$$\% \text{ Voltage regulation} = \frac{|V_{L(\text{N.L.})}| - |V_{L(\text{load})}|}{|V_{L(\text{load})}|} \times 100 \qquad (8\text{-}50)$$

Since we usually wish to maintain constant potential on our load apparatus, good performance calls for a *small* value of voltage regulation. The specific value

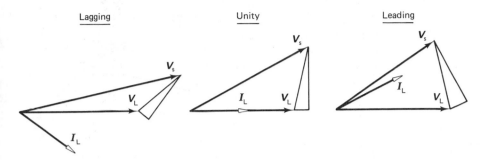

Figure 8-22 Phasor diagrams showing the $V_S - V_L$ relations of a short line model at different power factors.

depends upon the magnitude of the load current and the power factor of the load. This may be seen from the illustration of Figure 8-22 based on the short line model for the line.

It may be seen from the diagrams that the voltage regulation for these sample cases is greatest for the current lagging case and least for the current leading case. Indeed the regulation might be zero or even numerically negative for the current leading case. Note that, for the case of the short line model the load voltage and the source voltage are equal with no load current ($IZ = 0$).

If the shunt admittance is considered with one of the more precise models, we may find that the load end voltage, V_L is actually larger than the source voltage, V_S at zero load current. This is illustrated in Figure 8-23 which shows a nominal pi circuit at no load and the corresponding phasor diagram. The possible rise in load end voltage magnitude with constant source voltage when the load is removed is known as the *Ferranti effect*. If this rise in voltage level is enough to be undesirable, shunt inductive reactors may be used on the line to counter the effect. We follow with an example of voltage regulation computations with the three different line models used.

Examples
Suppose we have a 345-kV line with the following parameters:
$$Z = 0.1 + j0.6 \quad \Omega/\text{mile}$$
$$Y = 0 + j8.0 \times 10^{-6} \quad S/\text{mile}$$
$$Z_0 = \sqrt{\frac{0.1 + j0.6}{0 + j8 \times 10^{-6}}} = 275.74\underline{/-4.73°} \ \Omega$$
$$= 274.80 - j22.74 \ \Omega$$
$$= \sqrt{(0.1 + j0.6)(0 + j8 \times 10^{-6})} = 0.00221\underline{/85.27°} \ \Omega$$
$$= 0.000182 + j0.00220 \ \Omega$$

We will consider a case where the line delivers 500 MW at unity power factor to the load. With a load voltage of 345 kV, or 199 kV line-to-neutral, the current will be $500 \times 10^6 / \sqrt{3} \times 345 \times 10^3$ amperes. We will consider line lengths of 30, 150, and 300 miles. We will compute the source end voltage, the load end voltage, and the percent voltage regulation.

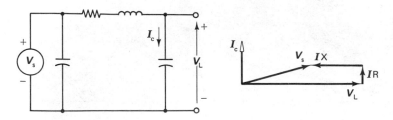

Figure 8-23   An illustration of Ferranti effect.

Each of the three line models will be used so that a comparison of results may be made. The results of the computations are given in Table 8-3.

TABLE 8-3   **Voltage Regulation for Example Cases**

	l	V_S	$V_{L(N.L.)}$	$V_{L(load)}$	% Regul.
Equiv.	30	201.83	202.27	199.19	1.55
Pi	150	214.45	226.57	199.19	13.74
Model	300	231.78	292.60	199.19	46.90
Nominal	30	201.83	202.27	199.19	1.55
Pi	150	215.26	227.54	199.19	14.23
Model	300	240.32	306.20	199.19	53.
Short	30	202.25	202.25	199.19	1.54
Line	150	224.73	224.73	199.19	12.82
Model	300	270.17	270.17	199.19	35.6

All voltages are in kV, line-to-neutral.

Much can be said about the adequacy of the three-line models by study of the results above. We note:

1. The nominal pi model is quite satisfactory for lines of 150 miles and below — it compares favorably with the (exact) equivalent pi representation.
2. The short line model is adequate for lines up to 30 miles in length.
3. The short line model is incapable of showing the Ferranti effect because shunt capacitance is not included.

8.13 LINE LOADING

Thermal limits

The voltage level of a line is limited by such things as insulation, conductor clearance, electromagnetic field effects on nearby objects, and radio and audio noise owing to ionization near the conductor surface.

The current handling ability of a line is determined by the conductor cross-sectional area and the environment in which the conductor operates. An overhead line with bare conductors can handle more current without overheating than can the same size conductor in a cable. Unfortunately the materials that form good electrical insulators are also good thermal insulators, thus preventing the heat from the I^2R losses from escaping to the external environment. Thermal considerations have a major effect on the load capacity of a cable line. Thermal limits for overhead lines are usually less significant. A design that limits the I^2R losses for efficiency of transmission tends to reduce the thermal problems of the overhead line.

The thermal limits of given conductors may be found from manufacturer's data and from the publications of the American National Standards Institute (ANSI). If we know the current rating of the conductor, the thermal loading limit of a line is

$$S = \sqrt{3} \ V_{l\text{-}l}I_{(\text{max thermal})} \qquad (8\text{-}51)$$

Maintenance of synchronism

As a first approximation, we often represent a line as a series inductive reactance between the voltages, V_L and V_S, at the load and source ends. The inductance of a line was found to vary as $\ln(D/r)$, and the factors that control the separation distance, D, and the conductor radius, r, work together to keep the logarithm of the ratio within narrow limits of being the same for all lines. The reactance, X_L, of most single conductor lines at 60 Hz is near 0.8 ohm per phase per mile, thus the total inductive reactance of a line is determined closely when we know the line length.

A review of Section 5.8 will remind the reader that there is a maximum power which can be transmitted between two voltage sources through an inductive reactance. In terms of the notation of this chapter

$$P = \frac{V_S V_L}{X} \sin \delta \qquad (8\text{-}52)$$

and

$$P_{\text{max}} = \frac{V_S V_L}{X} \qquad (8\text{-}53)$$

where δ is the phase angle between V_S and V_L and X is the total inductive reactance between the two voltages. Any attempt to pass more power than P_{max} will result in loss of synchronism (instability). If we find that the level of P_{max} is too low to meet the needs of transmission, we may increase the voltage or decrease the reactance, perhaps by cancelling some of the reactance by series capacitors.

The limit of Eq. (8-53) cannot be approached very closely because Eq. (8-52) applies only between constant voltages if phase angle, δ, is the variable. There is always impedance "behind" the busses, so the X total is greater than that of the line and the constant voltages are those "behind" the total reactance, and it is these voltages that must not have a phase difference, δ, of more than 90°. A study of these matters is undertaken in power system analysis courses under the heading of *"Steady-State Stability."*

Surge Impedance Loading

It is interesting to observe a special case where a line is loaded by being terminated in an impedance of Z_0 per phase such that

$$I_L = V_L/Z_0$$

If we insert this value of I_L into equations (8-42) and (8-43) and then take the quotient of these two equations we find

$$Z_S = \frac{V_S}{I_S} = Z_0 \qquad (8\text{-}54)$$

which tells us that the impedance "looking into" the source end will be equal to Z_0, no matter how long the line!

We sometimes use a load of this nature as an index to the relative magnitude of load on a given line. In this connection it is common to consider the special case of a lossless line ($R = 0$ and $G = 0$). For a lossless line

$$Z_0 = \sqrt{\frac{R + j\omega L}{G + j\omega C}} = \sqrt{\frac{L}{C}} \qquad (8\text{-}55)$$

The characteristic impedance, Z_0, thus reduces to a purely real number which may be represented by a resistance. This special value of Z_0 is known as the *surge impedance*. Lines are often assumed lossless for investigating the traveling waves impressed on lines by lightning surges, which may help to explain the origin of the name.

If a power line is terminated in the surge impedance, then from Eq. (8-54) we see that the impedance looking into the source end will also be equal to the surge impedance; that is, a pure resistance equal to $\sqrt{L/C}$. The physical significance of this case is that we have unity power factor at both ends of the line and the reactive voltamperes of the series inductance, $I^2 X_L$ and the shunt capacitance $V^2 B_C$ exactly balance each other. If we draw more current than that of surge impedance loading, there will be an excess $I^2 X_L$ required and inductive reactive volt-amperes, Q, must be supplied by the source. Conversely, a current smaller than that of surge imped-ance loading will result in an excess of capacitive reactive voltamperes (numerically negative) and negative Q must be supplied by the source.

Surge impedance loading is thus a pivot point in considering whether a load is relatively light or heavy for a given line. The actual value of this significant load is given by

$$P = \frac{|V|^2}{|Z_0|} = \frac{V^2}{Z_0} \text{ W} \qquad (8\text{-}56)$$

We note that, if V is the line-to-neutral voltage, P is the power per phase, and if V is the line-to-line voltage, P is the total three-phase power. It is usually convenient to express V in kilovolts, in which case P will be in megawatts (with Z_0 in ohms).

ADDITIONAL READING MATERIAL

1. Del Toro, V., *Electric Machines and Power Systems,* Englewood Cliffs, New Jersey: Prentice-Hall, Inc., 1985.

2. Elgerd, O. I., *Basic Electric Power Engineering,* Reading, Massachusetts: Addison-Wesley Publishing Company, 1977.

3. Elgerd, O. I., *Electric Energy Systems Theory: An Introduction,* New York: McGraw-Hill Book Company, 1982.

4. Gross, C. A., *Power System Analysis,* New York: John Wiley & Sons, 1979.

5. Nasar, S. A., *Electric Energy Conversion and Transmission,* New York: MacMillan Publishing Company, 1985.

6. Shultz, R. D., and R. A. Smith, *Introduction to Electric Power Engineering,* New York: Harper & Row, Publishers, 1985.

7. Stevenson, W. D., Jr., *Elements of Power System Analysis,* New York: McGraw-Hill Book Company, 1982.

8. Weeks, W. L., *Transmission and Distribution of Electrical Energy,* New York: Harper & Row, Publishers, 1981.

STUDY EXERCISES

1. Consider the 12.47-kV distribution line shown in Figure 8-4. The conductor is No. 4 copper, which has an outer diameter of 0.254 inch and a *GMR* of 0.00717 foot. Approximate carrying capacity is 200 A. $R = 1.5$ ohms per mile. The central conductor is elevated 8 inches over those on the side. The line is completely transposed.
   a. Find Z in ohms per mile.
   b. Find Y in siemens per mile.
   c. If 200 amperes is the thermal limit what is the total load capacity?

2. Consider again the coaxial cable of Figure 8-12, which was described in the example of Section 8.8. Suppose we have a 15-mile length of this cable with 34 kV impressed from line-to-ground at the source end. With a zero-impedance short-circuit at the load end the short-circuit current will return through the sheath. We find that the conductor resistance is 0.07 ohm/mile and that of the sheath is 0.5 ohm/mile. The *GMR* of the inner conductor is 0.035 foot. It is assumed that the thickness of the sheath is small compared to its inner radius. Frequency $= 60$ Hz.
   a. If we neglect shunt capacitance what is the Z_π of the cable (short line approximation)?
   b. What is the fault (short-circuit) current in amperes if the 34-kV voltage is maintained at the source end with the fault in place?
   c. If the shunt capacitance is included, using the nominal pi model, how much does the answer to part (b) change?

3. Suppose that we have a complex number value $\gamma l = 0.8\underline{/70°}$, what are $\cosh \gamma l$ and $\sinh \gamma l$?

4. Consider a cable with $R = 0.2$ ohm/mile, $X_L = 0.3$ ohm/mile, and $B_C = 3.5 \times 10^{-4}$ S/mile. (Per-phase values of three-phase system)
   a. What are Z_0 and γ, the characteristic impedance and the propagation constant?
   b. For three cases of line length, 10, 30, and 100 miles, find and tabulate the elements of the three-line models (short line, nominal π, equivalent π). Note the similarity and difference from the example of Table 8-2.

5. Consider again the cable with the constants of Problem 4. Suppose that the cable operates with 34.5 kV to ground (line-to-neutral) at the load end when delivering 300 A at unity power factor to a load. For each of the line lengths and line models of Problem 4 (9 cases) find V_S, the source voltage required (assumed held constant when the load is removed). Also find V_L when the load is removed, and the percent voltage regulation. Tabulate the results in a form similar to Table 8-3 for the overhead line example.

6. Consider a 200-mile, 230-kV, three-phase line with $Z = 0 + j0.8$ ohm/mile and $Y = 0 + j5.4 \times 10^{-6}$ S/mile—losses are to be neglected. The line is terminated at the load end in the surge impedance and the voltage V_L is maintained at 230 kV, line-to-line by proper adjustment of the source voltage V_S. We will take $V_L = (230/\sqrt{3})\underline{/0°}$ as the phasor representing the load-end voltage of the reference phase.

  a. What is I_L of the reference phase?

  b. What is the total complex power at the load end?

  c. What are the phasor voltage V_S and phasor current, I_S, at the source end?

  d. What is the total complex power at the source end?

  e. Repeat parts (b) through (d) if the magnitude of the load current is increased 25 percent over that of part (a).

  f. Repeat parts (b) through (d) if the magnitude of the load current is decreased 25 percent under that of part (a).

SOLUTIONS TO STUDY EXERCISES

1. a.

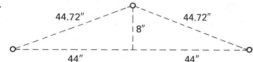

$$GMD = \sqrt[3]{\frac{(44.72)^2 88}{12^3}}$$

$$= 4.67 \text{ feet}$$

$$X_L = 0.2794 \log \frac{4.67}{0.00717} = 0.7862 \ \Omega/\text{mile}$$

$$Z = 1.5 + j0.7862 \qquad \Omega/\text{mile} \quad \longleftarrow \quad a$$

b. $B_C = \dfrac{14.63}{\log \dfrac{4.67 \times 2 \times 12}{0.254}} = 5.532 \ \mu \text{ siemen/mile}$

$$Y = 0 + j5.532 \times 10^{-6} \text{ siemen/mile} \quad \longleftarrow \quad b$$

c. $S = \sqrt{3} \times 12.47 \times 200 = 4319 \text{ kVA} \quad \longleftarrow \quad c$

2. a.

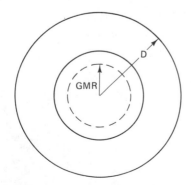

$$\lambda = 2 \times 10^{-7} \ln \frac{D}{GMR}$$

$$X_L = 0.2794 \log \frac{1/12}{0.035}$$

$$= 0.1053 \ \Omega/\text{mile}$$

$$X_{tot} = 15 \times 0.1053 = 1.5795 \ \Omega$$

$$R_{tot} = 15(0.07 + 0.5) = 8.55 \ \Omega$$

$Z = \underline{8.55 + j1.5795\ \Omega}\quad\longleftarrow\quad$ a

b. $I = \dfrac{34 \times 10^3\underline{/0°}}{8.55 + j1.5795} = \underline{3910\underline{/-10.5°}\ A}\quad\longleftarrow\quad$ b

$I_C = 34 \times 10^{-3}\underline{/0°}(j8.49 \times 10^{-4})(15/5)(1/2)$
   $= j43.3\ A$

$I_{tot} = 3910\underline{/-10.5^5} + j43.3$
$|I_{tot}| = 3902.8$
Change is only $\underline{7.2\ A}\quad\longleftarrow\quad$ c

3. The method of evaluation of hyperbolic functions of complex number arguments will depend strongly upon the type of calculator or computer assistance which is available to the student. It will be assumed that the student has available as a minimum a calculator with exponential and trig functions.

Method a — Note that:

$$\cosh(\alpha + j\beta) = \frac{\varepsilon^{(\alpha+j\beta)} + \varepsilon^{-(\alpha+j\beta)}}{2}$$

$$\sinh(\alpha + j\beta) = \frac{\varepsilon^{(\alpha+j\beta)} - \varepsilon^{-(\alpha+j\beta)}}{2}$$

and

$$\varepsilon^{(\alpha+j\beta)} = \varepsilon^{\alpha}\varepsilon^{j\beta}$$
$$\varepsilon^{-(\alpha+j\beta)} = \varepsilon^{-\alpha}\varepsilon^{-j\beta}$$

In the present case
   $\gamma l = 0.8\underline{/70°} = 0.2735 + j0.7517$
therefore
   $\alpha = 0.2735,\ \beta = 0.7517$
(Note that α is in *nepers* and β is in *radians*)
   $\varepsilon^{(\alpha+j\beta)} = 0.9604 + j0.8978$
   $\varepsilon^{-(\alpha+j\beta)} = 0.5556 - j5194$

$$\cosh(\gamma l) = \frac{0.9604 + j0.8978 + 0.5556 - j0.5194}{2}$$

$$= 0.7580 + j0.1892\quad\longleftarrow$$

$$\sinh(\gamma l) = \frac{0.9604 + j0.8978 - 0.5556 + j0.5194}{2}$$

$$= 0.2023 + 0.7087\quad\longleftarrow$$

Method b — Use the series:

$$\cosh(\gamma l) = 1 + \frac{(\gamma l)^2}{2!} + \frac{(\gamma l)^4}{4!} + \cdots$$

   $1 \longrightarrow 1$
   $(\gamma l)^2/2! \longrightarrow -0.2451 + j0.2056$
   $(\gamma l)^4/4! \longrightarrow 0.00296 - j0.0168$

$$\Sigma \longrightarrow \quad \underline{0.7578 + 0.1889} \longleftarrow$$

$$\sinh(\gamma l) = (\gamma l) + \frac{(\gamma l)^3}{3!} + \frac{(\gamma l)^5}{5!} + \cdots$$

$$(\gamma l) \longrightarrow \quad 0.2736 + j0.7517$$
$$(\gamma l)^3/3! \longrightarrow \quad -0.0739 - j0.0427$$
$$(\gamma l)^5/5! \longrightarrow \quad \underline{0.0027 - j0.00047}$$

$$\Sigma \longrightarrow \quad 0.2024 + j0.7086 \longleftarrow$$

4. a.

$$Z_0 = \sqrt{\frac{0.2 + j0.3}{0 + j3.5 \times 10^{-4}}} = 32.10\underline{/-16.85°}\ \Omega$$

$$= \underline{30.72 - j9.30} \longleftarrow \quad a$$

$$\gamma = \sqrt{(0.2 + j0.3)(0 + j3.5 \times 10^{-4})}$$
$$= 0.01123\underline{/73.15°}\ S = \underline{0.003255 + j0.01075} \longleftarrow \quad a$$

b. Short Line: $Z_\pi = Zl$, $Y_\pi/2 = 0$

l	Z_π
10	$2 + j3$
30	$6 + j9$
100	$20 + j30$

Nominal Pi: $Z_\pi = Zl$ (same as short line)
$$Y_\pi/2 = Yl/2$$

l	$Y_\pi/2$
10	$0 + j17.5 \times 10^{-4}$
30	$0 + j52.5 \times 10^{-4}$
100	$0 + j175 \times 10^{-4}$

Equivalent Pi: $Z_\pi = Z_0 \sinh(\gamma l)$
$$Y_\pi/2 = \frac{1}{Z_0} \frac{\cosh(\gamma l) - 1}{\sinh(\gamma l)}$$

l	Z_π	$Y_\pi/2$
10	$1.993 + j2.997$	$0.0000012 + j0.001752$
20	$5.812 + j8.921$	$0.0000281 + j0.00529$
100	$13.4594 + j27.0013$	$0.00126 + j0.01911$

Comment: Note by comparison with the example given in the text for a typical overhead line that the approximations used for a cable for a short, medium, or long line must be different because of the much higher B_C and its effect on the constants of the circuit models.

5.

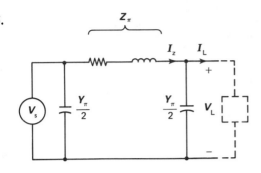

For equivalent and nominal pi section:

$V_S = V_{L(load)} + Z_\pi I_Z$ where $I_Z = V_{L(load)} Y_\pi/2 + I_L$

$V_{L(N.L.)} = V_S \left[\dfrac{2/Y_\pi}{Z_\pi + 2/Y_\pi} \right]$ (voltage divider ratio)

For short line model where $Y_\pi/2 = 0$:

$V_S = V_{L(load)} + Z_\pi I_L$, $V_{L(N.L.)} = V_S$

	l	V_S	$V_{L(N.L.)}$	$V_{L(load)}$	% Reg
Equiv.	10	34931	35116	34500	1.78
Pi	30	34822	36522	34500	5.86
Model	100	28005	48316	34500	40.05
Nomin.	10	34933	35118	34500	1.79
Pi	30	34876	36586	34500	6.04
Model	100	30746	52111	34500	51.05
Short	10	35111	35111	34500	1.77
Line	30	36400	36400	34500	5.50
Model	100	41488	41488	34500	20.25

Comments:

We see that 100 miles of this cable is an impractical length.

We see that the short line model is not very accurate beyond the 10-mile length.

We see that the nominal pi model is reasonably good up to the 30-mile length.

6. **a.** $Z_0 = \dfrac{j0.8}{j5.4 \times 10^{-6}} = 384.9\underline{/0°}$ (resistive)

$I_L = \dfrac{(230/\sqrt{3})\underline{/0°}}{384.9\underline{/0°}} = \dfrac{132.8\underline{/0°}}{384.9\underline{/0°}} = 0.345\underline{/0°}$ kA

b. $S_L = 3 \times 132.8 \times 0.345 = 137.4\underline{/0°}$ MVA
$= 137.4 + j0$ MVA

c. $V_S = V_L \cosh(\gamma l) + I_L Z_0 \sinh(\gamma l)$
$I_S = (V_L/Z_0) \sinh(\gamma l) + I_L \cosh(\gamma l)$
$\gamma = \sqrt{j0.8 \times j5.4 \times 10^{-6}} = 0 + j0.002078$
$\gamma l = j0.002078 \times 200 = j0.4157$

$\cosh(j0.4157) = \cos(0.4157) = 0.9148 + j0$

$\sinh(j0.4157) = j \sin(0.4157) = 0 + j0.4038$

$V_S = 132.8(0.9148 + j0) + 0.345(384.9)(0 + j0.4038)$

$\quad = 132.8\underline{/23.8°}$

$I_S = (132.8/384.9)(0 + j0.4038) + 0.345(0.9148 + j0)$

$\quad = 0.345\underline{/23.8°}$

d. $S_S = 3 \times 132.8\underline{/23.8°} \times 0.345\underline{/-23.8°} = 137.4 + j0$

e. $I_L = 1.25 \times 0.345\underline{/0°} = 0.431\underline{/0°}$

By same equations as in (c) except with new I_L

$V_S = 138.75\underline{/28.8°}$

$I_S = 0.418\underline{/19.45°}$

$S_S = 3(138.75\underline{/28.8°})(0.418\underline{/-19.45°}) = 171.78 + j28.55$

f. $I_L = 0.75 \times 0.345\underline{/0°} = 0.259\underline{/0°}$

$V_S = 127.96\underline{/18.3°}$

$I_S = 0.274\underline{/30.5°}$

$S_S = 3(127.96\underline{/18.3°})(0.274\underline{/-30.5°}) = 103.07 - j22.2$

Comments:

Note that, with surge impedance loading: V_S and I_S are in phase and equal in magnitude to V_L and I_L. $P_S = P_L$, which it must be since the line is lossless. Q_S is zero.

With load greater than surge impedance loading: I_S *lags* V_S and both magnitudes are different from the load end quantities. P_S still equals P_L but $Q_S > 0$; that is, the line must be supplied with positive reactive voltamperes from the source.

With load less than surge impedance loading: I_S *leads* V_S, and both magnitudes are different from the load end quantities. P_S still equals P_L but $Q_S < 0$; that is, the line must be supplied with negative reactive voltamperes from the source.

HOMEWORK PROBLEMS

1. A residence is supplied with single phase power from a three-wire 120/240-volt source as shown below.

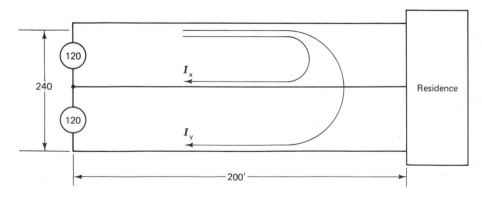

The 120/240-volt source is a transformer with the low-voltage winding center tapped. The transformer is on a pole behind the residence and the conductors run 200 feet to the residence. All three wires are No. 4 hard-drawn stranded copper with resistance $R = 1.518$ ohms/mile. The conductor *GMR* is 0.00717 feet. The conductor configuration is flat with spacing as shown below. Frequency = 60 Hz.

The loads on such a supply are conveniently studied in terms of two components, shown as I_x and I_y in the circuit diagram. I_x supplies power at 120 volts to lighting, small heaters, etc. I_y supplies power at 240 volts to water heaters, electric stoves, etc.
   a. If I_x is 60 amperes fed to a resistive load, what is the voltage at the residence end when the transformer voltage is maintained at 120 volts? (I_y is zero)
   b. If I_y is 120 amperes fed to a resistive load, what is the voltage at the residence end when the transformer maintains the 240-volt source fixed? (I_x is zero)

2. A certain all-aluminum stranded conductor has a cross-sectional area of 1113 kcmil.
   a. Given that the resistivity, ρ, for aluminum is 2.83×10^{-8} ohm-meter at 20 °C, what would be the resistance of 1000 feet of this conductor?
   b. If the manufacturer gives R_{dc} as 0.01558 ohm per 1000 feet at the same (20 °C) temperature, by what percentage does the spiraling of the conductor increase the resistance?
   c. What would be the R_{dc} of this same conductor at 50 °C?
   d. If the manufacturer gives the 60-Hz resistance as 0.0956 ohm/mile at 50 °C, what is the ratio of the ac resistance to the dc resistance?

3. A typical configuration for a 15-kV class distribution line is shown in the figure below.

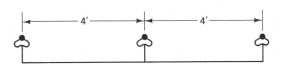

The series impedance of the line is found to be $(0.0490 + j0.1240)$ ohm/1000 feet. The frequency of the line is 60 Hz, and the line is completely transposed. Determine the *GMR* of the conductor.

4. A three-phase, sixty hertz, overhead transmission line has a symmetrical (delta) spacing as shown below.

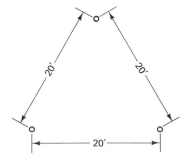

Manufacturer's data for the conductor (Cardinal) gives the *GMR* as 0.0402 foot and the outside diameter as 1.196 inches. $R_{ac} = 0.1082$ ohms/mile at 50 °C and 60 Hz. Find:

a. The inductance of the line in mH/phase/mile.
b. The inductive reactance in ohms/phase/mile.
c. The capacitance in μF/phase/mile.
d. The capacitive susceptance in μS/phase/mile.
e. The impedance, **Z**, in ohms/phase/mile.
f. The shunt admittance, **Y**, in siemens/phase/mile.

5. Repeat Problem 2 if the conductors are arranged in a flat configuration as below.

6. A three-phase line is found to have:

$$Z = 0.10 + j0.74 \quad \text{ohm/phase/mile}$$
$$Y = 0 + j5.7 \times 10^{-6} \quad \text{S/phase/mile}$$

as the series impedance and shunt admittance at 60 Hz.

a. What are the characteristic impedance, Z_0, and the propagation constant, γ?
b. If the line is 200 miles long, verify the following values for $\cosh(\gamma l)$ and $\sinh(\gamma l)$.

$$\cosh(\gamma l) = 0.9168 + j0.01108$$
$$\sinh(\gamma l) = 0.02538 + j0.4003$$

c. If this line delivers 400 MW (total) at 345 kV (*l-l*) and a power factor of 0.9, current lagging to a load, what are V_S and I_S at the source end? Use the exact method of Eqs. (8-42) and (8-43).
d. What is the complex power input, S_S, from the source?
e. What is the transmission efficiency, P_L/P_S, of the line?

7. Solve Problem 6(c) using the nominal pi model for the line.

8. Using the data and results of Problem 7, what is the percent voltage regulation of the line at the specified load if the source voltage is held fixed when the load is removed?

Chapter 9

Power

Electronics

- Introduction to power electronics.
- Characteristics of silicon-controlled rectifiers (thyristors).
- Analysis of ac to dc, dc to ac, ac to ac, and dc to dc circuits included.
- Importance of power factor, harmonics, quality of power in the design stressed.

9.1 INTRODUCTION

Power electronics is a relatively modern subject that emerged in the last two decades due to the continuous evolution of high-power semiconductor devices suitable for power or energy applications. The term *power electronics* was coined in the 1960s to distinguish it from classical electronics, which deals with devices and circuits that carry power in the milliwatt to a few watts range at microvolts to low-voltage level. The power portion of any power electronic circuit addresses a specific application, while the electronics part called *gate control circuitry* switches the devices in a controlled and logical manner to precisely achieve the desired function. Because of the magnitudes of power involved in the kilowatt to the megawatt range, the power electronic circuits are incorporated with proper heat sinks. The design of the heat sinks depends on the heat-dissipating characteristics of the power semiconductor devices and the computation of the power dissipated in the devices. Thus the design, fabrication, and testing of any power electronic circuit need a sound background in single- and three-phase power, analog and digital electronics, control system dynamics, magnetics, and thermodynamics.

The purpose of this chapter is to familiarize a novice with this new field and provide material for basic understanding of the subject. After a brief introduction to the various power semiconductor devices and their characteristics, commonly used power electronic conversion schemes will be introduced and analyzed. In so doing, a variety of important applications in the energy and power field will be mentioned. Some of these are: ac and dc motor control systems, static frequency conversion, high-voltage dc (HVdc) transmission systems mentioned in Chapter 8, and high power supply systems. Since the material presented here is a part of an energy-oriented text book, the electronic control portion of the circuit will not be covered here. Such material is usually covered in standard electronic courses or advanced power electronics courses taught at the senior or graduate level.

9.2 POWER SEMICONDUCTOR DEVICES AND CHARACTERISTICS

The following devices are often used in various power electronic applications: (1) diodes, (2) bipolar transistors, (3) MOS field-effect transistors (MOSFET), (4) diacs, (5) silicon-controlled rectifiers (SCR), (6) triacs, and (7) gate turn-off (GTO) SCR. The principle and performance characteristics of the first four solid-state devices are covered in standard electronic courses and text books. Therefore, this section will cover the steady-state characteristics and operation of the last three controllable devices, which for most applications operate as electronic or solid-state *switches*.

Thyristors

A "thyristor" is a generic name that applies to any of three or four junction devices. The most popular or commonly used thyristor is a silicon-controlled rectifier or SCR. It should be clarified, however, that in the literature, the terms *thyristor*

**TABLE 9-1   Maximum Voltage and Current Ratings
of Power Semiconductor Components**

Device (Single Unit or Cell)	Maximum Voltage Rating (kV)	Maximum Current Rating (A)
Diode	6	7,500
Bipolar Transistor	1.0	30
	0.3	100
MOS FET	0.1	50
	1.0	5
SCR	5.0	3,000
Triac	1.0	2,000
GTO SCR	2.0	1,000

and *SCR* are used synonymously. Table 9-1 shows the maximum voltage and current ratings of SCRs and other power electronic components that are currently available in the market. If higher ratings are desired, one should take recourse to a *valve* configuration, in which several units or cells are assembled in series/parallel arrangements. This approach is commonly adopted particularly for the bulk power system applications mentioned above.

Figure 9-1 shows the various forms of the thyristor representation. As the name "silicon-controlled rectifier" implies, the SCR is a diode or rectifier with control provided by an appropriate pulse of current sent through the gate. The static or steady-state *V-I* characteristic curves are given in Figure 9-2. It should be evident from these curves, that the thyristor operates as a switch.

Basically, the SCR operates in the following manner: When a positive anode-to-cathode voltage, V_{AK} is applied to the SCR, it will start conducting provided a gate

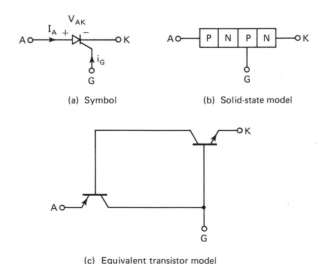

(a) Symbol   (b) Solid-state model

(c) Equivalent transistor model

Figure 9-1   SCR representations.

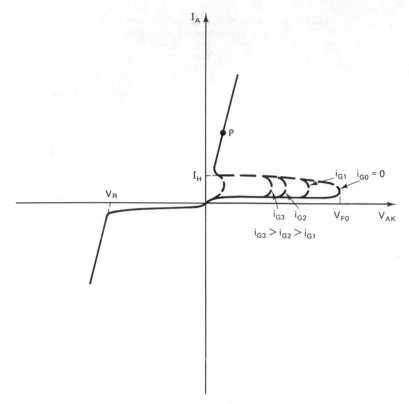

Figure 9-2   Volt-amp characteristics of SCR.

signal pulse, i_G, of sufficient amplitude and duration is also simultaneously applied. The positive-biased voltage at which it conducts depends on the strength of the gate current, as implied by the family of curves shown in Figure 9-2. The higher the gate current, the less the anode-cathode voltage required to turn-on the SCR. Once the SCR is turned on, the gate signal need not be applied any longer to keep the device conducting. At this point, the gate signal can, therefore, be terminated, and the SCR remains in the conducting state when V_{AK} is very small, as long as sufficient I_A is flowing. The steady-state or equilibrium point at which the SCR conducts, such as P, depends upon the resistance (or impedance) connected to it. While it is conducting, the forward anode-to-cathode drop is very small, usually of the order of a few volts, depending on the SCR turn-on speed.

Suppose the thyristor is conducting; it will turn off if the anode-to-cathode voltage is reduced to zero or negative, which is equivalent to applying a reverse-biased voltage. This phenomenon is called *natural commutation,* which is possible only when alternating voltage sources such as sinusoidal voltages are applied to the thyristor. The intent here is to reduce the thyristor current to below the holding current, I_H, to assure commutation. One must make sure not to apply a reverse-biased voltage more than V_R to avoid the avalanche breakdown of the thyristor, a situation which also happens with a diode. If the forward-biased voltage exceeds

V_{FO}, then the thyristor continues to conduct even without the presence of a gate signal.

A second method by which a thyristor can be turned off is called *forced commutation*, which implies the need for an external circuit to force the current through it below I_H, or to reduce V_{AK} below zero. A simple form of such a scheme is given in Figure 9-3. This series—capacitor commutation scheme works in the following manner: When the main SCR is turned-on, it starts charging the capacitor, C. When the capacitor voltage reaches the dc supply value, the conditions are ideal to turn off the main SCR; namely, the current through it is below its holding value and the anode-to-cathode potential is about to be reverse-biased. If now a suitable pulse is applied to the commutation SCR, it will turn on, while in this process the main SCR is turned off. The capacitor discharges its voltage through the commutation SCR and the resistor connected in series with it. When capacitor is completely discharged, conditions are ripe to turn on the main SCR again. In other words, the forced commutation of the main SCR is achieved by employing a second SCR.

Like any other device, thyristors need protection against the following situations:

1. overload
2. overcurrent
3. overvoltage
4. excess dv/dt
5. excess di/dt

Circuit breakers with thermal protection or expulsion fuses can protect a thyristor against overload conditions. Because of the fast speeds at which this device operates, usually a fast fuse of compatible rating is used to protect it against overcurrents. The SCR may be protected against overvoltages, due to switching or lightning, by connecting a surge arrester such as a metal oxide varistor across it.

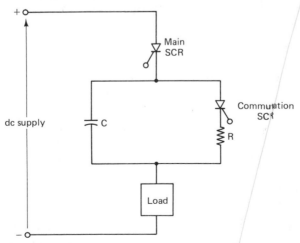

Figure 9-3   A simple forced-commutation circuit.

Because of the SCR's steep dynamic turn-on and turn-off characteristics, it needs both dv/dt (dV_{AK}/dt) and di/dt (dI_A/dt) protection. While the excessive dv/dt is not a serious factor, it can result in premature turn-on of the thyristor even without a gate signal applied to it. The protection against excessive dv/dt is more often than not provided by an RC snubber circuit shown in Figure 9-4.

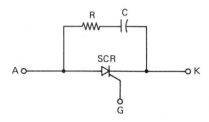

Figure 9-4 RC snubber circuit.

The most popular method to protect an SCR against high di/dt, which could create hot spots and burn the junction, is to provide an inductor connected in series with the SCR. The reactor is designed to go into magnetic saturation for currents greater than its rated value so that only a minimal voltage develops across the reactor.

The reader is referred to the suggested material listed at the end of the chapter to gain a more comprehensive understanding of this device.

Triac

A triac can be conceived as two SCRs connected back-to-back. However, it is manufactured as one integrated device with only one gate terminal. Its schematic symbol is shown in Figure 9-5. The two unmarked terminals play the role of anode and cathode, depending upon the relative polarities of the applied source. A proper (positive or negative) pulse of signal should be applied to the gate at the appropriate time to fire the triac. Figure 9-6 shows the V-I characteristic family of the device. As in the case of the SCR (Figure 9-2), the larger the gate signal, the less the forward breakdown voltage required to turn on the triac. In the same vein, it will turn off once the current falls below the holding current, I_H.

Figure 9-5 Triac representation.

In the absence of the gate terminal, the triac becomes a diac, an uncontrolled two-way rectifier, which is equivalent to two diodes connected back-to-back. The V-I characteristic of the diac is the envelope characteristic, with $i_{GO} = 0$ in Figure 9-6.

GTO Thyristor

A Gate Turn-Off (GTO) SCR is a device that can be turned off under load by applying a relatively large negative current pulse to the gate even if the anode current

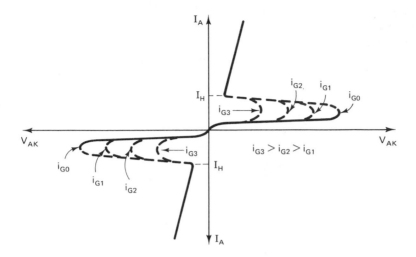

is more than I_H. Though these devices have existed for the last twenty-five years, they did not possess the degree of reliability for use in circuits where forced commutation is otherwise required. What this means is that these devices do not need an external commutation circuit. GTO thyristors are usually represented schematically as in Figure 9-7.

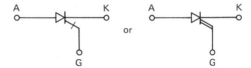

Figure 9-7   GTO SCR representations.

9.3 AC AND DC CONVERTERS

Converters with ac input and dc output are classically known as controlled rectifiers, when thyristors or any other controlled devices perform the power switching function to realize variable output dc voltage from constant input ac voltage source. The control is obtained by changing the firing angle α of the thyristors, which as stressed before, act as switches in the conversion scheme to be discussed here. The input could be either a single-phase or a three-phase source. These converters can be configured for half-wave or full-wave rectification, the latter providing better performance, as we shall observe later. Since the input is ac, natural commutation can be achieved. However, forced commutation can also be employed if more versatile control of the output is desired. These converters are very often utilized for dc motor control systems employed in process industries such as steel rolling mills. Other applications include portable tools, standard power supplies, and HVdc transmission systems. In this section, we will explore the basic principles underlying the operation of this type of converter and analyze simple schemes to gain a rudimentary understanding of this important conversion scheme.

Single-Phase Schemes

A simple form of a half-wave rectifier feeding a resistive load, such as an electric heater or incandescent lamp, is shown in Figure 9-8(a). If the input source $v_i(\omega t) = V_m \sin(\omega t)$ as drawn in Figure 9-8(b), the corresponding output voltage

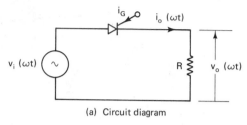

(a) Circuit diagram

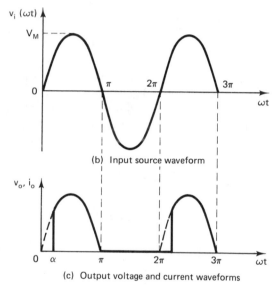

(b) Input source waveform

(c) Output voltage and current waveforms

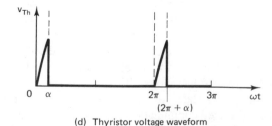

(d) Thyristor voltage waveform

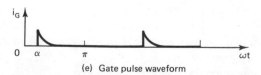

(e) Gate pulse waveform

Figure 9-8   Single-phase half-wave controlled rectifier scheme.

and the resulting current wave are indicated in Figure 9-8(c) for a particular triggering angle α. The anode-cathode voltage across the thyristor is depicted in Figure 9-8(d). The triggering current pulse of suitable magnitude and duration required to turn on the SCR at $\omega t = \alpha$ is shown in Figure 9-8(e). If $\alpha = 0$, we realize an uncontrolled half-wave rectifier employing a diode.

For this scheme, the average value of the output voltage V_{dc} is:

$$V_{dc} \triangleq \frac{1}{2\pi} \int_\alpha^\pi V_m \sin(\omega t)\, d(\omega t)$$

$$= \frac{V_m}{2\pi} [-\cos(\omega t)]_\alpha^\pi$$

$$= \frac{V_m}{2\pi} [1 + \cos \alpha] \tag{9-1}$$

and the corresponding current I_{dc} is given by:

$$I_{dc} = \frac{V_{dc}}{R} = \frac{V_m}{2\pi R} [1 + \cos \alpha] \tag{9-2}$$

If $\alpha = 0$, then $V_{dc} = V_m/\pi$ and this value goes to zero if $\alpha = \pi$. Therefore, by varying α from 0 to π, one can completely vary the dc output voltage from zero to the maximum value of (V_m/π). The corresponding power dissipated in R will be

$$P_{dc} = \frac{V_{dc}^2}{R} = I_{dc}^2 R = \frac{V_m^2}{4\pi^2 R} [1 + \cos \alpha]^2 \tag{9-3}$$

One important observation we can make at this point, with respect to controlled power electronics circuits, is the nonsinusoidal nature of resulting waveforms shown in Figure 9-8(c). The output voltage and current in this circuit contains various harmonics that can be determined by performing a Fourier series analysis. We shall delve into this aspect later.

The rms value of the output voltage, is given by:

$$V_{rms} \triangleq \left[\frac{1}{2\pi} \int_\alpha^\pi (V_m \sin(\omega t))^2\, d(\omega t) \right]^{1/2}$$

$$= \frac{V_m}{2\sqrt{\pi}} \left[(\pi - \alpha) + \frac{\sin 2\alpha}{2} \right]^{1/2} \tag{9-4}$$

Consequently,

$$I_{rms} = \frac{V_{rms}}{R} = \frac{V_m}{2\sqrt{\pi}R} \left[(\pi - \alpha) + \frac{\sin 2\alpha}{2} \right]^{1/2} \tag{9-5}$$

We will now define a measure called *Ripple Factor* (r) for determining the smoothness of the output voltage.

$$r(\%) \triangleq \frac{\text{rms value of all ac components of } v_0}{V_{dc}} \times 100 \tag{9-6}$$

This can be shown to reduce to:

$$r = \left[\frac{V_{rms}^2}{V_{dc}^2} - 1 \right]^{1/2} \times 100 \tag{9-7}$$

One can quickly observe that the output voltage, v_0 resulting from the scheme is far from ideal. It has in fact, no resemblance to the dc output waveform desired. However, the scheme still provides a component of dc output. In the following example, we will gain a better understanding of this factor.

Example
We want to vary the intensity of light of an incandescent lamp connected to a 120-V, 60-HZ ac source by means of the scheme discussed above. We want to control the power dissipated in the lamp from 0 to 60 Watts. Investigate P_{dc} as a function of α. Also draw the variation of V_{dc}, V_{rms} and r as a function of the triggering angle.

Solution From equation (9-3), with $\alpha = 0$

$$60 = \frac{\{\sqrt{2}\,(120)\}^2}{4\pi^2 R}(1 + 1)^2$$

$$= \frac{2(120)^2}{\pi^2 R}$$

or,

$$R = \frac{2(120)^2}{60\pi^2} = 48.63 \ \Omega$$

Now for any other value of α

$$P_{dc} = 15[1 + \cos \alpha]^2 \tag{9-8}$$

From equation (9-1)

$$V_{dc} = \frac{\sqrt{2}\,(120)}{2\pi}[1 + \cos \alpha]$$

$$= 27.01[1 + \cos \alpha]$$

$$V_{rms} = \frac{\sqrt{2}\,(120)}{2\sqrt{\pi}}\left[(\pi - \alpha) + \frac{\sin 2\alpha}{2}\right]^{1/2} \tag{9-9}$$

$$= 47.87\left[(\pi - \alpha) + \frac{\sin 2\alpha}{2}\right]^{1/2} \tag{9-10}$$

The plots for P_{dc}, V_{dc}, V_{rms} and r are given in Figure 9-9.

Because of excessive harmonic content in the output waveforms, this is not the best scheme, though it is the simplest. Most of the control is gained in the first half domain of α from 0 to 90°, and the scheme becomes very insensitive in the latter half of the α-domain.

A more versatile scheme, which can reduce the ripple is the full-wave scheme employing a bridge circuit shown in Figure 9-10. It can also be effectively used with nonideal input sources with dc offset. In other words, the half-wave or one-pulse scheme is rendered impractical if the input source has a dc component.

The full-wave scheme is also known as two-pulse rectifier in the sense that during the positive-half cycle of the source voltage, thyristors 1 and 3 can be turned on by applying proper pulse signals simultaneously to their gates at any desired triggering angle α. Similarly, a second pulse can be applied 180° later to the gates of the other two thyristors 2 and 4 during the negative-half-cycle of the input voltage. Since two gate pulses are required per cycle, the two-pulse connotation came into

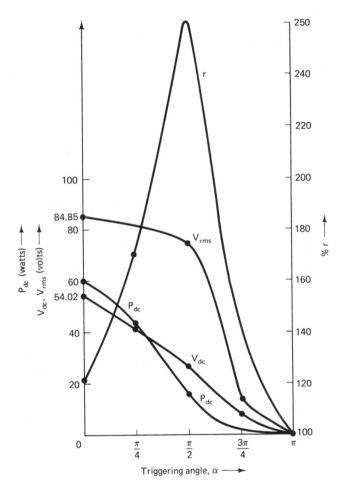

Figure 9-9   Plots for single-phase controller example.

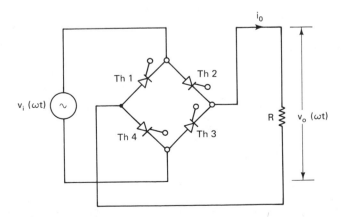

Figure 9-10   Schematic of a single-phase, full-wave (bridge) controlled rectifier.

vogue. The input and output waveforms are drawn in parts (a) and (b) of Figure 9-11. At any given time, since two nonconducting SCRs operate in parallel, the entire v_i is applied across them as shown in Figure 9-11(c) and (d).

We will now obtain the expressions for the average and the rms values of v_o as in the half-wave case:

$$V_{dc} = \frac{1}{\pi} \int_\alpha^\pi v_o(\omega t)\, d(\omega t)$$

$$= \frac{1}{\pi} \int_\alpha^\pi V_m \sin(\omega t)\, d(\omega t)$$

$$V_{dc} = \frac{V_m}{\pi}[1 + \cos \alpha] \tag{9-11}$$

and

$$V_{rms} = \left[\frac{1}{\pi} \int_\alpha^\pi V_m^2 \sin^2(\omega t)\, d(\omega t)\right]^{1/2}$$

$$= \frac{V_m}{\sqrt{2\pi}}\left[(\pi - \alpha) + \frac{\sin 2\alpha}{2}\right]^{1/2} \tag{9-12}$$

Comparing equations (9-1) and (9-11) quickly reveals that for a given α, V_{dc} in the full-wave case is twice that in the half-wave case, as might be expected. Also, V_{rms} for the full-wave rectifier is $\sqrt{2}$ times the value for the half-wave case.

Example

In Figure 9-10, $R = 10$ ohms, and $v_i = \sqrt{2}(240) \sin(\omega t)$ volts. For $\alpha = 30°$, find the following:

a. The average output current, I_{dc}
b. The power dissipated in the load, P_{dc}
c. The rms output current, I_{rms}
d. The average and rms thyristor currents
e. The power factor of the source

Solution   For the conditions stipulated, the load is purely passive and resistive.
a.   Using equation (9-11)

$$I_{dc} = \frac{V_{dc}}{R} = \frac{V_m}{\pi R}[1 + \cos \alpha]$$

$$= \frac{\sqrt{2}(240)}{\pi(10)}[1 + \cos 30°]$$

$$= 20.16 \text{ A}$$

b. $P_{dc} = I_{dc}^2 R = (20.16)^2(10) = 4{,}064.21$ watts
From these numerical values, the reader can realize that the situation described here is akin to controlling an electric range or a dryer at home from a 240-volt, single-phase supply.

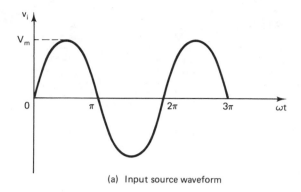

(a) Input source waveform

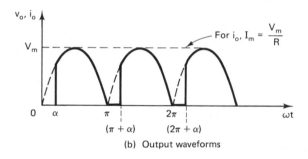

(b) Output waveforms

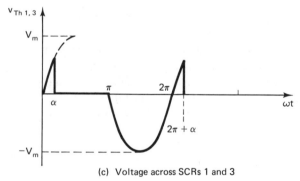

(c) Voltage across SCRs 1 and 3

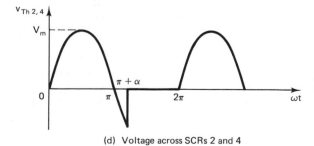

(d) Voltage across SCRs 2 and 4

Figure 9-11 Waveforms for single-phase, full-wave controlled rectifier scheme supplying resistive load.

c. From equation (9-12)

$$I_{rms} = \frac{V_{rms}}{R} = \frac{V_m}{\sqrt{2\pi R}}\left[(\pi - \alpha) + \frac{\sin 2\alpha}{2}\right]^{1/2}$$

$$= \frac{\sqrt{2}\,(240)}{\sqrt{2\pi}\,(10)}\left[\left(\pi - \frac{30°}{180°}\right) + \left(\frac{\sin 60°}{2}\right)\right]^{1/2}$$

$$= 25.0\ A$$

d. Average thyristor current $= \dfrac{I_{dc}}{2} = 10.08\ A$

rms thyristor current $= \dfrac{I_{rms}}{\sqrt{2}} = 17.68\ A$

e. Power factor of the source $\triangleq$

$$\frac{\text{Real power delivered to the load}}{\text{Input apparent power}}$$

$$= \frac{4,064.21}{240 \times 25} = 0.6774\ \text{lagging} \tag{9-13}$$

Observation. Even with a purely resistive load, the power factor seen by the source is not unity. In fact, with power electronic circuits, the input power factor can become very small depending upon α and other system parameters.

Inductive Loads

We have thus far considered the load to be purely resistive, which is the easiest type to analyze. Power electronic circuits involving inductive loads and active sources lead to more complex output waveforms and require the numerical solution of nonlinear transcendental equations for output currents. To expose the reader to this aspect, we will now briefly analyze a full-wave controlled rectifier scheme feeding an inductive load. It is illustrated in Figure 9-12.

We can draw the following inferences from the output waveforms shown in Figure 9-12(b) and (c).

1. The fact that the load is inductive in nature makes i_o lag behind v_o. Therefore, when v_o goes to zero at $(\omega t) = \pi$, i_o is still nonzero, and to facilitate the current flow through the load, a free-wheeling diode (D) is usually connected across the dc circuit as shown in Figure 9-12(a).

2. For small L/R ratio, i_o is discontinuous and is obviously nonsinusoidal.

3. The output current can be made smoother and continuous by inserting an inductor of suitable value in series with the load circuit. The curves corresponding to this situation are shown in Figure 9-13. For this case of continuous current operation, we can analyze the harmonic content in the output wave forms by a Fourier series:

Let

$$v_o \triangleq V_{dc} + \sum_{n=1}^{\infty}(a_n \sin(n\omega t) + b_n \cos(n\omega t)) \tag{9-14}$$

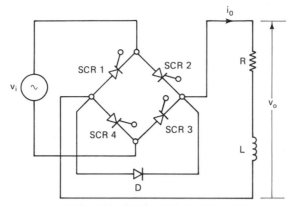

(a) Single-phase full-wave controlled rectifier
    scheme feeding an inductive load

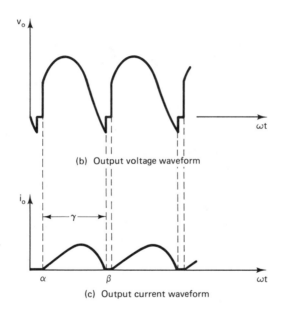

(b) Output voltage waveform

(c) Output current waveform

Figure 9-12   Single-phase, full-wave controlled rectifier scheme feeding inductive load.

$$= V_{dc} + \sum_{n=1}^{\infty} V_{nm} \cos(n\omega t - \theta_n) \qquad (9\text{-}15)$$

where V_{nm} = Maximum value of nth harmonic

$$\text{Component} = \sqrt{a_n^2 + b_n^2} \qquad (9\text{-}16)$$

$$\theta_n = \text{Arc tan}(a_n/b_n) \qquad (9\text{-}17)$$

In equations (9-14) to (9-17)

$$a_n = \frac{2}{\pi} \int_{\alpha}^{\pi+\alpha} v_o \sin(n\omega t)\, d(\omega t) \qquad (9\text{-}18)$$

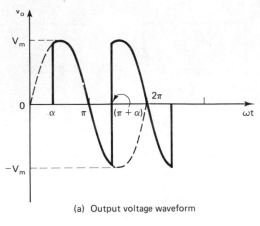

(a) Output voltage waveform

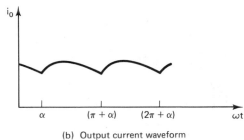

(b) Output current waveform

Figure 9-13 Single-phase, full-wave controlled rectifier scheme feeding inductive load (continuous output current case).

In this case

$$a_n = \frac{2}{\pi} \int_\alpha^{\pi+\alpha} V_m \sin(\omega t) \sin(n\omega t) \, d(\omega t)$$

$$= \frac{2V_m}{\pi} \left[\frac{\sin(n+1)\alpha}{n+1} - \frac{\sin(n-1)\alpha}{(n-1)} \right] \tag{9-19}$$

Similarly,

$$b_n = \frac{2}{\pi} \int_\alpha^{\pi+\alpha} v_o \cos(n\omega t) \, d(\omega t) \tag{9-20}$$

$$= \frac{2}{\pi} \int_\alpha^{\pi+\alpha} V_m \sin(\omega t) \cos(n\omega t) \, d(\omega t)$$

$$= \frac{2V_m}{\pi} \left[\frac{\cos(n+1)\alpha}{(n+1)} - \frac{\cos(n-1)\alpha}{(n-1)} \right] \tag{9-21}$$

The average output voltage (or the dc term in the series) is given by

$$V_{dc} = \frac{1}{\pi} \int_\alpha^{\pi+\alpha} V_m \sin(\omega t) \, d(\omega t)$$

$$= \frac{2V_m}{\pi} \cos \alpha \tag{9-22}$$

Compare this expression with equation (9-11) for the resistive load case. For this case, V_{rms}, with $n = 1$ in equations (9-16), (9-17), and (9-21), can be shown to be equal to $V_m/\sqrt{2}$. The ripple factor r can still be defined by equation (9-7).

A quick examination of Figures 9-11, 9-12, and 9-13 reveals that the frequency of output waveforms is twice that of the source frequency. This is to be expected, since the full-wave scheme is a two-pulse converter. The voltage output, therefore, contains only even order harmonics.

A similar series for output current can be expressed as:

$$i_o = I_{dc} + \sum_{n=1}^{\infty} I_{nm} \cos(n\omega t - \theta_n - \phi_n) \qquad (9\text{-}23)$$

where

$$I_{nm} = \text{Maximum or Peak value of } n\text{th}$$
$$\text{harmonic current component} = V_{nm}/Z_n \qquad (9\text{-}24)$$
$$Z_n = n\text{th harmonic impedance of the load}$$
$$= (R^2 + (n\omega L)^2)^{1/2}$$
$$\phi_n = \text{Arc } \tan(n\omega L/R) \qquad (9\text{-}25)$$

Three-Phase Schemes.   The single-phase, full-wave bridge scheme can be extended to accommodate a three-phase input source. Most popular and versatile is the 6-pulse converter, or controlled rectifier. Prior to the invention of thyristors, mercury arc rectifiers were employed to realize controlled rectification. The well-known Pacific ±400 kV, HVdc transmission line linking the northwest region at Celilo, Oregon, to Sylmar, near Los Angeles in the southwest region, adopted the mercury arc rectifiers. We will explore this important application later in this section after gaining a good understanding of this converter scheme.

The circuit of Figure 9-14 employs a Y-connected, three-phase source $v_i(\omega t)$, delivering dc output $v_o(\omega t)$ to resistive load through a thyristor bridge consisting of six controlled switches; hence it is called a 6-pulse converter.

The operation of the scheme can be described in the following manner: During $\omega t = \pi/3$ to $2\pi/3$, v_{ab} is the most positive voltage relative to other line-to-line voltages and conditions are congenial for thyristors 1 and 6 to conduct if suitable pulses are applied, or are already applied, to their respective gates, as shown in Figure 9-14(d). The two thyristors will conduct during the following 60° at which time Th 6 will commutate naturally, and if a gate pulse is applied at this moment to the Th 2, output can be delivered to load the next 60° through Th 1 and Th 2. If this process is continued with two thyristors conducting at any given time, we will essentially realize a 6-pulse controlled rectifier. The output voltage v_o and load current i_o are illustrated in Figure 9-14(c) for a typical case. The (d) part of the same figure shows the logical sequence in which the thyristors conduct, and the gate pulses need to be applied to obtain continuous output waveforms.

A close observation of v_o clearly shows that its fundamental frequency of variation is six times that of the input source. Therefore, the harmonic of the output voltage will be multiple orders of 6, a clear advantage with a polyphase source. In addition, the ripple in v_o will also be much lower than in the single-phase scheme

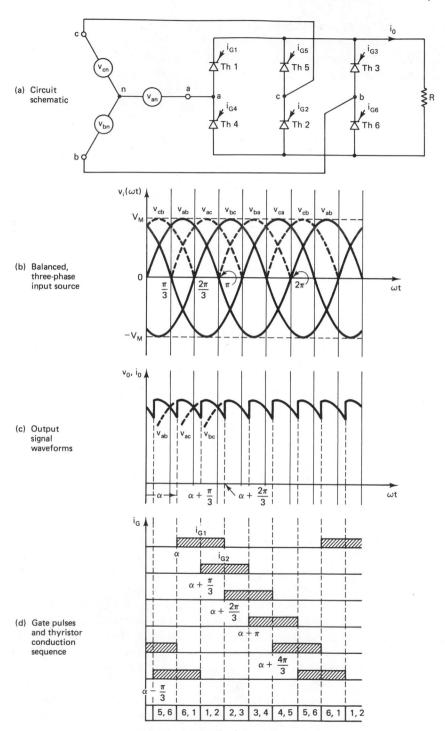

Figure 9-14   Three-phase, full-wave controlled rectifier scheme.

we discussed earlier in this chapter. Consequently, we can obtain higher quality (or less ripple) output.

The average value of $v_o = V_{dc}$ is given by:

$$
\begin{aligned}
V_{dc} &= \frac{3}{\pi} \int_{\alpha+(\pi/3)}^{\alpha+(2\pi/3)} v_o(\omega t)\, d(\omega t) \\
&= \frac{3}{\pi} \int_{\alpha+(\pi/3)}^{\alpha+(2\pi/3)} v_{ab}(\omega t)\, d(\omega t) \\
&= \frac{3}{\pi} \int_{\alpha+(\pi/3)}^{\alpha+(2\pi/3)} V_M \sin(\omega t)\, d(\omega t) \\
&= \frac{3V_M}{\pi} \cos\alpha
\end{aligned}
\tag{9-26}
$$

where V_M is the maximum line-to-line voltage. In a similar manner, the rms value of the load voltage is:

$$
\begin{aligned}
V_{rms} &= \left[\frac{3}{\pi} \int_{\alpha+(\pi/3)}^{\alpha+(2\pi/3)} v_o^2\, d(\omega t) \right]^{1/2} \\
&= V_M \left[\frac{1}{2} + \frac{3\sqrt{3}}{4\pi} \cos(2\alpha) \right]^{1/2}
\end{aligned}
\tag{9-27}
$$

The harmonic coefficients a_n and b_n for this scheme can be obtained by

$$
a_n = \frac{6}{\pi} \int_{\alpha-(\pi/3)}^{\alpha} v_o \sin(n\omega t)\, d(\omega t)
\tag{9-28}
$$

and

$$
b_n = \frac{6}{\pi} \int_{\alpha-(\pi/3)}^{\alpha} v_o \cos(n\omega t)\, d(\omega t)
\tag{9-29}
$$

$$
n = 6, 12, 18, \ldots.
$$

A further analysis of V_{dc} from Eq. (9-26) reveals that V_{dc} is maximum when $\alpha = 0$, corresponding to the uncontrolled case. It goes to zero when $\alpha = \pi/2$. Then for $\pi/2 < \alpha \leq \pi$, V_{dc} actually becomes negative. This corresponds to the "inverter" operation, which is possible, provided there is adequate series inductance in the load circuit. Looking from a different point of view, if the dc part of the circuit were a source, one could make the circuit operate as an inverter to realize three-phase, ac output from a dc input source. We will discuss the inverter or, more appropriately, dc-to-ac converter scheme in the following section.

The ripple factor, r, can still be obtained from Eq. (9-7), with V_{dc} and V_{rms} defined by Equations (9-26) and (9-27), respectively.

Before we proceed with a practical application for the three-phase controlled rectifier scheme, we will work an example to gain a better understanding of its basic principle of operation.

Example

The three-phase six-pulse controlled rectifier scheme can be adopted to realize a 125-V dc variable supply voltage from a constant voltage, constant frequency ac source such as 480 V or 208 V. However, in order to reduce the ripple, one prefers a 12-pulse, or even a 24-pulse, rectifier scheme, which requires twice, or four times, the number of thyristors we have used in the circuit of Figure 9-14.

Perform a detailed analysis of this 6-pulse circuit if it delivers dc power to a 5-ohm resistor at 125 V from a 208-V, 60-Hz three-phase ac supply (Figure 9-15).

Solution

a. For $V_M = \sqrt{2}\,(208)$ volts and $V_{dc} = 125$-V we need to determine α using equation (9-26).

$$V_{dc} = 125 = \left\{ \frac{3[\sqrt{2}\,(208)]}{\pi} \cos \alpha \right\}$$

or

$$\cos \alpha = \frac{125\pi}{(3)\,(\sqrt{2})\,(208)} = 0.4450$$

or

$$\alpha = 63.58°$$

b. $I_{dc} = \dfrac{V_{dc}}{R_L} = \dfrac{125}{5} = 25\text{-A.}$

c. $P_{dc} = V_{dc} I_{dc} = 125 \times 5 = 3.125\text{-kW.}$

Observation 1. Evidently, proper heat sinks will be needed to dissipate the heat in the thyristor.

d. V_{rms} from equation (9-27)

$$= \sqrt{2}\,(208) \left[\frac{1}{2} + \frac{3\sqrt{3}}{4\pi} \cos(127.15°) \right]^{1/2}$$

$$= 147.16\text{-V}$$

e. Average thyristor current $= \dfrac{I_{dc}}{3}$ (9-30)

This is true since each thyristor conducts for 1/3 cycle.

$$= 25/3 = 8.34 \text{ A}$$

f. rms value of current in each thyristor

$$= \frac{I_{rms}}{\sqrt{3}}$$

$$= \frac{V_{rms}}{\sqrt{3}\,R_L}$$

$$= \frac{147.16}{(\sqrt{3})\,(5)} = 16.99\text{-A}$$

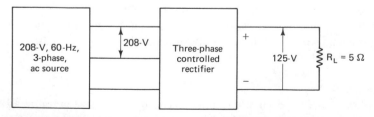

Figure 9-15 Block diagram for example on three-phase controlled rectifier scheme delivering resistive load.

g. Ripple factor, $r = \left[\left(\dfrac{V_{rms}}{V_{dc}} \right)^2 - 1 \right]^{1/2} \times 100$

$\qquad = \left[\dfrac{(147.16)^2}{(125)^2} - 1 \right]^{1/2} \times 100$

$\qquad = 42.10\%$

Observation 2. The ripple is very high, thus warranting a 12-pulse or even a 24-pulse scheme for this example system to mitigate the ripple.

h. Input power factor $= (P_{input}/S_{input})$

$\qquad\qquad\qquad\quad \cong P_{dc}/S_{input}$

$\qquad\qquad\qquad\quad \cong \dfrac{3,125}{\sqrt{3}\,(208)\,(\sqrt{3})\,(16.99)}$

$\qquad\qquad\qquad\quad = 0.2947$ current lagging

Observation 3. The power factor is extremely low as seen by the source from the source side.

Inductive Loads.   Besides supplying dc power to resistive loads, these can be adopted as dc motor drives in process industries, as converters at the terminals of a HVdc transmission line, for portable hand tools and many more applications. In most of these cases the load is no longer fully resistive. A general dc load includes in addition an active source (EMF) and an inductance as we emphasized while discussing single-phase schemes. In these situations, the output voltage and current are no longer in phase. The output current could be continuous or discontinuous, depending upon the value of α and other circuit parameters. Figure 9-16 illustrates continuous i_o from a passive inductive load.

HVDC Transmission System Application

We introduced the concept of energy transmission by dc in Chapter 8. It serves as an asynchronous link to interconnect two polyphase ac lines of the same frequency, or two different frequencies, to provide higher stability and efficiency. A schematic of such a system is given in Figure 9-17.

One such line is the famous Pacific NW/SW Intertie, which is approximately 900 miles (or about 1,450 km.) long, interconnecting two ac high-voltage systems

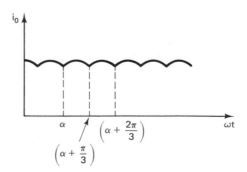

Figure 9-16 Continuous output waveform for inductive load.

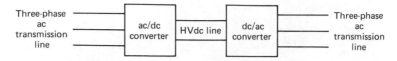

Figure 9-17   HVdc transmission system schematic.

between Celilo, Oregon and Sylmar, California, near Los Angeles. It was originally designed to operate at ±400 kV, utilizing an array of three 133-kV mercury-arc valve groups. This bipolar line with 800 kV between the lines had the capacity to transmit 1,800 amperes or about 1,440 MW of power. By proper control of the valves, the real power could be varied from zero to 1,440 MW in each direction. Putting it differently, the power flow direction implied the converter at one end, serving as a controlled rectifier, and the converter at the other end as an inverter. Their roles could be interchanged to facilitate the flow of energy in the opposite direction.

In 1978, the current capacity of the line was increased from 1,800 amperes to 2,000 amperes. Very recently, this line was further upgraded to ±500 kV with the installation of two 100-kV SCR valve groups on both the positive and negative sides, and at both ends of the line. This latest modification also needed the addition of a new smoothing reactor in series with the line and the replacement of new filters to isolate unwanted harmonics in the dc line.

The first solid-state HVdc line is the Eel River Project in New Brunswick, Canada. It started functioning in 1972. It is rated at 320 MW with four three-phase, ac-to-dc bridge converters connected back-to-back on each side of the line. Each of these bridges is rated at 40 kV between the positive pole (or the negative pole) and ground, and is designed to carry 2000 A. In each bridge, there are six SCR valves with each valve having 200 SCRs.

To understand the principle of the converter operation, or equivalently, the power flow control in the line, let us consider the Pacific Intertie again. A simplified circuit diagram of the line is shown in Figure 9-18. If it is desired to keep the voltage at the Celilo converter station at the rated value of 400 kV, the firing angle α^c_{top} of the top set of the SCR must be controlled so that:

$$V_{dc} = 400 = \left\{ \frac{3[\sqrt{2}\,(500)]}{\pi} \cos(\alpha^c_{top}) \right\} \qquad \text{by Eq. (9-26)}$$

or

$$\alpha^c_{top} = \cos^{-1}\left[\frac{400\pi}{(3)\,(\sqrt{2})\,(500)} \right] = 53.67°$$

The firing angle of the bottom SCR valve should be adjusted so that $V_{dc} = +400$ kV can be obtained from the ac side.

Or

$$\alpha^c_{bottom} = 53.67°$$

which is equal to α^c_{top}.

If it is desired to send 1,600 MW from the north to the south, which happens in the summer, the dc voltage from one pole to the ground at the Sylmar converter

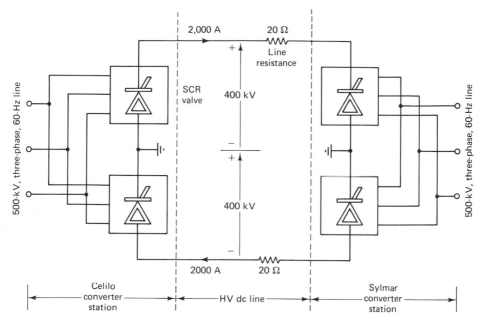

Figure 9-18   Pacific HVdc transmission intertie representation diagram.

station ought to be at $400 - 2 \times 20 = 360$ kV. This situation warrants that the converter at the Sylmar station operate as an inverter. The α^s_{top} at the Sylmar end must now be adjusted such that:

$$+360 = -\frac{3[\sqrt{2}(500)]}{\pi} \cos(\alpha^s_{top})$$

(The negative sign on the right side takes care of the reversal of power flow.)
    or

$$\alpha^s_{top} = 122.22°$$

We can deduce that α^s_{bottom} will also have to be =

$$\alpha^s_{top} = 122.22°$$

    In summary, the power flow magnitude and direction are essentially controlled by proper adjustment of the firing angles of the valves at both converter stations.

9.4 DC-TO-AC CONVERTERS

These inverters convert dc power to ac power at desired voltage and frequency. While we have already established that these are employed at the output side of a HVdc transmission line, other applications for these converters include:

1. Uninterruptible power supplies (UPS) for computers, which require high quality power supply,
2. Induction motor drives, which demand variable input ac voltage supply,

3. Space power systems,

4. Induction heating, and others.

In the following material, we will discuss the basic single-phase and three-phase circuits employed in one or more of the applications mentioned above to understand their principle of operation and the quality of the output waveforms.

Single-Phase Inverter

A simple scheme to obtain constant ac output voltage at a desired frequency, or range of frequencies, is illustrated in Figure 9-19. This circuit employs two thyristors, Th 1 and Th 2, which can be made to conduct alternately by applying suitable gate signals to each of them, a commutating capacitor C, and an output transformer with center tap in the primary winding. The circuit essentially operates in the following sequential manner:

Assume Th 1 is turned on by the application of i_{G1} to its gate, the input voltage V_{dc} will now be applied across the top half of the primary winding of the transformer. Since the bottom half of the primary is magnetically linked to the top winding, V_{dc} will be induced in the lower winding also. This implies a voltage $2V_{dc}$ will build across the capacitor and will remain at this value until Th 2 is turned on by injecting i_{G2} through its gate. Since the voltage across the capacitor cannot change instantaneously, a reverse-biased voltage will be momentarily applied across Th 1, causing it to turn off. At this stage, a negative voltage V_{dc} appears across the bottom half of the transformer primary winding. This induces a negative V_{dc} across the top half of the primary winding, resulting in $-2V_{dc}$ across the capacitor. This situation will continue until Th 1 is turned on again, completing one cycle of operation of the circuit. The various salient waveforms of this circuit are also shown in Figure 9-19.

We can make the following important observations on the operation of this circuit:

1. The output frequency of variation $f = 1/T$ Hertz; where T is duration between two successive gate pulses applied to each thyristor.

2. The output waveform is far from sinusoidal in shape. Hence, proper filters are required to extract the fundamental component from the almost rectangular output voltage v_o.

3. If the load is an ac induction motor, i_o will be different from v_o (Figure 9-19(b)). The reader is encouraged to go through standard power electronics or electronic drives material listed at the end of this chapter to gain an understanding on such circuits involving active sources and inductive parameters.

4. The capacitor C aids in achieving the forced commutation of the two thyristors at proper instants. It should be chosen along with n and R_L such that each thyristor has adequate time to turn off. At this stage, compare the forced commutation process attained here with the simplest circuit described in Figure 9-3.

5. With the advent of high-power transistors and GTO thyristors, a power electronic designer has better choices of devices along with the standard thyristors for circuits such as the one described here.

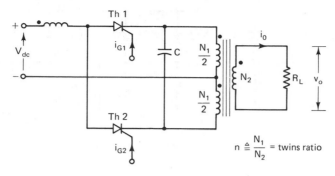

(a) Circuit diagram

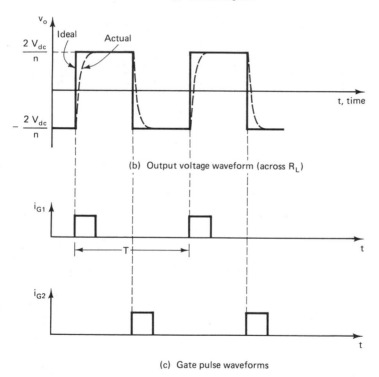

(b) Output voltage waveform (across R_L)

(c) Gate pulse waveforms

Figure 9-19   Single-phase dc-to-ac converter.

Three-Phase Inverter

Figure 9-20 illustrates a commonly used scheme employing six electronic switches (Th 1 to Th 6) and 6 free-wheeling diodes (D1 to D6). this figure does not explicitly show the commutation circuits to turn off the switches, or gating circuits.

The operation of this circuit is similar to the single-phase inverter scheme described earlier in this section, except each thyristor must be turned on at equal intervals of time in the natural sequence of 1 to 6, as shown in Figure 9-20. The time interval between two successive pulses applied to each thyristor's gate will dictate

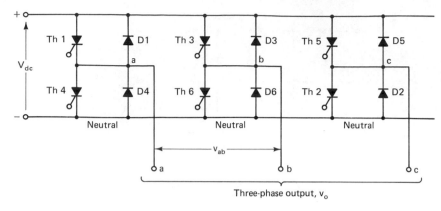

(a) Circuit schematic

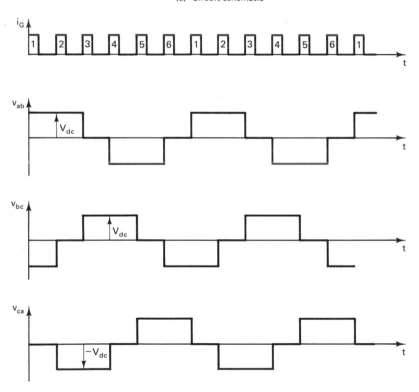

(b) Gate current and output voltage waveforms

Figure 9-20   Three-phase dc-to-ac converter.

the frequency of output of the ac waveform. The figure also shows the output waveforms v_{ab}, v_{bc}, and v_{ca}, which have a high harmonic content. We could improve the quality of the output voltage by doubling or even quadrupling the number of switching devices to realize a 12-pulse or a 24-pulse converter. We have to pay a higher price to achieve higher quality of power supply.

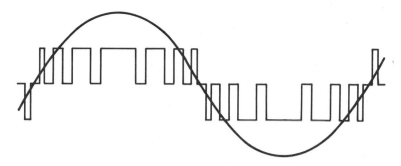

Figure 9-21   Pulse width modulation process.

There is yet another way to improve the quality of the 6-pulse inverter described above. Instead of a single gate pulse, one could design an efficient control logic for gating circuitry, whereby each gating pulse frequency and width is varied in such a manner as to approach sinusoidal output voltage. Such a scheme is called *pulse width modulation* (PWM) process, which is commonly adopted for modern ac drives. The PWM process is conceptually depicted in Figure 9-21.

9.5 AC-TO-AC CONVERTERS

The primary purpose of this type of converter is to vary the rms value of the output voltage applied to an ac load from a constant input source. The control is achieved by means of changing the firing angle of the switching device; hence the name *phase angle* or simply, *phase control*. Since the entire power circuit is ac, the commutation process can be accomplished naturally. In other words, no external commutation circuits are required to deactivate the thyristors.

Important applications for this type of controller vary from the simplest such as light dimmers to the sophisticated ac drives for induction motors. Other applications can be found in tap changing mechanisms for power transformers, and so forth.

As in the case of the converters discussed in the earlier two sections, we will briefly discuss both single-phase and three-phase schemes:

Single-Phase Full-Wave Controller

Consider the following full-wave controller feeding a resistive load from a standard single-phase supply. Figure 9-22 shows such a situation. It also shows the output voltage and current waveforms and the instants at which the gating pulses are applied. One could use a single triac in the place of the two thyristors.

The analysis of this scheme is similar to the full wave ac-to-dc converter circuit we discussed in Section 9.3. The average value of the output voltage is obviously zero. If one of the thyristors is replaced by a diode, V_{dc} across the load is not zero. The effective value of v_o is given by:

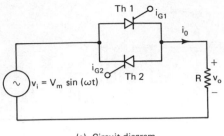

(a) Circuit diagram

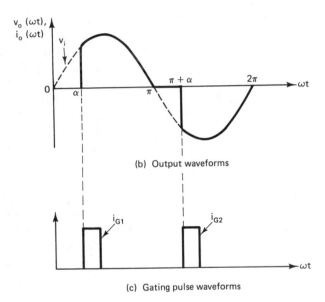

(b) Output waveforms

(c) Gating pulse waveforms

Figure 9-22   Single-phase, full-wave, ac-to-ac converter scheme.

$$V_{rms} = \left[\frac{1}{\pi} \int_{\alpha}^{\pi} v_o^2 \, d(\omega t)\right]^{1/2}$$

$$= \frac{V_m}{\sqrt{2\pi}}\left[(\pi - \alpha) + \frac{\sin 2\alpha}{2}\right]^{1/2} \tag{9-12}$$

which is exactly the same as in the case of a full-wave controlled rectifier. The rms value of each thyristor voltage can be shown to be $V_{rms}/\sqrt{2}$. We can observe from Figure 9-22 that v_o has significant odd harmonic components, which can be evaluated by using standard Fourier series. At this point let us consider an example to gather further knowledge on this scheme.

Example

In the circuit of Figure 9-22, let $v_i = \sqrt{2}\,(240)\sin(\omega t)$ and $R = 24\Omega$. For $\alpha = 45°$ or $\pi/4$ radians, determine:

a. V_{rms}, I_{rms} and real power absorbed by the load
b. The power factor as seen by the source
c. The rms value of third harmonic component of v_o

Solution
a.   By equation (9-12)

$$V_{rms} = \frac{\sqrt{2}\,(240)}{\sqrt{2}\,\pi}\left[\left(\pi - \frac{\pi}{4}\right) + \frac{\sin\left(\dfrac{\pi}{2}\right)}{2}\right]^{1/2}$$

$$= 228.84 \text{ volts}$$

Therefore

$$I_{rms} = \frac{228.84}{24} = 9.535 - A$$

Real power absorbed, $P_{dc} = V_{rms}I_{rms}$ (why?) $= I_{rms}^2 R_L = 2{,}181.97$ watts
b.   Input apparent power $S = (240)\,(240/24)$
$$= 2{,}400 \text{ VA}$$

Therefore, input power factor

$$= \frac{P_{dc}}{S} = \frac{2{,}181.97}{2{,}400} = 0.9092 \text{ lagging}$$

c.   The third harmonic component of v_o can be computed as follows:

$$a_3 = \frac{2}{\pi}\int_\alpha^\pi v_o \sin(3\omega t)\,d(\omega t) \tag{9-30}$$

which will reduce to:

$$a_3 = \frac{V_m}{\pi}\left[\frac{\sin 4\alpha}{4} - \frac{\sin 2\alpha}{2}\right] \tag{9-31}$$

For $\alpha = \pi/4$

$$a_3 = \frac{\sqrt{2}\,(240)}{\pi}\left[\frac{\sin\pi}{4} - \frac{\sin\left(\dfrac{\pi}{2}\right)}{2}\right]$$

$$= -54.02 \text{ volts}$$

Similarly,

$$b_3 = \frac{2}{\pi}\int_\alpha^\pi v_o \cos(3\omega t)\,d(\omega t) \tag{9-32}$$

which can be shown to yield:

$$b_3 = \frac{V_m}{\pi}\left[\frac{\cos(4\alpha) - 1}{4} - \frac{\cos(2\alpha) - 1}{2}\right] \tag{9-33}$$

For $\alpha = \pi/4$
$$b_3 = 0 \quad \text{(verify)}$$

Therefore,

$$V_{3\,rms} = \frac{V_{3m}}{\sqrt{2}} = \frac{(a_3^2 + b_3^2)^{1/2}}{\sqrt{2}} = \frac{54.02}{\sqrt{2}} = 38.20 - V$$

or the per unit value of the third harmonic

$$= \frac{V_{3\,rms}}{V_{rms}} = \frac{38.20}{129.11} = 0.1669$$

$$= 16.69\%$$

Viewed differently, the third-harmonic component will be $= 38.20/240 = 0.1592$ or 15.92 percent of the input rms value. This value is fairly significant, and the designer must devise means of mitigating this particular harmonic content.

Three-Phase Full-Wave Converter

The schematic diagram of a Y-connected three-phase converter often adopted for induction motor drives is drawn in Figure 9-23.

The basic principle of operation is very similar to the single-phase scheme presented in Figure 9-22. The logical sequence at which the gate signals are applied is exactly the same as in the three-phase, controlled rectifier scheme presented in Figure 9-14 and the accompanying material explained in Sec. 9.3.

9.6 DC-TO-DC CONVERTERS

This type of converter is intended for developing variable dc output voltage, usually required for the speed control of dc motors, from a constant dc source, such as a battery. This converter scheme, classically known as a *chopper*, employs one or more electronic switches, devices for commutating the switches at required instants, and, of course, gating circuits. The simple forced commutation scheme presented in Figure 9-3, though not versatile, can perform the functions of the dc-to-dc converter.

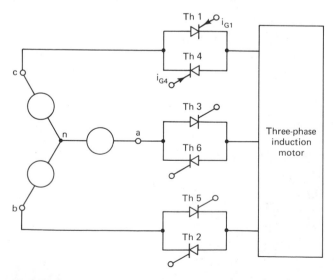

Figure 9-23   Three-phase, ac-to-ac Converter Scheme (Y-connected).

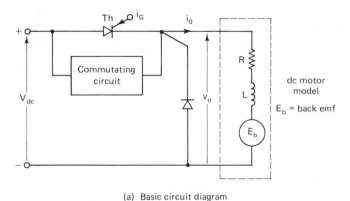

(a) Basic circuit diagram

(b) Gating current waveform

(c) Output voltage waveform

(d) Output current waveform

Figure 9-24   A simple dc-to-dc converter scheme.

The basic principle of operation calls for controlling the ratio of the turn-on-time (t_{on}), of each device to its periodic time, T, as illustrated in Figure 9-24. We can easily observe from this figure that:

$$V_o = \left(\frac{t_{on}}{T}\right)V_{dc} = \left(\frac{t_{on}}{t_{on} + t_{off}}\right)V_{dc} \qquad (9\text{-}34)$$

The load voltage V_o can be varied in one of the following ways:

1. Varying t_{on} while keeping T constant, which implies adopting the pulse width modulation (PWM) principle introduced in Figure 9-21 of Section 9-4,

2. Keeping t_{on} constant, while varying T, which means employing the frequency modulation principle of switching, or

3. A judicious combination of 1 and 2.

We can also observe from Figure 9-24(d) that i_o is not an ideal dc waveform, which may require additional conditioning if warranted.

9.7 SUMMARY

The primary objective of this chapter is to enable the reader to gain a rudimentary knowledge on the rapidly emerging and expanding field of power electronics. To obtain this objective, the principles of operation of power electronic switching devices and converter schemes have been introduced. The problems demanding a designer's attention have been brought out; some of these problems are gating circuitry design, commutating circuitry design, harmonics generation, and means of reducing their effects with proper filter design. The four commonly used converter schemes, namely, ac-to-dc (controlled rectifier), dc-to-ac (inverter), ac-to-ac (phase controller), and dc-to-dc (choppers) are introduced. Several applications encountered in energy and power system applications have either been briefly described, or introduced. In an introductory level of material, it is very difficult, if not impossible, to offer detailed coverage of the material presented here. The reader is therefore strongly recommended to refer to the following reference books:

ADDITIONAL READING MATERIAL

1. Chapman, S.J., *Electric Machinery Fundamentals,* New York: McGraw-Hill Book Company, 1985.

2. Datta, S. K., *Power Electronics and Controls,* Reston, Virginia: Reston Publishing Company, Inc., 1985.

3. Dewan, S. B., and A. Straughen, *Power Semiconductor Circuits,* New York: John Wiley & Sons, Inc., 1975.

4. Institute of Electronic and Electrical Engineers, IEEE Tutorial Course, *Power Electronics Applications in Power Systems,* 78 EH0135-4-PWR, Piscataway, N.J.: IEEE Service Center, 1978.

5. Kusko, A., *Solid-State dc Motor Drives,* Cambridge, Mass.: MIT Press, 1969.

6. Motto, J. W. Jr., ed., *Introduction to Solid-State Power Electronics,* Youngwood, Pennsylvania: Westinghouse Electric Corporation, 1977.

7. Nasar, S. A., *Electric Machines and Electromechanics,* New York: Schaum's Outline Series in Engineering, McGraw-Hill Book Company, 1981.

STUDY EXERCISES

1. The current output waveform, shown in Figure 9-12(c), resulting from a controlled rectifier scheme can be approximated as a triangular function drawn as follows:

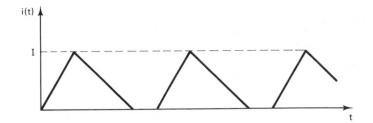

a. Find I_{dc}, the mean value of this periodic function.

b. Find I_{rms}, the root-mean-square of this current waveform.

2. In dealing with electronic or power electronic circuits, it is often desirable to design and fabricate power supply modules to derive variable dc output voltage from a standard ac, single-phase source. Draw a schematic of such a power supply module to realize 0 to 15-V dc from a 120-V, 60-Hz, ac supply. Specify the ratings for each SCR.

3. We have mentioned in Section 9-3 that in dealing with controlled rectifier circuits with inductive loads, we need to solve transcendental equations, which are nonlinear in nature. In order to appreciate and gain a better understanding of this fact, analyze the simplest case of the following circuit with $\alpha = 0$, by finding the average value of the output voltage.

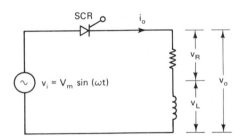

4. We presented the six-pulse HV dc transmission scheme in Figure 9-18. As we emphasized in the text, a 12-pulse scheme can improve the quality of the dc output voltage. Draw a schematic of this improved scheme, showing only one end of the converter station.

5. We have thus far totally concentrated on SCRs as power electronic switches. However, we mentioned in the beginning that power transistors are also available for high power application. Draw a full-bridge dc-to-ac converter scheme for 100-W to 1-kW applications in the 120-V, 400-Hz range employed in the aerospace industry.

SOLUTIONS TO STUDY EXERCISES

1. a. $i(t) = \left(\dfrac{4}{T}\right)It;\quad 0 \leqslant t \leqslant \dfrac{T}{4}$

$i(t) = -I\left(\dfrac{4t - 3T}{2T}\right);\quad \dfrac{T}{4} \leqslant t \leqslant \dfrac{3T}{4}$

$i(t) = 0;\quad \dfrac{3T}{4} \leqslant t \leqslant T$

$$I_{dc} = \frac{1}{T}\left\{\int_0^{T/4}\left(\frac{4}{T}\right)It\,dt + \int_{T/4}^{3T/4} - I\left(\frac{4t - 3T}{2T}\right)dt + 0\right\}$$

$$= \frac{1}{T}\left\{\left[\frac{2It^2}{T}\right]_0^{T/4} - \left[\frac{I}{2T}(2t^2 - 3tT)\right]_{T/4}^{3T/4}\right\}$$

$$= \frac{I}{8} + \frac{I}{4} = \underline{0.375\ I} \longleftarrow \quad a$$

b. $$I_{rms} = \sqrt{\frac{1}{T}\left\{\int_0^{T/4}\frac{16}{T^2}I^2t^2\,dt + \int_{T/4}^{3T/4}\frac{I^2}{4T^2}(4t - 3T)^2\,dt\right\}}$$

$$= \sqrt{\frac{1}{T}\left\{\left[\frac{16I^2}{3T^2}t^3\right]_0^{T/4} + \left[\frac{I^2}{4T^2}\left(\frac{16t^3}{3} - 12t^2T + 9tT^2\right)\right]_{T/4}^{3T/4}\right\}}$$

$$= \sqrt{\frac{1}{T}\left\{\frac{I^2T}{12} + \frac{I^2T}{6}\right\}}$$

$$= \sqrt{I^2/4} = \underline{0.5\ I} \longleftarrow \quad b$$

2.

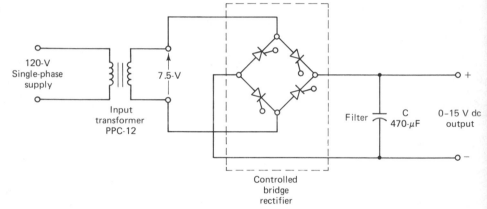

120-V
Single-phase
supply

7.5-V

Input
transformer
PPC-12

Filter   C   470-μF

0–15 V dc
output

Controlled
bridge
rectifier

The salient ratings of the SCR selected are as follows: These are usually provided by manufacturers in the form of commercial data sheets.

	Specification		Rating Chosen
1.	Maximum repetitive peak reverse voltage = $2 \times \sqrt{2} \times 7.5 = 21.21$-V. Choose the next available rating, V_{RRM}	=	100-V
2.	Assume that the maximum RMS current drawn by the load is 20-A. Then each SCR has to withstand at least 10-A. Therefore, maximum on-state RMS current of each SCR, I_T (RMS)	=	40-A
3.	Maximum on-state average current of each SCR = $40\ (2/\pi)$, I_T(AV.)	=	25-A
4.	Maximum one-cycle, non-repetitive surge current, I_{TSM}	=	1,255-A
5.	Minimum critical rate-of-rise of off-state voltage, dv/dt	=	200-V/μs

6. Maximum non-repetitive rate-of-rise of turn-on
   current, di/dt =   200-A/μs
7. Junction temperature range, T_J = $-40°$ to 125 °C
8. Maximum dc gate current rquired to trigger
   SCR, I_{GT} =   100-mA
9. Maximum dc gate voltage required to trigger
   SCR, V_{GT} =   2.5-V

3. The input and output waveforms are drawn below:

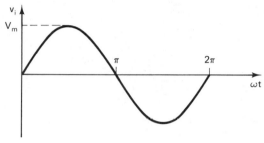
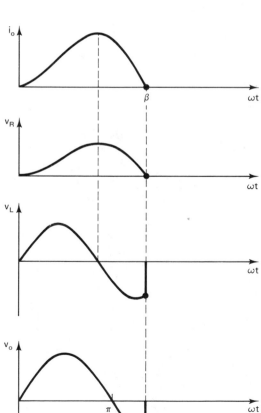

Analysis

By Kirchhoff's voltage law, during the time the SCR conducts

$$v_o = v_R + v_L = v_i$$

or

$$Ri(t) + L\frac{di(t)}{dt} = V_m \sin(\omega t)$$

Solving this first-order differential equation results in

$$i(t) = \frac{V_m}{Z}[\varepsilon^{-(Rt/L)} \sin \phi + \sin(\omega t - \phi)] \qquad \text{(Verify)}$$

where

$$Z = [R^2 + (\omega L)^2]^{1/2}$$

and

$$\phi = \text{Arc} \tan(\omega L/R)$$

It should be noted that the SCR continues to conduct up to $\omega t = \beta > \pi$ until all the energy stored in L is dissipated through R. For a given set of parameters and input conditions, β can be found by solving the transcendental equation for $i(t) = 0$ with $\omega t = \beta$. That is

$$[\varepsilon^{-(R\beta/\omega L)} \sin \phi + \sin(\beta - \phi)] = 0$$

Output Voltage, V_{dc}

The average value of the output voltage, V_{dc} can be obtained by considering the original equation

$$Ri + L\frac{di}{dt} = V_m \sin(\omega t)$$

or

$$i = \frac{V_m \sin(\omega t)}{R} - \frac{L}{R}\frac{di}{dt}$$

$$= \frac{V_m \sin(\omega t)}{R} - \frac{\omega L}{R}\frac{di}{d(\omega t)}$$

Integrate both sides with limits 0 to β and averaging over 2π yields

$$I_{dc} = \frac{1}{2\pi}\int_0^\beta \frac{V_m \sin(\omega t)}{R} d(\omega t) - \frac{1}{2\pi}\int_0^\beta \frac{\omega L}{R}\frac{di}{d(\omega t)} d(\omega t)$$

$$= \frac{V_m}{2\pi R}(1 - \cos \beta) - 0 \text{ (Why?)}$$

then

$$V_{dc} = I_{dc}R = \frac{V_m}{2\pi}(1 - \cos \beta)$$

4.

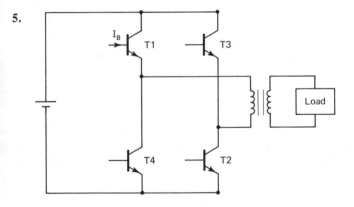

The use of a Y/Y transformer on one side together with a Y/Δ transformer essentially provides the 30° phase shift required between each adjacent phase. Each of the two 6 pulse converter blocks is basically the same as the scheme presented in Figure 9-14.

5.

The switching sequence involves simultaneous energizing of T1 and T2 first, followed by T3 and T4 and repeating the sequence at the required frequency. This circuit is recommended for higher frequencies to avoid the output transformer saturation. The output waveform is similar to Figure 9-19(b). The control of switching the transistors is achieved by varying each transistor's base current, I_B.

HOMEWORK PROBLEMS

1. The continuous current function drawn in Figure 9-12(c) can be linearized as in the following diagram.

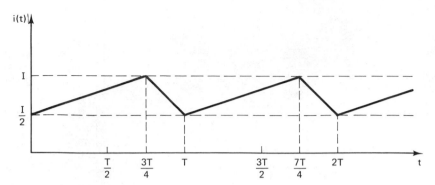

a. Determine the mean value of $i(t)$.

b. Find the rms value of $i(t)$.

c. Compare the results with the corresponding values of Study Exercise 1.

2. In the single-phase, bridge-controlled-rectifier scheme shown in Figure 9-10, $R = 10\Omega$, $v_i = \sqrt{2}(240)\sin(\omega t)$ volts. For $\alpha = 90°$, find the following:

a. I_{dc}

b. P_{dc}

c. I_{rms}

d. The average and rms thyristor currents

e. The power factor of the source

Compare your answers with the corresponding values of the example following equation (9-12). Sketch the various voltage waveforms similar to Figure 9-11.

3. For the six-pulse rectifier scheme presented in Figure 9-14, $R = 5\Omega$, $V_M = \sqrt{2}(480)$-V. For a delay angle of $\alpha = 45°$, calculate the following:

a. I_{dc}

b. P_{dc}

c. The average and rms thyristor currents

d. Ripple factor, r

e. The power factor at the source. Sketch the output current waveform for the given conditions.

4. Repeat Study Exercise 3 for $\alpha = 30°$ case. This corresponds to a controlled rectifier scheme feeding an inductive load with non-zero delay angle.

5. Power transistors (both bipolar and MOSFET) are being increasingly utilized for ac to dc applications. Draw a single-phase, full-bridge scheme to derive 200-V dc from 120-V ac. Specify the most important ratings of each bipolar transistor.

6. We state in the text that the Pacific HVdc line has been upgraded to realize ±500-kV dc voltage and 2,000-A current rating. Under the newer conditions, it is possible to transmit 2,000 MW through the line in either direction.

a. If we keep the dc voltage at Celilo at 500-kV, what should be α_{top}^c?

b. What is the rated dc voltage at Sylmar Station for the given operating conditions? Assume that the dc line resistance has not changed significantly due to reconductoring.

c. Determine α_{top}^s under these conditions.

7. Consider the single-phase ac-ac converter scheme given in Figure 9-22. For all conditions remaining the same except for $\alpha = 60°$, determine:

a. V_{rms}, I_{rms} and real power absorbed by the load.

b. The power factor at the source.

   c. The rms value of the third harmonic component of V_o.

   d. Compare these results with the respective values given in the example following Eq. (9-12) and comment accordingly.

8. The dc-to-dc converter scheme presented in Figure 9-24 is referred to as a Type A chopper. If $E_b = 0$ and $L = 0$,

   a. Find V_o as a function of time, circuit parameters and input source voltage.

   b. Express the rms value of V_o in a similar form as in (a) above.

   c. Repeat (a) above for output current.

   d. Repeat (b) above for output current.

Appendix A

Vector

Products

Many students have been introduced to the concept of a *vector* in courses preceding this course in electric energy. Classical mechanics and electromagnetics are often presented using vector notation. Some students who may not have been exposed to vector methods may have found some of the expressions in this book rather obscure. Two of the operations of vector mathematics are simple and yet of much help in presenting some of the ideas of the book. A brief review (or introduction) is in order.

A *vector* is a multipart number often representing a physical quantity having both magnitude and direction. In two dimensions two numbers suffice to describe a vector. In a Cartesian coordinate system we have an x-component and a y-component. In a three dimensional situation we would have a z-component as a third number. Numerous other coordinate systems exist, but the Cartesian system suffices for our present purposes.

The two-dimensional vectors and complex numbers have much in common — the rule for addition, for example. The product operation is different, however. Multiplication of vectors is done in two forms different from that of complex numbers — the *dot* product and the *cross* product, and both of these forms have appeared in this book.

THE DOT PRODUCT

Suppose that we have two vectors, $\vec{A}$ and $\vec{B}$. One form of multiplication is called the *dot product* and is symbolized as $\vec{A} \cdot \vec{B}$. We form the dot product by taking the product of the magnitudes of the two vectors times the cosine of the angle between them. For example,

$$C = \vec{A} \cdot \vec{B} = +AB \, \cos < \frac{\vec{B}}{\vec{A}} \tag{A-1}$$

An example of the dot product from mechanics is the work done by a force vector, $\vec{F}$, moving through a displacement vector, $\vec{X}$ where work, W, is given by

$$W = \vec{F} \cdot \vec{X} = FX \, \cos < \frac{\vec{X}}{\vec{F}}$$

If by chance the force and displacement vectors are colinear; that is, the angle between the two is zero, we have the simpler relation, $W = FX$. If, on the other hand, the force acts perpendicular to the displacement, no work is done since

$$W = \vec{F} \cdot \vec{X} = FX \, \cos 90° = 0$$

In our consideration of magnetic circuits, we find a relation such as

$$d\mathscr{F} = \vec{H} \cdot d\vec{l}$$

or the summation of the $d\mathscr{F}$ elements as

$$\mathscr{F} = \int \vec{H} \cdot d\vec{l}$$

which is called a *line integral,* since we sum along a path determined by the sequence of increments $d\vec{l}$.

Our labor is simplified if we follow a path (as in the case of the toroid of Chapter 3) where $\vec{H}$ and $d\vec{l}$ are colinear, thus resulting in

$$d\mathscr{F} = H \, dl \quad \text{and} \quad \mathscr{F} = \int H \, dl$$

If the integral is taken around a closed path as in Ampère's law we symbolize that fact by means of a small circle on the integral sign as in

$$\mathcal{F} = Ni = \oint \vec{H} \cdot d\vec{l}$$

It should be noted that the dot product of two vectors results in a *scalar;* that is, a magnitude without direction, as in the preceding examples of work and magnetomotive force, $\mathcal{F}$.

THE CROSS PRODUCT

The other form of multiplication of two vectors, as $\vec{A}$ and $\vec{B}$ for example, is symbolized as $\vec{A} \times \vec{B}$. We form the cross product by taking the product of the two magnitudes times the *sine* of the angle between them as

$$|\vec{C}| = |\vec{A} \times \vec{B}| = AB \, \sin < \frac{\vec{B}}{\vec{A}} \qquad \text{(A-2)}$$

The symbols modifying C in Eq. (A-2) indicate two things: that the cross product is a *vector,* which has direction, and that the expression given is that of the magnitude of the cross product vector. The direction is *normal* (perpendicular) to the plane determined by the two factors, $\vec{A}$ and $\vec{B}$. The sense is given along the normal direction by the right-hand rule which says:

Let the fingers of the right hand pass from the first vector to the second; the thumb then points in the direction of the cross product vector.

An example of the cross product occurs in Chapter 4 where we have the Eq. (4-15)

$$d\vec{f} = i \, d\vec{l} \times \vec{B}$$

for the force on a conductor segment in a magnetic field. Also we have the expression

$$\vec{f} = i\vec{l} \times \vec{B}$$

for the force on a straight conductor in a uniform field.

It should be noted that the current, i, is *not* a vector but a scalar. The direction of the current is contained in the vector $\vec{l}$.

The direction of the cross product is dependent on the sequence of the two factors. If the sequence of the factors is reversed, the direction of the cross product is reversed; $\vec{A} \times \vec{B}$ is opposite in direction from $\vec{B} \times \vec{A}$. It must be noted, therefore, that the cross product formulation gives us the direction of the magnetic force on a current-carrying conductor only if we write the factors of the equations in the proper sequence.

Appendix B

Glossary of Terms, Symbols, Units, and Abbreviations

Symbol	Name	Units	Abbreviation
a	Number of armature paths		
A	Area	sq. mtrs	m^2
B	Magnetic flux density	tesla	T
B	Susceptance	siemen	S
C	Capacitance	farad	F
E	Voltage	volt	V
$\mathcal{F}$	Magnetomotive force	ampere (turns)	NI
G	Conductance	siemen	S
H	Magnetic field intensity	amp-turn/mtr	NI/m
I	Current	ampere	A
J	Polar moment of inertia	kilogram-m^2	kg-m^2
l	Length	meter	m
L	Inductance	henry	H
λ	Flux linkages	weber-turn	Wb
n	Rotational speed	rev/min	r/min
N	Number of turns		
ω	Angular velocity	radians/sec	rad/sec
p	Power—instantaneous	watt	W
p	Number of poles		
P	Average power	watt	W
$\mathcal{P}$	Permeance (magnetic)	henry	H
μ	Permeability (magnetic)	henry/meter	H/m
Q	Charge	coulomb	C
Q	Reactive voltamperes	voltamperes	var
r	Radius	meter	m
$\mathcal{R}$	Reluctance (magnetic)	henry^{-1}	H^{-1}
R	Resistance	ohm	Ω
s	Slip	per-unit	
S	Slip	rev./min	r/min
S	Complex power	voltampere	VA
t	Time	second	sec
T	Time constant	second	sec
V	Voltage	volt	V
w	Energy density	joule/m^3	J/m^3
W	Energy	joule	J
X	Reactance	ohm	Ω
Y	Admittance	siemen(mho)	S
Z	Impedance	ohm	Ω
ϕ	Flux (magnetic)	weber	Wb
ϕ	Phase		

Appendix C

Inductance of a Long, Straight Wire

Consider a long, straight wire of circular cross section with radius r and carrying a current I. The current, I, may be thought of as the instantaneous value of a time-varying current or as either the maximum or rms value of a sinusoidal current if the interpretation of any derived quantities takes the nature of the current into account. We show a cross-sectional view of the conductor in Figure C-1.

An analysis of the field effects is conveniently split into two parts: an inductance component owing to the field inside the conductor and an inductance component owing to the field external to the conductor.

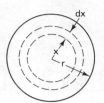

Figure C-1   Cross-sectional view of a long, straight conductor.

INDUCTANCE COMPONENT DUE TO INTERNAL FLUX

The component of inductance due to the internal flux is conveniently derived using the stored energy concept given by Eq. (8-6)

For a linear medium ($\mu = \mu_o = 4\pi 10^{-7}$) the energy density in a magnetic field is given by

$$w = \frac{1}{2}\mu_o H^2 \quad J/m^3 \tag{C-1}$$

Consider a small concentric ring inside the wire, with radius x, thickness dx and length $l = 1$ m.

If the total current is I (uniformly distributed) the current inside the ring is

$$I_x = I\frac{\pi x^2}{\pi r^2} \tag{C-2}$$

and a line integral around a concentric path of radius x to apply Ampère's law gives

$$\oint \vec{H} \cdot \vec{ds} = I_x \tag{C-3}$$

or

$$H\oint ds = I\frac{x^2}{r^2} \tag{C-4}$$

$$H 2\pi x = I\frac{x^2}{r^2} \tag{C-5}$$

$$H = \frac{Ix}{2\pi r^2} \tag{C-6}$$

The energy stored in this ring is

$$dW = \frac{1}{2}\mu_o H^2 \, dV \tag{C-7}$$

where the volume, dV is given by

$$dV = 2\pi x \, dx \quad m^3 \tag{C-8}$$

Hence

$$dW = \frac{1}{2}\mu_o\left(\frac{Ix}{2\pi r^2}\right)^2 2\pi x \, dx$$

$$= \frac{\mu_o I^2}{4\pi r^4}x^3 \, dx \tag{C-9}$$

Summing

$$W = \int_{x=0}^{x=r} dW = \left[\frac{\mu_o I^2}{4\pi r^4} \frac{x^4}{4} \right]_o^r \tag{C-10}$$

$$= \frac{\mu_o}{16\pi} I^2 \tag{C-11}$$

Since $W = (1/2)LI^2$

$$L_{int} = \frac{\mu_o}{8\pi} = \frac{1}{2} \times 10^{-7} \quad H/m \tag{C-12}$$

It is at first surprising to note that the inductance per unit length is a constant, independent of conductor radius!

INDUCTANCE COMPONENT DUE TO EXTERNAL FLUX

The second component of inductance of a long, straight conductor is that due to the flux linkages that are completely external to the conductor. By computing these flux linkages in any region, say from x_1 to x_2 in Figure C-2 we may use Eq. (8-5) to evaluate an inductance component. The values of x_1 and x_2 of interest are determined by the particular configuration at hand in a given problem. The results of this exercise are used in Chapter 8 to evaluate certain lines.

From symmetry, the flow lines of the H vector are concentric circles and we may easily evaluate a line integral around one of these circles. From Ampère's law

$$\oint \vec{H} \cdot \vec{ds} = I$$

and for the path around the circle of x radius

$$H_x 2\pi x = I$$

thus

$$H_x = \frac{I}{2\pi x} \tag{C-13}$$

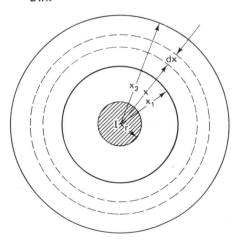

Figure C-2   A conductor of radius r, which carries a current I. The concentric rings define a region wherein magnetic flux linkages will be computed.

and for a region of permeability μ_o (air) we have

$$B_x = \mu_o H_x = \frac{\mu_o I}{2\pi x} \qquad Wb/m^2 \qquad\qquad (C\text{-}14)$$

In the ring of thickness dx and length 1 meter (into the paper) we have an increment of magnetic flux

$$d\phi = \vec{B} \cdot \vec{dA} = \frac{\mu_o I}{2\pi x} dx \qquad Wb/m \qquad\qquad (C\text{-}15)$$

Since this flux component links the circuit just once we also have

$$d\lambda = d\phi = \frac{\mu_o I}{2\pi x} dx \qquad Wb/m \qquad\qquad (C\text{-}16)$$

If we sum the $d\lambda$ increments from x_1 to x_2, we have as the flux linkages contributed in this region

$$\lambda_{12} = \int_{x_1}^{x_2} d\lambda = \int_{x_1}^{x_2} \frac{\mu_o I}{2\pi x} dx$$

$$\lambda_{12} = \frac{\mu_o I}{2\pi} \ln(x_2/x_1) \qquad\qquad (C\text{-}17)$$

Since μ_o is $4\pi 10^{-7}$, Eq. (C-17) may be written

$$\lambda_{12} = 2 \times 10^{-7} I \ln(x_2/x_1) \qquad\qquad (C\text{-}18)$$

Last, we recognize from Eq. (8-5) that

$$L = \frac{\lambda}{I}$$

and hence

$$L = 2 \times 10^{-7} \ln(x_2/x_1) \qquad H/m \qquad\qquad (C\text{-}19)$$

or

$$L = 0.7411 \log(x_2/x_1) \qquad mH/mile \qquad\qquad (C\text{-}20)$$

or

$$= 0.4605 \log(x_2/x_1) \qquad mH/km \qquad\qquad (C\text{-}21)$$

These results may then be used as in Chapter 8 to find the inductance of lines of various configurations. Note that in the above equations, ln means a natural logarithm and log means a logarithm to base ten.

Appendix D

Electric Field of a Long,

Straight Wire

Consider a long, straight wire of radius r, which has a charge of Q coulombs per meter of length. As a result of the charge, there is an electric field around the wire. The flow lines of the E vector are directed radially outward from the wire as illustrated in Figure D-1 below.

The potential difference between two points in space is found from the integral

$$V = \int \vec{E} \cdot \vec{dx} \qquad \text{(D-1)}$$

It will be noted that the equipotential surfaces form cylindrical shells around the conductor at the center. It will be convenient for purposes of computing line capaci-

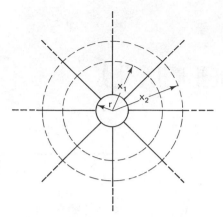

Figure D-1   Electric field around a long, straight wire.

tance in Section 8.8 to derive an expression for the potential difference between two equal potential surfaces of radii x_1 and x_2.

From Gauss's law we can write an equation for D, the displacement vector as

$$D = \frac{Q}{2\pi x} \quad C/m^2 \tag{D-2}$$

This vector is directed radially outward from symmetry considerations.

The E vector (electric field intensity) may be found from the relation

$$E = \frac{D}{\varepsilon_0} \tag{D-3}$$

where ε_0 is the permittivity of free space — which is essentially the same as that of air and approximately equal to 8.85×10^{-12} farads per meter

If we combine Eqs. (D-2) and (D-3) we have

$$E = \frac{Q}{2\pi\varepsilon_0 x} \quad V/m \tag{D-4}$$

We may now evaluate the potential difference between surfaces at x_1 and x_2 by using Eqs. (D-1) and (D-2) and noting that the E vector and dx are colinear when we follow a path radially outward between two equipotential shells.

$$V_{12} = \int_{x_1}^{x_2} E\,dx = \int_{x_1}^{x_2} \frac{Q}{2\pi\varepsilon_0 x}\,dx$$

$$= \frac{Q}{2\pi\varepsilon_0} \ln(x_2/x_1) \qquad \text{volts} \tag{D-5}$$

Since x_1 region is more positive than x_2 for a reference positive charge, Q, the potential difference, V_{12}, is of the nature of a *drop* in voltage, which is consistent with the double subscript notation used in earlier chapters.

It will be noted that there is a strong resemblance between Eqs. (D-5) and (C-19), which latter equation applied to the magnetic field. An important difference between the cases is the fact that there is no field *inside* the conductor, so there is no relation analogous to that of the internal flux linkages of the magnetic case.

Index